Low-Temperature District Heating Implementation Guidebook

Final Report of IEA DHC Annex TS2

Implementation of Low-Temperature District Heating Systems

Helge Averfalk
Theofanis Benakopoulos
Isabelle Best
Frank Dammel
Christian Engel
Roman Geyer
Oddgeir Gudmundsson
Kristina Lygnerud
Natasa Nord
Johannes Oltmanns
Karl Ponweiser
Dietrich Schmidt
Harald Schrammel
Dorte Skaarup Østergaard
Svend Svendsen
Michele Tunzi
Sven Werner

Edited by Kristina Lygnerud and Sven Werner,
Halmstad University 2021

FRAUNHOFER VERLAG

Disclaimer notice (IEA DHC):
This project has been independently funded by national research resources and performed within the International Energy Agency Technology Collaboration Programme on District Heating and Cooling (IEA DHC). Any views expressed in this publication are not necessarily those of IEA DHC. IEA DHC can take no responsibility for the use of the information within this publication, nor for any errors or omissions it may contain. Information contained herein have been compiled or arrived from sources believed to be reliable. Nevertheless, the authors or their organizations do not accept liability for any loss or damage arising from the use thereof. Using the given information is strictly your own responsibility.

Disclaimer Notice (Authors):
This publication has been compiled with reasonable skill and care. However, neither the authors nor the IEA DHC Contracting Parties make any representation as to the adequacy or accuracy of the information contained herein, or as to its suitability for any particular application, and accept no responsibility or liability arising out of the use of this publication. The information contained herein does not supersede the requirements given in any national codes, regulations or standards, and should not be regarded as a substitute.

Copyright:
All property rights, including copyright, are vested in IEA DHC. In particular, all parts of this publication may be reproduced, or transmitted in any form or by any means, electronic, mechanical, photocopying, recording or otherwise only by crediting IEA DHC as the original source. Republishing of this report in another format or storing the report in a public retrieval system is prohibited unless explicitly permitted by the IEA DHC programme manager in writing.

Image Source (Front Cover):
Major front cover picture: Maryna Miliushchanka, Fraunhofer IEE
Additional front cover pictures: Sven Werner, Halmstad University

Citation:
Please refer to this report as:
Averfalk H et al, Low-Temperature District Heating Implementation Guidebook. IEA DHC Report, 2021

International Energy Agency Technology Collaboration Programme on District Heating and Cooling

www.iea-dhc.org

Table of Contents

	Preface		6
	Executive Summary		7
	Definitions of expressions and terms		14
1	**Path to future district heating**		**17**
	1.1	Past, present, and future of district heating	17
	1.2	Four generations of district heating	18
	1.3	District heating within the fossil energy society	21
	1.4	Transition challenges	21
	1.5	District heating within the renewable energy society	23
	1.6	Economics of low-temperature heat distribution	24
	1.7	Previous and current works on low-temperature heat distribution	26
	1.8	Major conclusions from this introduction	27
	1.9	Organisation of this guidebook	27
	1.10	Literature references in Chapter 1	28
2	**Economic benefits of low-temperature district heating**		**29**
	2.1	More geothermal heat extracted	30
	2.2	Less electricity used in heat pumps	32
	2.3	More waste heat extracted	33
	2.4	More heat obtained from solar collectors	34
	2.5	More heat recovered from flue gas condensation	36
	2.6	More electricity generated in combined heat and power plants	37
	2.7	Higher heat storage capacities	39
	2.8	Lower heat distribution loss	41
	2.9	Ability to use plastic pipes instead of steel pipes	41
	2.10	Non-economic benefits of lower temperatures	43
	2.11	Summary of economic benefits	43
	2.12	Major conclusions concerning economic benefits	45
	2.13	Literature references in Chapter 2	46
3	**Lower temperatures inside buildings**		**47**
	3.1	Temperature requirements and heat demands in buildings	47
	3.2	Building installations and their impact on district heating temperatures	50
	3.3	Customers with high supply temperatures requirements	53
	3.4	Substation malfunctions and faults giving higher temperatures	54
	3.5	Use of data to identify heating system improvements	57
	3.6	Actions to lower the temperatures in space heating systems	59
	3.7	Actions to lower the temperatures in domestic hot water systems	64
	3.8	Actions to lower the temperatures in ventilation systems	67
	3.9	How to design new building installations for lower temperatures	68
	3.10	What to consider when existing buildings are connected to district heating	70
	3.11	Major conclusions on lower temperatures in buildings	71
	3.12	Literature references in Chapter 3	72
4	**Lower temperatures in heat distribution networks**		**75**
	4.1	Tracking malfunctioning substations with high return temperatures	75
	4.2	Identification of unintentional circulation flows	79

	4.3	Addressing bottlenecks in network sections	80
	4.4	Successful cases of temperature reduction in existing systems	81
	4.5	Subnetworks	84
	4.6	Cascading solutions	85
	4.7	Installing heat pumps to address subsection demands	86
	4.8	Increased decentralised supply	87
	4.9	Digitalisation opportunities	89
	4.10	Design criteria for new systems	91
	4.11	New innovative supply and distribution concepts	92
	4.12	Lessons learned for obtaining lower system temperatures	97
	4.13	Major conclusions concerning low-temperature systems	97
	4.14	Literature references in Chapter 4	98
5	**Applied study: Campus Lichtwiese at TU Darmstadt**		**100**
	5.1	Transferability of the applied study	100
	5.2	Background of the applied study	100
	5.3	Monitoring data for identification of temperature reduction opportunities	101
	5.4	Impact of district improvement measures on the return temperature	104
	5.5	Comprehensive building renovation	104
	5.6	Performance of hot water preparation	106
	5.7	Performance of space heating circuits	108
	5.8	Performance of ventilation heating circuits	110
	5.9	Reduction of the district heating supply temperature	113
	5.10	Recommendations for actions	113
	5.11	Energetic, ecologic and economic comparison of the proposed actions	114
	5.12	Major conclusions from the applied study	117
	5.13	Literature references in Chapter 5	118
6	**Competitiveness of low-temperature district heating**		**119**
	6.1	Traits of business models in early LTDH installations	119
	6.2	The cases	120
	6.3	Results from case reviews	122
	6.4	Questions to address when expanding to LTDH	124
	6.5	National context and potential for LTDH to increase district heating competitiveness	125
	6.6	Heat distribution characteristics in early implementations	130
	6.7	Concentration of heat demand	132
	6.8	Temperature levels	134
	6.9	Heat distribution costs	136
	6.10	Major conclusions concerning competitiveness	138
	6.11	Literature references in Chapter 6	138
7	**Practical implementation of low-temperature district heating**		**139**
	7.1	Introduction	139
	7.2	Successful implementation stories	140
	7.3	Transition of a small-scale system - City of Gleisdorf, Austria	140
	7.4	Housing estate Weihenbronn in Wüstenrot, Germany	142
	7.5	Mijnwater project in Heerlen, The Netherlands	143
	7.6	Cold district heating at FGZ Zürich, Switzerland	145
	7.7	Stadtwerk Lehen in Salzburg, Austria	146
	7.8	Smart City Reininghaus in Graz, Austria	147

	7.9	Excess heat recovery from data centre in Braunschweig, Germany	**148**
	7.10	Excess heat from research facilities in Brunnshög in Lund, Sweden	**149**
	7.11	Geosolar District Heating in "Feldlager" in Kassel, Germany	**150**
	7.12	Conversion area 'Lagarde' in Bamberg, Germany	**151**
	7.13	Ultra-low-temperature district heating in Bjerringbro, Denmark	**152**
	7.14	Temperature reductions in existing apartment building in Viborg, Denmark	**152**
	7.15	Automatic return temperature limitation in Frederiksberg, Denmark	**153**
	7.16	District LAB Experimental Facility in Kassel, Germany	**154**
	7.17	Summary and lessons learnt from the case studies	**155**
	7.18	Major conclusions from the case studies	**157**
	7.19	Literature references in Chapter 7	**158**
8	**Transition strategies**		**159**
	8.1	Adopted transition initiatives from five urban areas in Europe	**160**
	8.2	Visions	**161**
	8.3	Strategies	**162**
	8.4	Planning measures	**163**
	8.5	University campus systems as forerunners	**164**
	8.6	Major conclusions from transition strategies	**165**
	8.7	Literature references in Chapter 8	**165**
9	**Conclusions**		**167**
	9.1	Technological developments	**167**
	9.2	Non-technical aspects	**168**
	9.3	Policy implications	**168**
	9.4	Recommendations	**170**
	9.5	Main conclusion of this guidebook	**171**
10	**Annexes**		**172**
	10.1	Typical configurations for low-temperature heat distribution networks	**172**
	10.2	Net list of completed detailed descriptions of cases	**179**
	10.3	Gross list of low-temperature inspiration initiatives	**182**
	10.4	Location index	**197**
	10.5	Participant organisations in the TS2 annex	**198**
	10.6	Literature references in Chapter 10	**200**
Imprint			**201**

PREFACE

This guidebook aims to provide tangible information that will facilitate the implementation of low-temperature district heating (LTDH) systems. These systems provide renewable heat and low-temperature excess heat at a lower cost than high-temperature district heating systems. Through the increased use of low-temperature district heating systems, a significant transformation of basic district heating technology can be accomplished.

This technological transformation will be an efficient elimination of fossil fuel use for heating buildings. The replacement of fossil fuels is essential to reducing the ongoing global warming created by our massive carbon dioxide emissions from combustion of fossil fuels. Hence, this guidebook supports the required substitution of fossil fuel–based heating in buildings with heat supplied by renewables and recycled heat.

This guidebook aims to provide simple advice and recipes for obtaining lower network temperatures and other new features in existing and new district heating systems. This ambition is accomplished by summarising gained experiences from early adopters in various urban areas throughout Europe. Notably, lessons learned can be applied in new and expanding district heating systems. Current first-, second- and third-generation district heating systems contain features that are barriers to lower network temperatures. A vital step in the development of low-temperature systems involves the understanding of current barriers to avoid them in new and expanding systems.

While the readers of this guidebook include various professionals, district heating practitioners represent our target audience since they will ultimately execute this technological transformation. The guidebook is meant to inspire the practitioner with our technical overviews, demonstrations, case descriptions and examples of transition strategies. However, policy- and decision-makers in the energy sector will also be reassured by our conclusions concerning technology development, non-technical barriers, and policy implications and recommendations. Moreover, market managers should consider our examination of the competitiveness of low temperatures in heat distribution networks. Although most of our references from early adopters of low-temperature systems are from Europe, our advice and strategies can also be applied in the USA, Canada, Russia, China, Japan, Korea and Chile to make future district heating systems more competitive.

This guidebook has been written within the TS2 annex of the IEA technology collaboration programme concerning district heating and cooling (also known as the IEA DHC/CHP programme, www.iea-dhc.org). Started in 2018, the TS2 annex was called 'Implementation of Low-Temperature District Heating Systems' and was active until 2021. Several research groups from Austria, Denmark, Germany, Norway, Sweden, and the United Kingdom have been involved in this study. The annex was funded by a task-sharing effort since the work contribution from each partner was financed by national research financing schemes. Kristina Lygnerud from Halmstad University in Sweden coordinated the annex.

EXECUTIVE SUMMARY

Introduction

In many urban areas, district heating systems are used to move heat through pipes from available heat sources to buildings and processes that require heat. Major heat sources include recycled heat from various societal processes that have considerable amounts of residual heat (e.g. thermal power plants). Some heat is also obtained from renewable energy sources. However, fossil fuels have continued to be used as a primary energy source in varying proportions. In most systems, some boilers use fossil fuels to provide heat only for peak and redundancy purposes as a complement to the major heat sources.

Future district heating systems will have market conditions that differ from those of current systems. The high heat demands from customers will decline. On the supply side, renewables and heat recycling will replace current heat recycling that relies on processes having fossil fuels as primary energy sources. Hence, future systems will have to use enhanced district heating technology to accomplish decarbonisation. Lowering heat distribution temperatures is key since lower temperatures result in higher efficiency in heat supply.

During the last decade, the collective label 'fourth-generation district heating' (4GDH) has been used to describe these enhanced district heating systems. The overarching goal with these systems is to obtain fully decarbonised district heating systems. According to the 2014 definition, a 4GDH system should have the following abilities:

- To supply low-temperature district heating for space heating and hot water preparation
- To distribute heat with low grid losses
- To recycle heat from low-temperature sources
- To integrate thermal grids into a smart energy system
- To ensure suitable planning, cost and motivation structures

From these five abilities, it is evident that low-temperature district heating (LTDH) will be a key technology to obtain more efficient district heating systems in the future.

> Our definition of 4GDH in this guidebook applies to all new technological features and concepts using low temperatures, which are considered best available from 2020 onward. As experienced in previous technology generations, a wide diversity of technology choices in 4GDH is expected. Hence, cold district heating systems are also included in our definition of 4GDH. The corresponding technology comprises all heat distribution technologies that will utilise supply temperatures below 70 °C as the annual average. 4GDH technology is a family of many different network configurations for heat distribution. Notably, cold and warm networks are siblings in this family of configurations.

The transition from traditional district heating systems to completely decarbonised systems will support international, national, and local ambitions for decarbonisation by obtaining lower emissions of carbon dioxide. The corresponding EU ambitions are summarised in Figure 5 below from the introduction in the **first** chapter. The conclusion from this figure is that the rate of change up to 2030 should increase more than five times compared to the previous goal for 2020. This higher achievement calls for more intensive work concerning the decarbonisation of the European district heating systems.

The following introductory conditions were essential input for the content of this guidebook:

- The decarbonisation of energy systems will eliminate heat recovery from fossil-based combined heat and power (CHP) plants that currently dominate the heat supply to district heating systems. The proportion of non-combustible and low-temperature heat supply will then increase in future district heating systems since the levels of biomass and waste combustion will not be sufficient to replace the current use of fossil fuels.
- The requirement for high temperatures in existing district heating systems is an initial barrier to the implementation of many suitable low-temperature heat sources.
- The introduction of low-temperature heat distribution with supply temperatures below 70 °C will increase the profitability of implementing geothermal heat, heat pumps, industrial excess heat,

solar collectors, flue gas condensers, and heat storage options into district heating systems.
- The low-temperature heat distribution will be a key economic driver together with higher carbon prices for obtaining decarbonisation within the EU because of the larger reduction target of carbon dioxide emissions for 2030.

Economic benefits of low-temperature district heating

At lower distribution temperatures, the economic benefits of renewables and recycled heat are based on the following nine efficiency gains analysed in the **second** chapter:

- More heat extracted from geothermal wells since lower temperatures of the geothermal fluid can be returned to the ground.
- Less electricity used in heat pumps when extracting heat from heat sources with temperatures below the heat distribution temperatures since lower pressures can be applied in the heat pump condensers.
- More excess heat extracted since lower temperatures of the excess heat carrier will be emitted to the environment.
- More heat obtained from solar collectors since their heat losses are lower, thereby providing higher conversion efficiencies.
- More heat recovered from flue gas condensation since the proportion of vaporised water (steam) in the emitted flue gases can be reduced.
- More electricity generated per unit of heat recycled from steam combined heat and power (CHP) plants since higher power-to-heat ratios are obtained with lower steam pressures in the turbine condensers.
- Higher heat storage capacities since lower return temperatures can be used in conjunction with high-temperature outputs from high-temperature heat sources.
- Lower heat distribution losses with lower average temperature differences between the fluids in heat distribution pipes and the environment.
- Ability to use plastic pipes instead of steel pipes to save cost.

Additional benefits include a reduced risk of low-cycle fatigue for steel pipes (due to less variation in supply temperatures); smaller temperature drop in the flow direction, allowing for lower supply temperatures from heat supply plants (since less heat will be lost), and a lower risk of scalding during pipe maintenance (lethal accidents have occurred in high-temperature systems).

To quantify the cost reductions related to lower temperatures for various heat supply technologies, a key performance indicator called 'cost reduction gradient' (CRG) is used. Examples of CRG for different heat supply technologies are presented in Table 1.

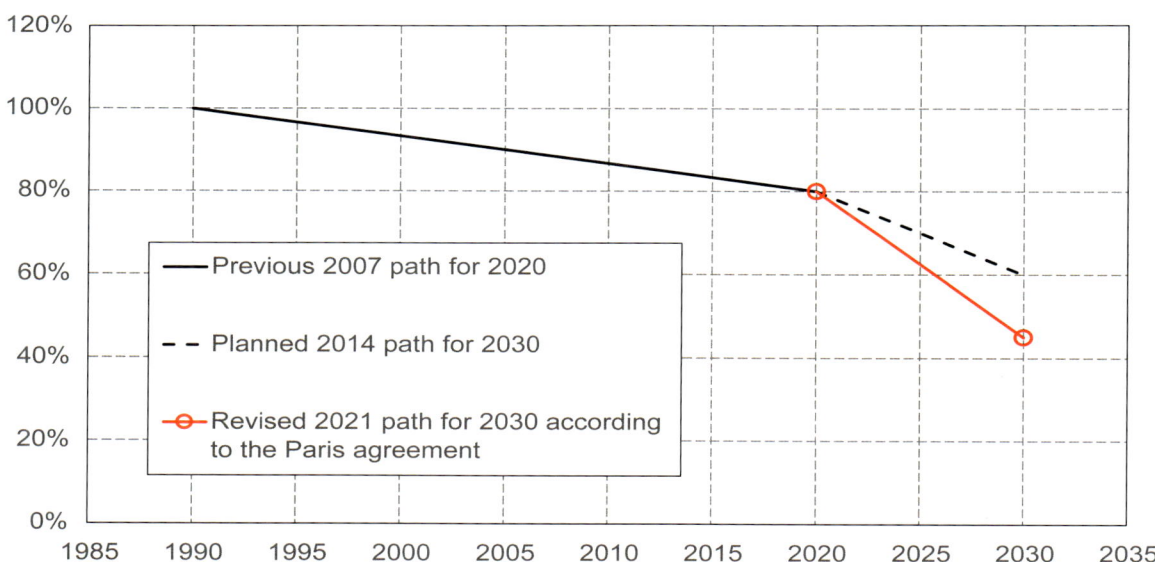

Figure 5. Three paths for reduction of the EU carbon dioxide emissions until 2020 and 2030.

Table 1. Overview of assessed economic effects, indicated with the cost reduction gradient (CRG) in euro/(MWh·°C), of reduced system temperatures.

Chapter section and the assessed heat supply technology, either by itself or dominating in a system	CRG in euro/(MWh·°C)	
	Investment cases when investment costs are reduced	**Existing cases when operating costs are reduced**
2.1 Low-temperature geothermal heat	0.45 – 0.74	0.67 – 0.68
2.2 Heat pump	0.41	0.63 – 0.67
2.3 Low-temperature waste heat	0.65	0.51
2.4 Solar thermal – flat plate collectors	0.35 – 0.75	Not available
2.4 Solar thermal – evacuated tube collectors	0.26	Not available
2.5 Biomass-boiler with flue gas condensation	Not available	0.10 – 0.13
2.6 Biomass-CHP with back-pressure turbine	Not available	0.10 – 0.16
2.6 Biomass-CHP with extraction turbine	Not available	0.09
2.6 Waste-CHP with flue gas condensation	Not available	0.07
2.7 Daily storage as tank thermal storage	0.01	0.07
2.7 Seasonal storage as pit thermal storage	0.07	0.07
2.8 Heat distribution losses	Not available	0 – 0.13

Table 1 reveals that traditional combustion processes in CHP plants without flue gas condensation have CRGs between 0.10 and 0.13 euro/(MWh·°C). In addition, corresponding CRGs for low-temperature heat sources, such as geothermal, heat pumps and waste heat, are between 0.5 and 0.7 euro/(MWh·°C). Hence, these new low-temperature heat sources have CRGs that are approximately five times higher than those for traditional heat supply.

Therefore, the cost savings for European low-temperature district heating systems is forecasted to be roughly 0.5 euro/(MWh·°C), which results in a total cost reduction potential of 14 billion euro per year, assuming the future annual EU district heat sales of 950 TWh and a temperature reduction of 30 °C. This cost reduction represents a net present value of more than 200 billion euro.

Lower temperatures inside buildings

The **third** chapter explains that lower district heating temperatures than those that are typically applied today can be used in new and existing buildings. In most cases, district heating temperatures are much higher than needed to comply with national temperature requirements to control the *Legionella* risk and typical comfort requirements for space heating.

The main obstacles to lower district heating temperatures are the occurrences of simple malfunctions and faults in the district heating substations, and which, therefore, need to be eliminated. Additionally, proper maintenance and automatic fault detection in building substations should be applied in district heating systems.

Legionella treatment is an area of focus when discussing low-temperature operation. The main risk of *Legionella* growth is often due to the poor design or operation of internal building installations. Alternatives to thermal treatment of *Legionella* are available, but in most existing buildings, district heating companies must rely on frequent wireless readings of energy meters to ensure the district heating supply temperature on entry to each customer is high enough to meet national requirements for the thermal treatment of *Legionella* in domestic hot water installations.

For the long term, energy renovations will enable lower district heating temperatures by reducing the heat demands in existing buildings and, therefore, reducing the space heating temperatures needed in commonly over-sized existing heating systems. The clever design of new heating installations (in existing and new buildings) using robust components will allow for lower temperatures in the future.

Longer thermal lengths in heat exchangers, *Legionella*-safe supply of domestic hot water and the automatic balancing of space heating systems are crucial for low-temperature heating. Current, best-available technologies in this regard include externally accessible flat stations for domestic hot water supply and smart return temperature thermostats with automatic balancing functionalities.

For the design of new space heating systems and domestic hot water installations, new standards are needed to address the use of low-temperature heating based on renewable heat sources in the future. Research is needed to develop more solutions to provide *Legionella*-safe and comfortable domestic hot water without the current high heat requirement.

Lower temperatures in heat distribution networks

The **fourth** chapter is dedicated to eleven measures that can be implemented on the system/distribution side to maintain low system temperatures and to reduce them further. Five major takeaways are identified.

Before investing in any improvement measure, it is necessary to compare customer supply temperatures requirements with the primary supply side temperatures. In some cases, the critical supply temperatures needed by customers and those provided by the district heating network do not match. This happens because the district heating operator does not know the exact supply temperature required by customers and, thus, ensures that the temperature supplied is never below that needed to guarantee comfort.

When replacing existing substations or when designing new ones, heat exchangers with longer thermal lengths should be preferred as these will enable to obtain low supply and return temperatures

Many systems have already started their transformation to lower temperatures, demonstrating the wide range of possibilities and proven solutions that exist. However, it is important to maintain focus to avoid undermining improvement efforts.

All temperature reduction experiences should be utilised; this means taking advantage of the lessons learned by forerunners and the knowledge transfer that occurs within the district heating community.

Low-temperature systems contribute to the reduction of greenhouse gas emissions. Targeted policy instruments and effective subsidies enable accelerated transformation. It is particularly important to raise awareness and sensitise political decision-makers to the necessity of low-temperature systems, especially given current energy policy frameworks hardly address their importance.

Darmstadt applied study

In the **fifth** chapter, the applied study of temperature reduction in the district heating network at TU Darmstadt's Campus Lichtwiese serves as a showcase of an existing district heating system and is equally applicable to many other district heating systems. The study shows that operational errors within the building heating infrastructure lead to considerable increases in the network temperatures and that problems in a few buildings can have a significant impact on the entire network. In addition, addressing the most critical issues helps to reduce network temperatures considerably, especially on the return side.

The study also reveals a major barrier: as long as the heat generation in a district heating system is realised via CHP plants and boilers, reduced network temperatures will not immediately improve cost because many benefits presented in Chapter 2 do not apply to a fossil-based energy system. Additionally, renewable heat sources, such as geothermal, solar thermal or local waste heat, are low-temperature heat sources, which are neither economically nor energetically feasible in high-temperature district heating systems. In a high-temperature district heating network, low-temperature renewable heat can only be integrated using a heat pump at low efficiencies, resulting in high electric energy demands. For an effective transition from fossil-based to renewable district heating, a transition from high-temperature district heating to low-temperature district heating is first necessary.

Competitiveness of low-temperature district heating

Increased operational efficiency from lower system temperatures and optimised technical configurations for low-temperature heat distribution are frequently discussed. Less discussed are the competitive advantages of low-temperature solutions. Therefore, the **sixth** chapter addresses how a stand-alone low-temperature district heating solution, or a combination of conventional district heating and low-temperature solutions can increase the overall competitiveness of the district heating business case. First, an overall business model perspective and a national viewpoint are presented. The discussion is concluded with a more detailed analysis of the heat distribution cost in the low-temperature district heating context.

From an overall perspective, traits in low-temperature district heating business models can be complementary to the conventional district heating model. The selling point of a combination for an existing district heating system or of a stand-alone solution in greenfield investments is that local resources are used, minimising the carbon footprint. In an era of increased digitalisation, engaging in dialogue and establishing long term relationships are valuable, an upside for the low-temperature district heating prosumer relationship.

The market maturity of low-temperature district heating is low, and as such, an emphasis is placed on ensuring functional, technical solutions rather than a simultaneous development of the business case. For future installations, tandem development is recommended.

Retaining heat distribution costs in district heating systems at feasible levels is vital to maintain competitiveness. The most significant component of the heat distribution cost is the specific capital cost that is higher in low heat density areas. Second is the cost of the heat distribution loss. But only the latter cost can be considerably reduced by low-temperature heat distribution, since only the ability to use plastic pipes can reduce the capital cost for LTDH.

Furthermore, LTDH should be able to be supplied (input at heating plants) with heat from low-temperature heat sources, which are expected to yield a lower heat generation cost. In low heat density areas, lower heat generation costs and lower heat distribution losses obtained by LTDH cannot completely compensate higher specific capital costs. Hence, it is impossible to increase the total competitiveness of district heating with LTDH in low heat density areas.

Practical implementation of low-temperature district heating

The introduction and application of new concepts and technologies, such as low-temperature district heating, often face concerns of feasibility and reliability. The earlier chapters discuss measures to be taken on a building and system level and show the various technical and economic benefits of low-temperature district heating. The **seventh** chapter highlights that many innovative low-temperature district heating systems have already been built and operate successfully.

Information from implemented, planned, proposed, and simulated cases have been digested from many early initiatives for lower temperatures in buildings and heat distribution networks. An overview of the locations for these initiatives are provided in Figure 75.

★ Cases analysed in detail and presented in Chapter 7
● Cases analysed in detail in this project, see Section 10.2
○ Cases described in the gross list of low-temperature initiatives, see Section 10.3

Figure 75. Locations of the regarded demonstrator cases.
© Fraunhofer IEE, own representation. Map taken from Eurostat: https://ec.europa.eu/eurostat

Particularly from the displayed case studies, the following main conclusions can be drawn:

- From a technical point of view, the large variety of system configurations shows the flexibility in the implementation and realisation of low-temperature district heating systems. For the operation, a sufficient monitoring and management system secures the success of the project. For the integration of multiple heat sources and more complicated systems, a wider digitalisation of the processes is needed.
- The regulatory boundary conditions are not always beneficial. So, for example, the integration

of geothermal heat requires a long (or too long) approval process, which can potentially derail the implementation. Furthermore, real cross-sectoral energy systems are not foreseen with today's rules, which makes a realisation complicated.
- The cases clearly show that a high connection rate and support from the customer could be gained when the system is owned by the municipality or a cooperative.
- From a business point of view with the above-mentioned ownership issue, interest rates might be lower, and long payback times are manageable. Some cases show that a transition to low-temperature district heating systems is economically feasible; some cases indicate the price level for the heat supply could be up to 10% lower compared to a conventional solution, even without accounting for future damage cost from global warming.

These conclusions prove that low-temperature district heating is a market-ready heat supply technology that can operate under various boundary conditions. Furthermore, case experiences show the need for digitalisation measures to secure a successful operation under the new boundary conditions, such as the integration of fluctuating renewable or waste heat sources or changed network (bidirectional) operation.

Transition strategies

The essence of conversion, transformation and transition is that all humans are capable of profound change. Required changes in our communities can be initiated, communicated, and implemented by three steps – (1) visions, (2) strategies and (3) planning measures. Changes to our energy system should also be identified at all levels in our global community. Although changes are necessary and inevitable in all areas, they are often accompanied by concerns. The **eighth** chapter shows how local transition strategies have addressed these apprehensions in some urban areas.

Adopted visions, strategies, and planning measures from five urban areas are presented when the heat distribution temperature issue is properly identified within the decarbonisation context. The three steps are vital to implementing district heating and cooling systems based on renewable or recycled heat or cold in every urban area. Additional university examples are briefly provided since some universities are forerunners in low-temperature district heating.

The major conclusions concerning visions, strategies, and planning measures within the local transition strategies are the following:

- Lower distribution temperatures are necessary in transition strategies for the decarbonisation of district heating systems.
- Cooperation with research organisations should be considered when new technologies are implemented in local district heating systems.
- University campuses are often forerunners with new technologies for heat distribution.

Complementary detailed information

The **tenth** chapter contains complementary detailed information on low-temperature district heating on the following:

- Various heat distribution configurations for lower heat-distribution temperatures
- A net list of completed descriptions of demonstration cases performed within this project
- A gross list of 165 low-temperature inspiration initiatives from Europe and North America that have been identified within this project
- Locations of all identified low-temperature initiatives that are mentioned in this guidebook
- An overview of the organisations participating in this project and corresponding dissemination activities

Conclusions from this guidebook

Conclusions about technological development, non-technical aspects, and policy implications, along with recommendations, are summarised in the **ninth** chapter.

Tangible proper technologies and methods are available for the implementation of low-temperature district heating. Early adopters have tested and implemented lower temperatures in existing and new heat distribution networks. Building owners can and should adopt the technology now for the utilisation of lower temperatures in the future. Reductions of specific heat demands will also facilitate lower temperatures. However, current technologies and methods can be further elaborated and refined by research and development.

The primary non-technical barrier to undertaking a low-temperature district heating investment is the resistance to change. One major factor that can explain

the limited interest in future-proof low-temperature district heating technology is that the risk of limited heat supply in 2050, since current fossil fuels will not be available, has not yet become apparent for most end users and heat providers.

The economic benefit of low-temperature district heating can reduce the levelized cost of heat from future district heating systems, but the savings in current systems is limited. Hence, this advantage is not now strong enough alone to encourage a transition towards more decarbonised district heating systems. Carbon pricing or other efficient policy drivers must be used as strong parallel economic drivers for incentivising decarbonisation. In addition, old institutional rules require proper revision for better alignment with low-temperature district heating.

Low-temperature district heating is easier to implement than many people fear, but adequate organisation is required. In the transition work, long term visions express the future direction for the decarbonisation, short and long term strategies identify what to do, and short and long term planning measures outline the steps to take.

The three main conclusions from this guidebook are the following:

- Low-temperature district heating together with expected national carbon pricing schemes are major economic drivers for the decarbonisation of the European district heating systems.
- The implementations of low-temperature district heating are possible since several early adopters have provided clear evidence for its suitability.
- However, the hurdles to start the transition are old habits and lock-in effects from application of current technology together with a lack of understanding of how to efficiently link stakeholders to each other.

DEFINITIONS OF EXPRESSIONS AND TERMS

The subject of low-temperature district heating has introduced new expressions and terms into the traditional district heating vocabulary. Within this guidebook, the following expressions and terms are used:

Expression	Definition
Bidirectional connection	Connection of a customer building that encompasses heat delivery to and heat supply from the customer with a combination of supply-to-return and return-to-supply connections
Bypass flow	Synonym for circulation flow
CHC configuration	Network configuration that is based on combined heating and cooling (see Section 10.1)
Circulation flow	Flow that bypasses a substation to maintain the supply temperature when no heat delivery appears in the substation (bypass and shortcut flows are synonyms)
Classic configuration	The traditional network configuration that uses one supply pipe and one return pipe by blending delivery and circulation flows into the return pipe (see Section 10.1)
Cold district heating	District heating based on a cold network
Cold network	Heat distribution network that in general require additional heating in customer substations since ultra-low supply temperatures are used
Combined heat and power	Synergy that utilises the excess heat from thermal power generation processes for heating purposes
Combined heating and cooling	Synergy that utilises the excess heat from cooling processes for heating purposes
Cost reduction gradient	Key performance indicator showing the reduction in heat delivery cost when applying lower heat distribution temperatures, expressed as cost reduction per heat delivered and per reduction of the temperature level
Delivery flow	Flow that passes through substations
Heat delivery	Heat delivered to customers in substations
Heat distribution loss	Heat lost from the distribution network, defined as the difference between heat supply and heat delivery
Heat recycling	General term for recovery of heat to be finally reused before released to the ambient temperature
Heat supply	Heat supplied into heat distribution networks
High supply temperature	Supply temperature beyond approximately 100 °C
High-temperature district heating	District heating based on high supply temperatures
Low supply temperature	Supply temperature below approximately 70 °C
Low-temperature district heating	District heating based on low supply temperatures
Medium supply temperature	Supply temperature between approximately 70 °C and 100 °C
Medium-temperature district heating	District heating based on medium supply temperatures
Modified classic configuration	Modification of the classic configuration by the introduction of a third pipe designed to accommodate circulation flow (see Section 10.1)
Multi-level configuration	Network configuration that provides more than one supply temperature (see Section 10.1)
Network configuration	General term for option to organise and manage the heat distribution network
Prosumer	Customer that can also supply heat into the heat distribution network

Expression	Definition
Return-to-return connection	Substation connection that adds heat to or subtracts heat from a return pipe
Return-to-supply connection	Substation connection that adds heat to a flow from the return pipe and delivers it to the supply pipe
Shortcut flow	Synonym for circulation flow
Substation	Installed unit that transfers and measures the delivered heat from the heat distribution network to a customer, or vice versa in the prosumer case
Supply-to-return connection	Traditional substation connection for heat delivery to a customer by the subtraction of heat from the heat distribution network
Supply-to-supply connection	Substation connection that adds heat to or subtracts heat from a supply pipe
Temperature level	The annual average temperature for supply and return flows at a heat supply plant
Total flow	The sum of delivery and circulation flows
Ultra-low configuration	Network configuration that uses ultra-low supply temperatures (see Section 10.1)
Ultra-low supply temperature	Supply temperature below approximately 50 °C
Warm district heating	District heating based on a warm network
Warm network	Heat distribution network with no additional heating in substations, except for customers with unusually high-temperature demands

Abbreviations

Within this guidebook, the following abbreviations are used:

1GDH	First generation of district heating
2GDH	Second generation of district heating
3GDH	Third generation of district heating
4GDH	Fourth generation of district heating
CDH	Cold district heating
CHC	Combined heating and cooling
CHP	Combined heat and power
COP	Coefficient of performance
CRG	Cost reduction gradient
DHC	District heating and cooling
DHC+	The European platform for district heating research
ETC-CPC	Evacuated tube collectors with compound parabolic concentrators
EU	European Union
EUDP	The Energy Technology Development and Demonstration Program performed by the Danish Energy Agency
FPC	Flat plate collectors
HVAC	Heating, ventilation, and air-conditioning
IEA	International Energy Agency
IEA-ECBCS	IEA technology collaboration programme called Energy Conservation in Buildings and Community Systems, renamed in 2013 to Energy in Buildings and Communities programme (IEA-EBC)
LTDH	Low-temperature district heating
RHC-ETIP	Renewable Heating & Cooling – the European Technology and Innovation Platform
TU	Technical University
ULTDH	Ultra-low temperature district heating
WDH	Warm district heating

1 PATH TO FUTURE DISTRICT HEATING

Author: Sven Werner, Halmstad University

In many urban areas, district heating systems move heat through pipes from available heat sources to buildings and other processes that require heat. Major heat sources include recycled heat from various societal processes that have considerable amounts of residual heat (e.g. thermal power plants). Some heat is also obtained from renewable energy sources. However, fossil fuels continue to be the primary energy source in varying proportions. In most systems, the direct use of fossil fuels in boilers only provides heat for peak and redundancy purposes as a complement to the major heat sources.

Heat distribution pipes should be short to limit heat losses and distribution costs. Pipes that are too long are not competitive with local heat generation units, such as heat pumps, biomass boilers, or electric boilers. Short pipes appear in areas with high heat demand within a certain land area (also known as high heat density areas), while longer pipes are required in areas with low heat densities.

District heating systems increase the energy efficiency of our global energy system because recycled heat can replace a large proportion of the primary energy supply. By using district heating systems, considerably smaller carbon footprints can be obtained since the primary energy supply for buildings in Europe is currently dominated by fossil fuels.

District cooling systems have a similar function to district heating systems. However, instead of delivering heat to buildings, these systems remove heat from buildings to obtain comfortable indoor climates during hot and warm days.

Future district heating systems will have market conditions that differ from those of current systems. The high customer heat demands will be reduced. On the supply side, renewables and heat recycling will replace current heat recycling from processes based on fossil primary energy sources. Hence, future systems will use enhanced district heating technology to accomplish decarbonisation. A vital component of this enhanced technology is the use of lower heat distribution temperatures, which will bring higher efficiency to the heat supply.

The following eight sections of this introduction will provide the overall context for low-temperature district heating (LTDH), the core subject of this guidebook:

1. Past, present and future of district heating
2. The main drivers for the four generations of district heating
3. A retrospective analysis of district heating within the fossil energy society
4. The transition with respect to future market conditions
5. The future of district heating within the renewable energy society
6. The economics of LTDH
7. A summary of previous works on LTDH
8. The major conclusions of this introduction

These eight steps are followed by an introduction to the content in each of the guidebook chapters.

1.1 Past, present, and future of district heating

Commercial district heating systems have been in operation for more than a century. One small medieval French district heating system (Chaudes-Aigues) has even been delivering geothermal heat since the 14th century. Currently, approximately 13% of all heat demand for buildings within the European Union (EU) is met by district heating, while the corresponding global proportion is approximately 8% (Werner, 2017). However, a major challenge for commercial district heating systems is that future buildings will require less heat than current buildings (Mathiesen et al., 2019).

Commercial district cooling systems have been in operation since the 1960s. These exist in the USA, Middle East, Europe, China, Japan and Singapore. Within the EU, district cooling systems supply just over 1% of all cooling delivery (Werner, 2016). However, since the total cooling delivery from all cold sources is only 6% of the total heat delivery from all heat sources, cooling delivery from European district cooling systems only corresponds to 2% of total heat delivery from district heating systems. However, the latter proportion is expected to increase with the increasing cooling demand (IEA, 2018), while heat demand is expected to decrease.

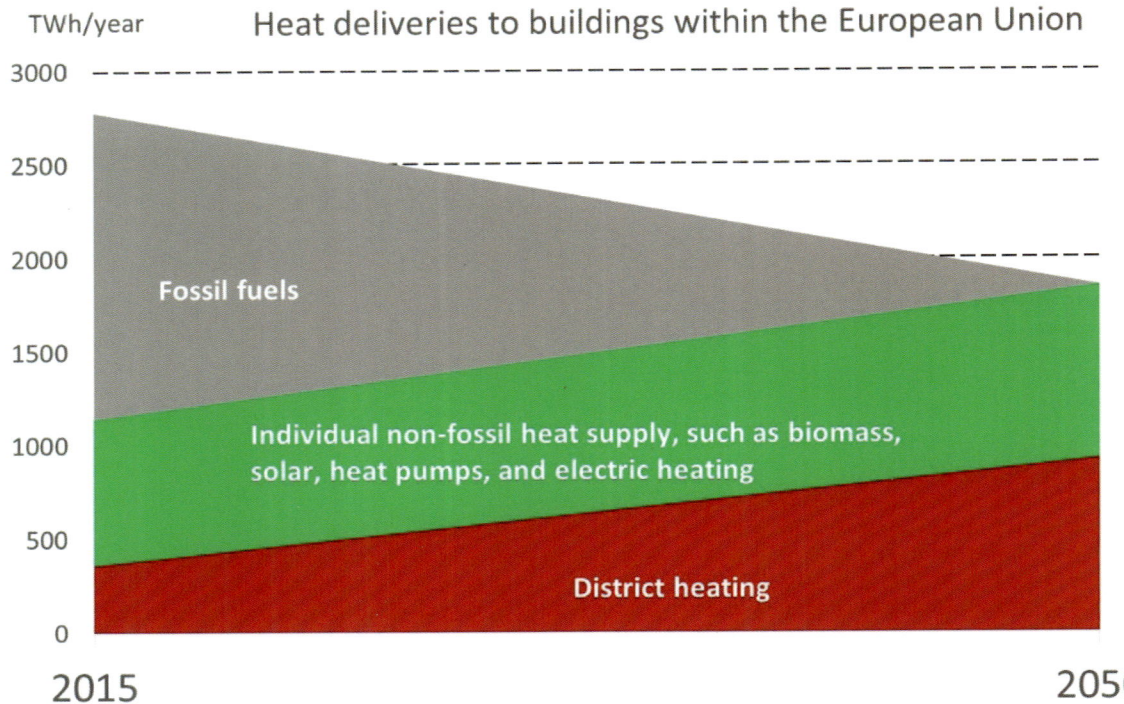

Figure 1. Possible transition from the current heat supply (expressed with the origins of the supply) to buildings within the EU being fully decarbonised by 2050, according to the Heat Roadmap Europe cluster project.

According to the final report from the Heat Roadmap Europe cluster project (Mathiesen et al., 2019), district heating and cooling can play an important role in the EU's goal for the reduced use of fossil fuels for heating buildings. A summary of this cluster project with respect to building heat delivery within the EU is presented in Figure 1.

Based on the information presented in Figure 1, approximately one-third of the current heat demand can be eliminated by energy efficiency measures within buildings. The current proportion of individual non-fossil heat supply (<30%) could increase to approximately 50% by 2050. Hence, this market share could nearly double. The corresponding possible development of district heating could also increase from 13% to approximately 50% by 2050, resulting in a market share roughly four times higher. The stronger development of decarbonised district heating compared to individual solutions can be explained by the large volumes of fossil fuels used in dense urban areas where district heating is more competitive. Thus, the use of fossil fuels for heating in European buildings can potentially be eliminated by 2050.

1.2 Four generations of district heating

Currently, district heating technologies refer to four generations reflecting different periods (of approximately 40 years) based on the best available technology at that time, as shown in Figure 2. Notably, these generations are defined by various characteristic parameters, such as the temperature level of the heat distribution. Additional parameters include various supply, distribution, and delivery options.

In the USA, the first commercial district heating systems were established in the 1880s based on the steam distribution results obtained by Birdsill Holly in his pioneering Lockport experiments in 1876–1877. This first-generation district heating (1GDH) technology was recognised as the best available technology between 1890 and 1930. Today, steam technology is considered outdated for district heating. However, steam distribution is still used in two major district heating systems: the Manhattan system in New York and the city-wide system in Paris. These two urban areas are extremely dense, resulting in highly favourable conditions for low heat distribution costs. Paris has the best conditions for efficient heat distribution in the entire EU (Persson & Werner, 2011). Hence, both systems can still afford to use the first generation of outdated district heating technology as part of their business activities.

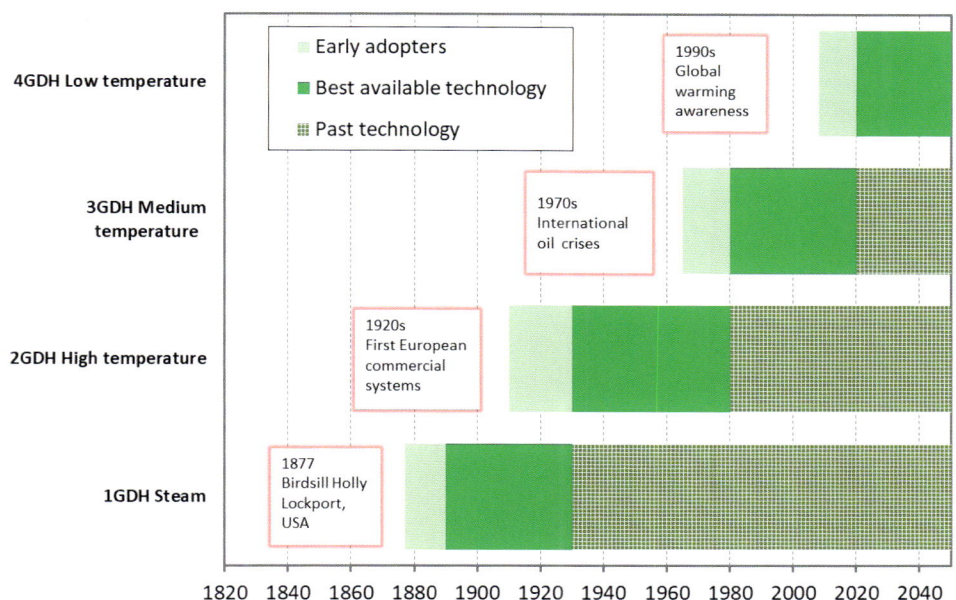

Figure 2. Overview of the time intervals defining the four generations of district heating systems with temperature level as one typical characteristic parameter.

In Europe, the first commercial district heating systems were introduced in Germany during the 1920s, using the 1GDH technology of steam distribution. However, several German engineers questioned early the choice of steam as a heat carrier and advocated for water as a heat carrier to increase system efficiency. These engineers became early adopters of second-generation district heating (2GDH) when their new ideas were implemented into new district heating systems. However, rather high supply temperatures were applied (above 100 °C), with a large difference between supply and return temperatures due to the use of distribution pipes with small diameters. The 2GDH technology was also applied in the USSR when district heating was introduced in the 1930s and expanded in the 1950s. The Russian experiences and methods were later transferred to China, where district heating systems were first introduced in the 1950s and 1960s. This 2GDH technology was recognised as the best available technology between 1930 and 1980.

Two international oil crises in the 1970s created greater interest in Europe for the use of district heating systems as a general tool to reduce dependence on imported fuel oil for heating, particularly in Denmark, Sweden, and Finland. Engineers in these three Nordic countries advocated lower supply temperatures (below 100 °C) to improve system efficiency. Simultaneously, other productivity gains were obtained using pre-insulated pipes with prefabricated substations. This third-generation district heating (3GDH) technology is recognised as the best available technology since about 1980 and is currently utilised in all European district heating systems. 3GDH technology is also used in Russia and China for the expansion of existing systems.

The first three generations of district heating have one common denominator: all heat was supplied into the distribution network from various supply plants. The supply temperature in distribution networks was high enough to satisfy local heat demands. Therefore, no additional heat had to be supplied at the customer level to meet customer temperature demands. Thus, this concept of district heating contained supply guarantees for both heat delivery and available capacity. These system solutions can be labelled as 'warm district heating' (WDH).

However, the heat distribution in these three generations was never an absolute monoculture. Several variants appeared between countries and regions. Various technical choices in Berlin, Moscow, Stockholm, Odense and Beijing created diversity within each technology generation.

The emerging awareness of global warming in the 1990s – with the creation of United Nations Framework Convention on Climate Change in 1992 and the corresponding Kyoto protocol in 1997 – created a renewed interest in district heating systems as a tool for replacing fossil fuels with renewables and various low-temperature heat sources.

At the 'District Energy Futures' workshop in Reykjavik on August 29, 2008, the expression 'fourth-generation district heating' (4GDH) was introduced. The workshop was hosted by the IEA-DHC technology collaboration programme in cooperation with Annex 49 of the then IEA-ECBCS collaboration programme. The purpose of this new expression was to distinguish the various ways of moving heat in cities when traditional high-temperature heat distribution was discussed along with new low-temperature heat distribution.

Early adopters of low temperatures in warm district heating systems became the engineers that designed several pilot solar district heating systems in Sweden, Denmark, Germany, and Austria. These experiences were aggregated into the Marstal system in Denmark when major seasonal heat storage was also introduced into a European town-wide district heating system. The development of the Marstal system was supported by two Sunstore projects, which were financed by European framework research programmes.

Cold district heating (CDH) systems can be used as a complement to warm district heating systems. In these cases, an additional decentralised heat supply is required to cover typical customer temperature demands. Central heat sources can then be distributed with ultra-low supply temperatures in the distribution networks. The heat supply uses local temperature boosters such as boilers or heat pumps to meet temperature demands. Beyond the label 'cold district heating' in English, these hybrid system solutions using both central and local heat supply are called 'kalte Fernwärme' in German, 'kold fjernvarme' in Danish, 'kall fjärrvärme' in Swedish, 'Anergienetze' in Swiss German and 'boucle d'eau tempérée' in French. Early adopters of cold networks appeared before 2010 in Switzerland, Germany, Italy, Norway and the Netherlands (Buffa, Cozzini, D'Antoni, Baratieri, & Fedrizzi, 2019).

Hence, the CDH label includes all district heating system solutions requiring an additional local heat supply to satisfy individual customer temperature demands. Cold district heating systems can be an alternative to traditional warm district heating systems when suitable excess heat sources with the required temperatures are not available in the neighbourhood. By applying low supply and return temperatures in a variant of the cold district heating network, the same temperatures used in district cooling systems can be obtained, allowing both heating and cooling to be supplied from the same distribution network. This synergy is called combined heating and cooling (CHC).

Based on the aforementioned points, some pros and cons of warm and cold district heating systems can be identified. Warm district heating systems are preferrable when excess heat is available at temperatures high enough to support the typical temperature demands of customer substations with no additional heat supply. Cold district heating systems can be applied when ambient heat sources with low temperatures are used as the main heat source. While heat losses will appear in warm district heating systems, they nearly vanish in cold district heating systems. Thus, investments are required for additional heat supply in customer substations with cold district heating systems, unlike warm district heating systems. More details on this subject are presented in Chapter 4.

Cold district heating systems are sometimes referred to as fifth-generation district heating (5GDH), as proposed by (Buffa et al., 2019). While this definition focuses almost exclusively on temperature level as the generation divider, the original generation divider (Figure 2) has always been the periods when each generation technology was used as the best available technology. While temperature level is an important parameter for describing each generation, it is not the main definition parameter.

> Our definition of 4GDH in this guidebook applies to all new technological features and concepts using low temperatures, which are considered best available from 2020 onward (Figure 2). As experienced in previous technology generations, a wide diversity of technology choices in 4GDH is expected. Hence, cold district heating systems are also included in our definition of 4GDH. The corresponding technology comprises all heat distribution technologies that will utilise supply temperatures below 70 °C as the annual average. 4GDH technology is a family of many different network configurations for heat distribution. Notably, cold and warm networks are siblings in this family of configurations.

1.3 District heating within the fossil energy society

District heating was primarily introduced for the more efficient use of fossil fuels. However, this driving force for district heating was weak during years with low fossil fuel prices, when the economic value of higher efficiency became low.

Higher oil prices after the two international oil crises in the 1970s created greater interest in district heating when higher efficiency increased its economic impact.

The main heat source for district heating systems became the heat loss from electricity generation with thermal power, also known as combined heat and power (CHP). However, heat losses from waste incineration and industrial processes were also used – but to a lower extent. The energy industry favoured the choice of heat from CHP since the synergy of heat recycling from CHP was retained within the energy industry. Cooperation on heat recycling from a third party required the synergy to be shared. Hence, heat recycling from various societal processes was key to district heating (economy of scope). The centralisation and scale (economy of size) of heat supply was also valuable but less economically advantageous than heat recycling.

For heat delivery, the customer heating systems required rather high temperatures to satisfy relatively high heat demand with standard radiators. Hence, high distribution temperatures in the district heating networks were necessary, but this reduced the ability to generate electricity in CHP plants, resulting in reduced electricity revenue from these plants. However, this traditional cost gradient was not high enough to implementing lower distribution temperatures in the networks.

The historical conditions for heat supply can be summarised as a high proportion of heat recovery from fossil CHP plants (almost a monoculture) and a low proportion of renewables or recycled heat from various heat-generating processes. The corresponding conditions for heat delivery included both high heat and temperature demands.

Generation temperatures were chosen based on the technology that was available when each generation was introduced. It was natural for American engineers in the 1880s to use steam distribution in 1GDH since water distribution was not possible without the availability of electricity-driven water pumps. The use of these pumps later became an option when German engineers introduced water-based heat distribution in 2GDH in the 1920s. The development towards 3GDH was facilitated by the lower cost of pre-insulated pipes and compact heat exchangers. More flow could be distributed, which could reduce the difference between supply and return temperatures in the networks.

The development of different district heating systems over several decades has resulted in many major long term district heating systems in Europe having various combinations of the different generations in their distribution networks. While Paris started with steam distribution in 1930, it expanded with secondary networks that use either 3GDH or 4GDH technologies. Munich, Geneva and Basel began with the 2GDH technology, but extensions are being performed with 3GDH technologies. Moreover, Vienna has a central 2GDH transmission network, but apply lower temperatures in more than 500 secondary 3GDH networks. The same situation will appear when new 4GDH network sections are implemented in conjunction with existing district heating systems.

1.4 Transition challenges

Global warming and the awareness of climate change call for new requirements for all parts of the existing energy system. Carbon dioxide emissions from fossil fuels must be reduced over the coming decades via the decarbonisation of the energy system. The European Commission has presented a heating and cooling strategy to reach decarbonisation in the EU (European Commission, 2016). Hence, decarbonisation is vital and expected for all new and existing district heating systems.

The direct consequence of decarbonisation for district heating systems will be that fossil CHP will not be available in the future. However, fossil CHP heat sources still dominate district heating systems in most countries. In the future, CHP plants can only be used for the combustion of biomass and waste, which will not be available to the same extent that fossil fuels are available today. Notably, this transition will change the heat sources for current district heating systems. Traditional thermal power plants will become less dominant in the electricity market and provide lower volumes of excess heat from electricity generation. Hereby, the proportion of non-combustible heat supply must be higher in the future.

In recent years, the construction industry has learned to create buildings with low heat demands. This technology has also been applied to existing buildings. Hence, heat demands are steadily decreasing. With standard sizes of heat emitting devices, lower heat demands will also provide lower temperature demands.

Lower heat demands will also increase the current heat distribution costs of district heating systems with lower heat densities. However, the heat density is high in most urban areas in Europe, which will, therefore, experience only a moderate impact of this cost increase (Persson, Wiechers, Möller, & Werner, 2019).

Typical examples of non-traditional heat supply include heat pumps, geothermal heat, solar collectors, flue gas condensation and low-temperature excess heat from industrial processes and concentrated electricity use. All these examples will benefit from lower temperatures since their cost sensitivity is higher than that obtained in traditional CHP plants.

With fossil energy, heat distribution networks were designed to adapt to the characteristics of fossil fuels, such as the ability to generate high temperatures. For renewable energy, these networks must be redesigned to adapt to the higher proportion of non-traditional heat supply.

The transition from fossil resources to renewable and recycled resources has been implemented to varying degrees in existing district heating systems. Figure 3 summarises this situation by comparing the average carbon dioxide emissions to the proportion of heat recycling and renewables in 47 countries with significant proportions of district heating systems. EU nations generally have lower carbon dioxide emissions than those in other regions. Notably, the lowest emissions are found in Iceland, Sweden and Norway. Figure 3 indicates that many countries have started on the transition by decarbonising their district heating systems, while some have not.

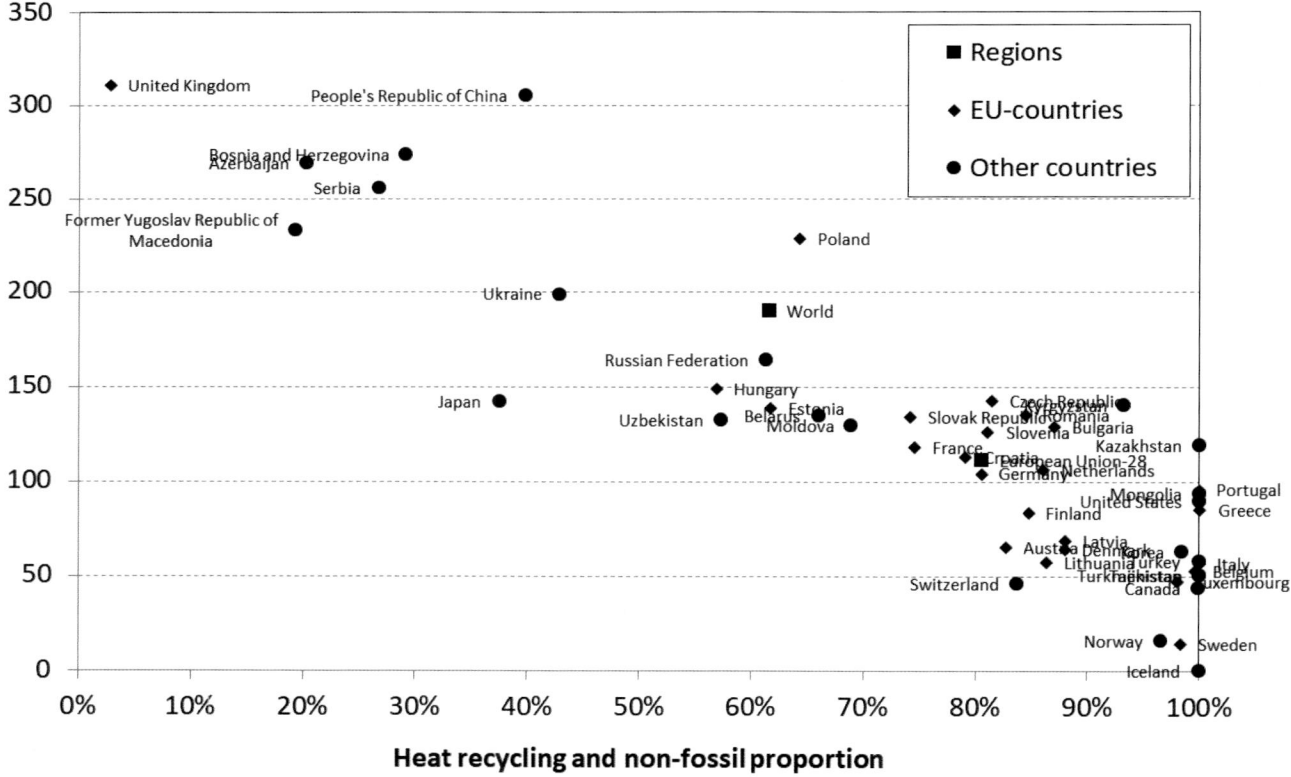

Figure 3. Estimated specific carbon dioxide emissions during 2014 from all district heating systems in 24 EU countries and 23 other countries. Source: (Werner, 2017)

1.5 District heating within the renewable energy society

The transition challenge for district heating systems needs to be addressed by meeting the new conditions for both heat delivery and heat supply. Heat delivery should meet lower heat and temperature demands in buildings, while heat supply will use higher proportions of solar, geothermal and biomass heat together with higher proportions of recycled heat from excess heat from both heat-generating processes and electricity use. Interaction with the electrical market will provide excess wind power for electric boilers and heat pumps, demand will remain for electricity from biomass and waste CHP. However, the total proportion of heat from CHP will be lower since traditional thermal power plants will be less dominant in the electricity market, resulting in lower volumes of excess heat from electricity generation.

The use of warm district heating systems will rely on heat sources that can provide temperatures high enough to meet customer temperature demands. These heat sources can be either heat recycling from excess heat or a primary energy supply from renewable resources, such as high-temperature solar collectors and geothermal heat.

The use of cold district heating systems will allow the use of district heating systems when no suitable warm excess heat is available in a neighbourhood. Central heat supply in these systems includes all heat sources that are warmer than ambient water, air or ground since the ambient temperature is the true reference temperature for all heating systems. Typical low-temperature heat sources in cold district heating systems can be low-temperature solar collectors; low-temperature excess heat from industrial, cooling, and sewage plants; ambient heat, such as ground heat; and as sea, lake, ground, and mine waters.

For new systems, the choice between warm or cold networks depends on several parameters. If the temperature of the heat source is high enough for warm district heating, a warm network is preferable. If an ambient heat source will be used, the choice depends on the heat density for the area to be served and the cost difference between large and small heat pumps. A cold network is better when the heat density is low since warm networks have high heat losses at this condition. Large heat pumps normally have lower specific investment costs than small heat pumps. However, small heat pumps manufactured in highly automated manufacturing processes can have low specific investment costs. The distribution cost will be somewhat higher for cold networks since wider pipes are required and more electricity is needed for the circulation of the higher water flow. On the other hand, the heat input for cold systems is almost cost-free since ambient and low-temperature heat can be used. In warm networks, the heat input often requires some form of payment.

Since the basic future heat sources will differ from today's sources, a different distribution of the economic drivers for district heating can be expected. The synergy of centralisation and scale (economy of size) may be more prominent, especially concerning the use of renewables in combination with heat storage. However, heat recycling (economy of scope) will remain significant.

Notably, district heating technology must be redesigned to meet the new conditions for heat delivery and supply. The general conditions for this redesign and the corresponding five expected abilities of 4GDH have been defined by (Lund et al., 2014):

1. To supply LTDH for space heating and hot water preparation
2. To distribute heat with low grid losses
3. To recycle heat from low-temperature sources
4. To integrate thermal grids into a smart energy system
5. To ensure suitable planning, cost and motivation structures

To obtain these abilities under future market conditions, low-temperature heat distribution is vital for 4GDH, as described in Figure 4, and is the main focus of this guidebook.

When redesigning district heating technology to use lower temperatures in new systems, the conditions for redesign are easier than those of existing district heating systems. Conditions will also vary depending on the age of the connected buildings.

Regarding existing district heating systems, some older connected buildings and their higher heat and temperature demands represent barriers to the general implementation of lower heat distribution temperatures. Lowering the temperature demands inside buildings is

The context for fourth generation district heating:

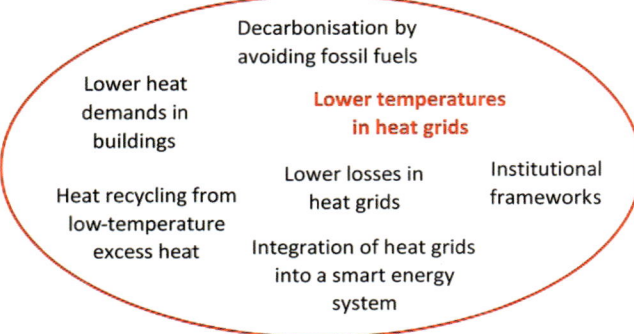

Figure 4. Abilities of fourth-generation district heating within the context of future market conditions.

possible at low or moderate costs. However, new buildings connected within these systems can and should be directly designed for future lower heat distribution temperatures.

The degree of freedom is higher for new district heating systems than for existing systems. However, the possibility of using lower temperatures in these systems can be hampered if old buildings with higher heat and temperature demands are connected. The combination of new systems and new buildings present a situation in which the best available technology for low-temperature heat distribution can be applied.

1.6 Economics of low-temperature heat distribution

The first step in the transition from fossil fuels to renewables and heat recycling is the internalisation of future damage costs from climate change into the prices of fossil fuels, a method called carbon pricing. The internalisation can be managed by adding carbon dioxide taxes to the market prices of fossil fuels. A cap-and-trade system can also be applied for emitted greenhouse gases. This approach has been taken in the European trading system since 2005 for carbon dioxide emitted from the largest emitters within the EU.

The second step, the primary interest in this guidebook, is the optimisation of the heat supply in district heating systems by implementing lower heat carrier temperatures in the heat distribution networks. In this step, the introduction of renewables and heat pumps becomes more profitable. Likewise, direct heat recycling from low-temperature sources without heat pumps also becomes more profitable.

For the first step, carbon pricing should be at a level that corresponds to the expected damage cost of using fossil fuels. According to the Stern Review (Stern, 2007), the future annual damage cost of climate change can be estimated to be between 5 and 20% of the annual global gross domestic product. Hence, proper carbon pricing should be between 75 and 300 USD per ton of emitted carbon dioxide equivalent. This estimation is based on a current global gross domestic product of 80–85 trillion USD per year and current annual greenhouse gas emissions of 55 billion ton of carbon dioxide equivalent. According to the World Bank Carbon Pricing Dashboard (World Bank, 2020), only three countries (Sweden, Switzerland and Finland) apply domestic carbon dioxide taxes within or close to this required carbon pricing interval. To date, the prices within the European Trading System for carbon dioxide emissions have also been considerably lower than the interval of proper carbon pricing for fossil fuel use estimated above. Hence, many countries are not really incentivised to replace fossil fuels to mitigate climate change. Effective incentives are needed such that the current energy system will reduce the emission of carbon dioxide from fossil fuels soon.

However, many European countries will likely soon apply higher domestic carbon dioxide taxes as an efficient tool for achieving the new revised EU 2030 reduction target of at least 55% for the EU carbon dioxide emissions. This emission reduction path is illustrated in Figure 5 together with the previous 2020 path and the pre-revision 2014 path for 2030. This new reduction target increases the required rate of reduction by more than five times compared to the previous 2020 target (as shown by the considerably steeper negative slope for the red line in Figure 5. Hence, the lower 2030 target will support the decarbonisation of the European district heating systems.

In the second transition step concerning low-temperature heat distribution, the economic benefits of renewables and recycled heat are based on the following nine major efficiency gains obtained at lower distribution temperatures:

1. More geothermal heat extracted from wells with temperatures between 60 °C and 100 °C since lower-temperature geothermal fluid can be returned to the ground.
2. Less electricity used in heat pumps when extrac-

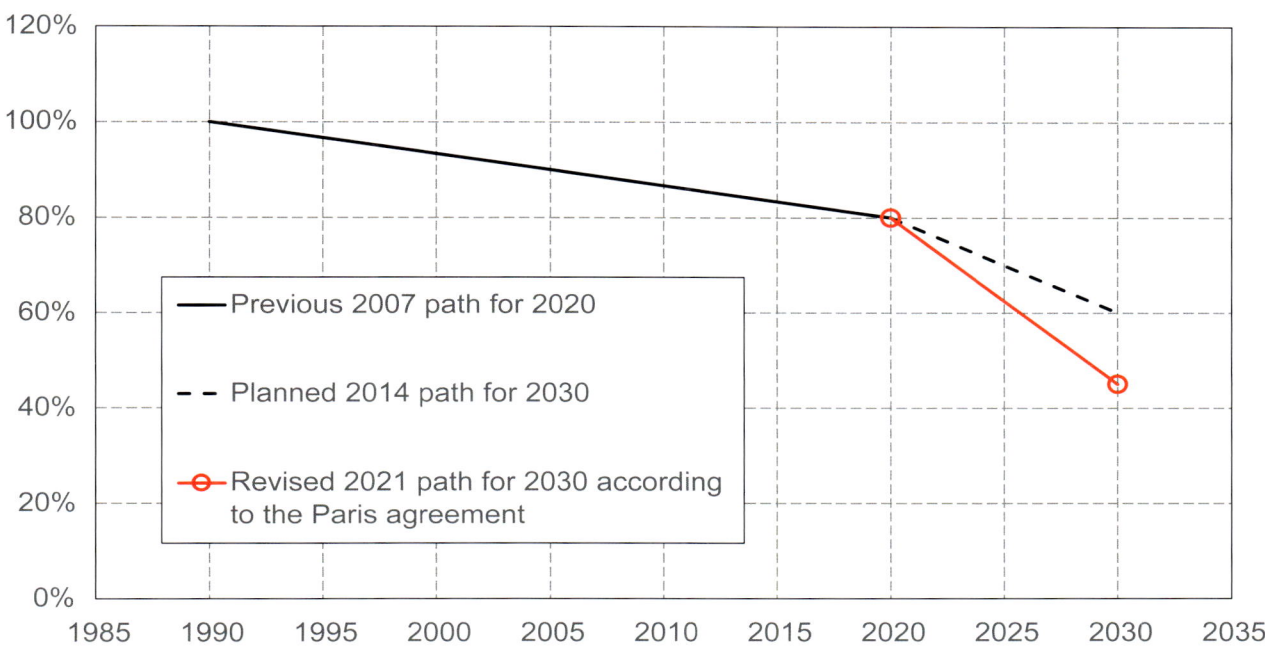

Figure 5. Three paths for the reduction of the EU carbon dioxide emissions through 2030.

ting heat from heat sources with temperatures below the heat distribution temperatures since lower pressures can be applied in the heat pump condensers.
3. More excess heat extracted from heat sources with temperatures between 60 °C and 100 °C since lower temperatures of the excess heat carrier will be emitted to the environment.
4. More heat obtained from solar collectors since their heat losses are lower, thereby providing higher conversion efficiencies.
5. More heat recovered from flue gas condensation since the proportion of vaporised water (steam) in the emitted flue gases can be reduced.
6. More electricity generated per unit of heat recycled from steam CHP plants since higher power-to-heat ratios are obtained with lower steam pressures in the turbine condensers.
7. Higher heat storage capacities since lower return temperatures can be used in conjunction with high-temperature outputs from high-temperature heat sources.
8. Lower heat distribution losses with lower average temperature differences between the fluids in heat distribution pipes and the environment.
9. Ability to use plastic pipes instead of steel pipes to save cost.

Additional benefits include a reduced risk of low-cycle fatigue for steel pipes (due to lower variation in supply temperatures); smaller temperature drop in the flow direction, allowing for lower supply temperatures from heat supply plants; and a lower risk of scalding during pipe maintenance (lethal accidents have occurred in high-temperature systems).

The status paper for 4GDH (Lund et al., 2018) concluded that these efficiency gains make up the primary driver for low-temperature heat distribution (summarised in Figure 6). When proper maintenance and minor additional investments are implemented in networks, substations and buildings, new low-temperature heat sources can either supply more heat at the same cost or the same amount of heat at a lower cost, owing to efficiency gains. The paper also estimated that the annual economic savings of lower temperatures for some future district heating systems were over four times higher than the estimated annual costs of the technology.

These results were further elaborated in (Averfalk & Werner, 2020), who defined and explored cost reduction gradients (CRGs) from the use of lower temperatures in heat distribution networks for various heat supply options. They concluded that the cost reduction gradients for new heat sources (e.g. heat pumps, solar collectors, geothermal heat and low-temperature excess heat) are approximately five times higher than those for traditional combustion in CHP plants. These economic benefits of LTDH make up an important part of this guidebook and are further explored in detail in Chapter 2.

1.7 Previous and current works on low-temperature heat distribution

In addition to this guidebook, other documents have been written over the past decade about the LTDH approach within the 4GDH concept and the associated technology options. Within the IEA-DHC technology collaboration programme, several research and development initiatives for LTDH have been undertaken. Early experiences with LTDH and its potential were highlighted by (Dalla Rosa et al., 2014), while the Transformation Roadmap project was based on experiences from

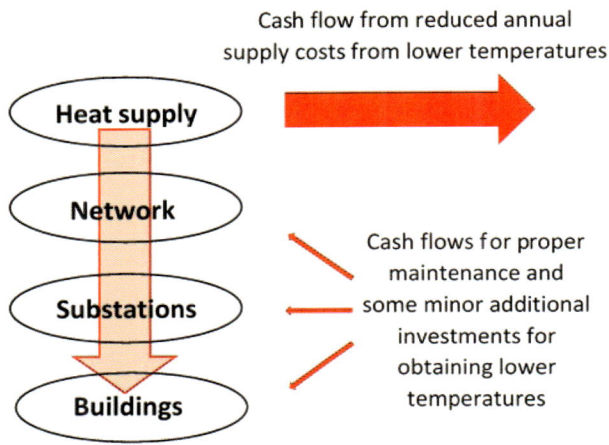

Figure 6. Economics of LTDH.

earlier generations (Averfalk et al., 2017). Another relevant IEA-DHC effort was the Low-Temperature District Heating for Future Energy Systems project. This project was reported on by (Schmidt & Kallert, 2017) and (Schmidt, 2018), who provided an early overview of pioneering LTDH implementations. Notably, participants from these three IEA-DHC projects have now merged into the group of participants creating this guidebook.

In Europe, DHC+ (the European platform for district heating research) presented a strategic research agenda in 2012 (Euroheat & Power, 2012) that outlined visions for 2020 and 2050. The goals of the EU on almost zero-carbon energy solutions by 2050 were shared by the DHC+, while the 4GDH/LTDH concept was addressed as a solution based on lower and more flexible distribution temperatures, assembly-oriented components, and flexible materials. The low-temperature concept was later confirmed in the new strategic research and innovation agenda from RHC-ETIP (Renewable Heating & Cooling – the European Technology and Innovation Platform) (RHC, 2020).

Several research projects within the European framework research and other programmes have been focusing on the development of low-temperature heat distribution. In earlier framework programmes, the Sunstore2 and Sunstore4 projects developed solar district heating systems in Denmark. Moreover, the Horizon 2020 programme contained LTDH-oriented projects, such as Storm, Flexynets, SDHp2m, Reuseheat, Cool DH, TEMPO, Related and Rewardheat. The regional Interreg programmes financed LTDH projects, such as Heatnet NWE, D2Grids and Lowtemp. Projects from other European research programmes include Life4HeatRecovery and Heatstore. The total turnover of these European research projects exceeds 100 million euro.

Major national initiatives have also been taken in Denmark, Germany, Austria, France and the Netherlands. In Denmark, the 4DH research centre was active between 2012 and 2018 with contributions from several Danish and international universities as well as numerous Danish district heating and industrial companies. The total turnover was nearly 10 million euro, with basic funding coming from Innovation Fund Denmark. Notably, researchers within this research centre wrote the 4GDH definition paper (Lund et al., 2014) and the 4GDH status paper (Lund et al., 2018). Many LTDH papers have also been published in scientific journals from the annual international 4DH conferences since 2015 (Østergaard, Lund, & Mathiesen, 2019). In Germany, the 'Wärmenetzsysteme 4.0' initiative provided 100 million euro between 2017 and 2020 for funding feasibility studies and pilot projects related to 4GDH/LTDH (Kühne, 2017). This support scheme will be extended to 800 million euro between 2021 and 2024 by the 'Bundesförderung für effiziente Wärmenetze' (Vollbrecht & Miller, 2020). In Austria, the ThermaFlex flagship project will use 8+ million euro between 2019 and 2022 to explore higher flexibility in new and extended district heating systems (Epp, 2019); (AEE INTEC, 2020). In France, the national heat fund can support the introduction of cold district heating systems. In the Netherlands, the WarmingUP project cluster organises several projects concerning LTDH.

1.8 Major conclusions from this introduction

The following four conclusions from this chapter are essential as conditions for the guidebook:

1. The decarbonisation of energy systems will eliminate heat recovery from fossil CHP plants that currently dominate the heat supply to district heating systems. The proportion of non-combustible and low-temperature heat supply will then increase in future district heating systems since future levels of available biomass and waste combustion will not be enough to replace the current use of fossil fuels.
2. The presence of high temperatures in existing district heating systems is an initial barrier to the implementation of many low-temperature heat sources.
3. The introduction of low-temperature heat distribution with supply temperatures below 70 °C will increase the profitability of implementing geothermal heat, heat pumps, industrial excess heat, solar collectors, flue gas condensation and heat storage into district heating systems.
4. Low-temperature heat distribution will be a significant economic driver to the expected extended use of higher carbon prices for obtaining decarbonisation within the EU because of the stronger 2030 reduction target for carbon dioxide emissions.

1.9 Organisation of this guidebook

The purpose of LTDH system development is to establish technology options that can be harmonised for European conditions to support the expansion of district heating in European countries that have little district heating presence. New 4GDH/LTDH technologies are expected to serve the same role that 3GDH had in expanding district heating among Nordic countries in the 1970s and 1980s.

In the following second chapter, the main economic benefits of lower temperatures in heat distribution networks are elaborated and summarised. This chapter is very important, since it contains the main economic arguments for LTDH.

The third chapter summarises the experiences of reduced temperatures inside buildings since the technical ability for buildings to cool down the heat distribution networks is the basic condition for obtaining LTDH. The presenting order of this information is from existing buildings to new buildings.

The fourth chapter describes the corresponding experiences in existing and new district heating networks for obtaining lower temperatures with the expected future heat sources. The presenting order of this information is from existing networks to new networks. Hence, temperature reductions in buildings and networks must be achieved to obtain the expected benefits of LTDH.

The technical possibilities in buildings and systems are further presented in the fifth chapter through examples and an in-depth applied analysis of the Lichtwiese campus system belonging to Technical University of Darmstadt in Germany. The purpose of this chapter is to provide evidence for the findings from the two preceding chapters about buildings and networks.

In the sixth chapter, the keywords are business models and competitiveness for LTDH. The initial essential research question is, 'Do district heating operators need to develop their business models when LTDH systems are implemented?' The competitiveness of LTDH is also explored by a benchmarking study of early LTDH projects concerning essential heat distribution characteristics and investment costs for both cold and warm heat distribution networks.

In the seventh chapter, early experiences from several pilot and demonstration projects performed in Europe are presented in the context of low-temperature heat distribution for existing and new buildings and networks. These brave early adopters can inform others of what is possible with reduced temperatures in district heating networks. These examples are also sorted from existing to new buildings and from existing to new networks.

Some examples of sustainable transition initiatives with proper LTDH are presented in the eighth chapter. The overview contains the visions, implementation strategies and planning measures for five urban areas in Europe. Moreover, some university campus systems are also presented as forerunners and early adopters of low-temperature heat distribution. Finally, conclusions about technological development, non-technical aspects, policy implications and recommendations are summarised in the ninth chapter.

In the tenth chapter, complementary detailed information is presented on various heat distribution configurations, a net list of the completed descriptions of demonstration cases, a gross list of identified low-temperature initiatives, locations of all identified LTDH initiatives appearing in this guidebook, and an overview of the organisations participating in this project.

Each chapter has its own reference list of literature sources at the end of the chapter, such that the references are located closer to the relevant chapters and a long reference list at the end of the guidebook is avoided.

1.10 Literature references in Chapter 1

AEE INTEC. (2020). THERMAFLEX - Thermal demand and supply as flexible elements of future sustainable energy systems. Retrieved from https://www.aee-intec.at/thermaflex-thermal-demand-and-supply-as-flexible-elements-of-future-sustainable-energy-systems-p238

Averfalk, H., & Werner, S. (2020). Economic benefits of fourth generation district heating. Energy, 193, 116727. doi:https://doi.org/10.1016/j.energy.2019.116727

Averfalk, H., Werner, S., Felsmann, C., Rühling, K., Wiltshire, R., Svendsen, S., . . . Quiquerez, L. (2017). Transformation Roadmap from High to Low Temperature District Heating Systems. IEA-DHC TC annex 11. Retrieved from https://www.iea-dhc.org/fileadmin/documents/Annex_XI/IEA-DHC-Annex_XI_Transformation_Roadmap_Final_Report_April_30-2017.pdf

Buffa, S., Cozzini, M., D'Antoni, M., Baratieri, M., & Fedrizzi, R. (2019). 5th generation district heating and cooling systems: A review of existing cases in Europe. Renewable and Sustainable Energy Reviews, 104, 504-522. doi:https://doi.org/10.1016/j.rser.2018.12.059

Dalla Rosa, A., Li, H., Svendsen, S., Werner, S., Persson, U., Rühling, K., . . . Bevilaqua, C. (2014). Toward 4th Generation District Heating: Experience and Potential of Low-Temperature District Heating. IEA-DHC Annex X report. Retrieved from https://www.iea-dhc.org/fileadmin/documents/Annex_X/IEA_Annex_X_Final_Report_2014_-_Toward_4th_Generation_District_Heating.pdf

Epp, B. (2019). Therma-Flex: Intelligente Wärmeversorgung der Zukunft. Euroheat & Power, 48(7-8), 47-48.

Euroheat & Power. (2012). Strategic Research Agenda - District Heating and Cooling. Retrieved from https://www.euroheat.org/publications/brochures/district-heating-cooling-strategic-research-agenda/

European Commission. (2016). An EU strategy on Heating and Cooling. Communication COM(2016)51. Retrieved from https://eur-lex.europa.eu/legal-content/EN/TXT/PDF/?uri=CELEX:52016DC0051&from=EN.

IEA. (2018). The Future of Cooling - Opportunities for energy-efficient air conditioning. International Energy Agency, Paris. Retrieved from https://www.iea.org/reports/the-future-of-cooling

Kühne, J. (2017). Fördervorhaben Wärmenetzsysteme 4.0. Euroheat & Power, 46(11), 14. Lund, H., Werner, S., Wiltshire, R., Svendsen, S., Thorsen, J. E., Hvelplund, F., & Mathiesen, B. V. (2014). 4th Generation District Heating (4GDH): Integrating smart thermal grids into future sustainable energy systems. Energy, 68, 1-11. doi:http://dx.doi.org/10.1016/j.energy.2014.02.089

Lund, H., Østergaard, P. A., Chang, M., Werner, S., Svendsen, S., Sorknæs, P., . . . Möller, B. (2018). The status of 4th generation district heating: Research and results. Energy, 164, 147-159. doi:https://doi.org/10.1016/j.energy.2018.08.206

Mathiesen, B. V., Bertelsen, N., Schneider, N. C. A., García, L. S., Paardekooper, S., Thellufsen, J. Z., & Djørup, S. R. (2019). Towards a decarbonised heating and cooling sector in Europe - Unlocking the potential of energy efficiency and district energy. Aalborg University. Retrieved from https://vbn.aau.dk/ws/portalfiles/portal/316535596/Towards_a_decarbonised_H_C_sector_in_EU_Final_Report.pdf

Persson, U., & Werner, S. (2011). Heat distribution and the future competitiveness of district heating. Applied Energy, 88(3), 568-576. doi:http://dx.doi.org/10.1016/j.apenergy.2010.09.020

Persson, U., Wiechers, E., Möller, B., & Werner, S. (2019). Heat Roadmap Europe: Heat distribution costs. Energy, 176, 604-622. doi:https://doi.org/10.1016/j.energy.2019.03.189

RHC. (2020). Strategic Research and Innovation Agenda for Climate-Neutral Heating and Cooling in Europe. Renewable Heating & Cooling - the European Technology and Innovation Platform. Retrieved from https://www.rhc-platform.org/content/uploads/2020/10/RHC-ETIP-SRIA-2020-WEB.pdf

Schmidt, D. (2018). Low Temperature District Heating for Future Energy Systems. Energy Procedia, 149, 595-604. doi:https://doi.org/10.1016/j.egypro.2018.08.224

Schmidt, D., & Kallert, A. (2017). Annex TS1 - Low Temperature District Heating for Future Energy Systems. IEA-DHC technology collaboration programme. Retrieved from https://www.iea-dhc.org/fileadmin/documents/Annex_TS1/IEA_DHC_Annex_TS1_Final_Report.pdf

Stern, N. (2007). The Economics of Climate Change - The Stern Review: Cambridge University Press.

Werner, S. (2016). European space cooling demands. Energy, 110, 148-156. doi:http://dx.doi.org/10.1016/j.energy.2015.11.028

Werner, S. (2017). International review of district heating and cooling. Energy, 137, 617-631. doi:https://doi.org/10.1016/j.energy.2017.04.045

Vollbrecht, K., & Miller, J. (2020). Konzept für Bundesförderung für effiziente Wärmenetze (Federal funding for efficient heating networks). Euroheat & Power, 49(10), 10-11.

World Bank. (2020). Carbon Pricing Dashboard. Retrieved from https://carbonpricingdashboard.worldbank.org/map_data

Østergaard, P. A., Lund, H., & Mathiesen, B. V. (2019). Developments in 4th generation district heating. International Journal of Sustainable Energy Planning and Management, 20, 1-4. doi:10.5278/ijsepm.2019.20.1

2 ECONOMIC BENEFITS OF LOW-TEMPERATURE DISTRICT HEATING

Authors: Roman Geyer, AIT; Sven Werner, Halmstad University; Christian Engel, Austroflex; Isabelle Best, Uni Kassel; Harald Schrammel, AEE INTEC; Karl Ponweiser, TU Wien

This chapter probes the motivating factors based on increased efficiency and capacity gains resulting from lower heat distribution temperatures, a major tool for lower cost decarbonisation.

Regarding heat generation, the chapter focuses on the most promising technologies for future district heating networks, as well as describing effects on heat storages and distribution networks. Accordingly, this second chapter provides an analysis beyond heat distribution loss, which constitute a single minor aspect. Section 2.11 summarises the economic benefits of lower system temperatures highlighted in Figure 7.

Regarding heat supply, two fundamentally different cases exist for cost reductions. Lower heat supply costs can be obtained by either lowering investment costs for new heat supply plants or lowering the operation costs of existing district heating systems.

In the case of lowering investment costs for a new system, a smaller new supply plant operating at a lower temperature can supply the same heat output as a somewhat larger plant operating at a higher temperature. This produces lower annuity for the lower investment cost, indicating that the new plant itself is responsible for reduced costs.

In the case of lowering the operation costs of an existing system, lower temperatures increase an existing supply plant's efficiency, enabling it to increase its heat output and, thus, replace the heat supply of a more expensive plant. This produces lower annual operation costs across the entire district heating system; that is, cost reduction is obtained outside of the newly installed more efficient plant.

To quantify these cost reductions, a key performance indicator is the 'cost reduction gradient' (CRG). Defined graphically in Figure 8, CRG is calculated by dividing the reduction of the levelized cost of heat (LCOH) for a supply technology or overall system by the achieved temperature reduction. The difference between a higher LCOH (reference case) and a lower LCOH (assessment case) is the LCOH Benefit. In the assessment case, a lower LCOH is the result of lower temperature levels. The CRG describes the economic benefits in terms of reduced cost per degree Celsius temperature reduction and per MWh of the reference heat volume. The higher the CRG value in euro/(MWh·°C), the greater the cost reduction sensitivity for the heat supply technology or the district heating system.

Given temperature reduction should consider reductions in the heat distribution network's supply and return temperatures, it should be expressed as the

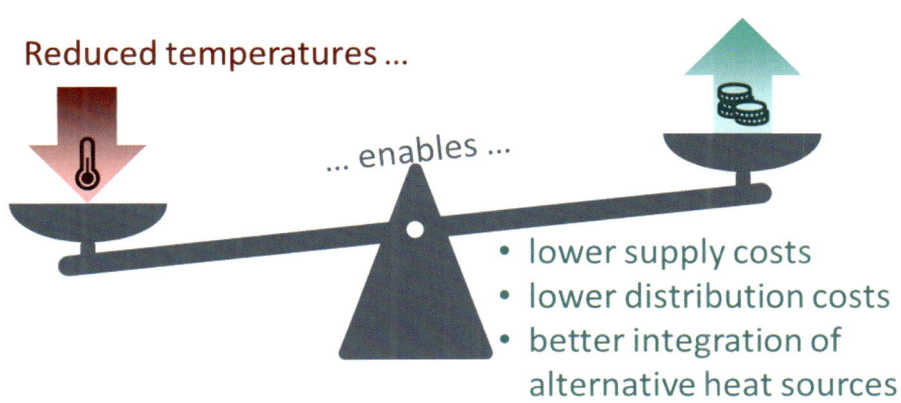

Figure 7. Overview of positive effects on reduced system temperatures and the scope of the assessments.

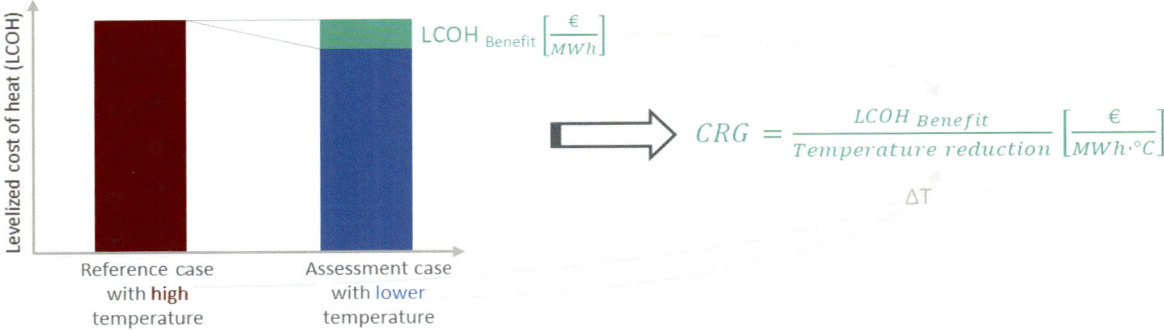

Figure 8. Concept of the energy-economic assessments with the key performance indicator 'cost reduction gradient' (CRG) in euro/(MWh·°C) (Geyer, 2020)

change in the annual average network temperature. This temperature reduction can be initially achieved through lower return temperatures, enabling lowering supply temperatures by maintaining the greatest possible flow in the network. This strategy is important because the efficiency in many heat supply plants is associated with lower supply temperatures. However, economic benefits are sometimes associated with lower return temperatures. Such cases are further discussed when they arise.

The reference heat volume in district heating systems refers to the heat delivered to customers, a system's ultimate service. When considering only a certain heat supply technology, that technology's heat output should be adjusted for heat distribution loss to facilitate comparisons with CRG values obtained by other technologies or systems.

For all of the heat supply technologies listed in this chapter, energy-economic assessments for reference and assessment cases have been performed. Both systems and separate heat supply technologies have been assessed. Thermodynamic models have been used to investigate capacity and efficiency improvements. Based on the models, simulation runs were applied for different temperature levels. For reference plants, cost data were derived from extant literature, with indicative cost estimations from manufacturers used to calculate the LCOH determined for both reference and assessment cases. A default heat distribution heat loss of 10% has been used to estimate CRG values for separate heat supply technologies.

It is possible to obtain CRG estimations by assessing either a specific heat supply technology or an entire district heating system. When a specific heat supply technology is assessed, a particular example load case is examined with different temperature levels, an approach used by (Geyer, 2020). When assessing an entire system, operations during one year are simulated, and the aggregated economic benefits from all temperature-dependent supply plants were estimated together, an approach used by various researchers (Dahlberg & Werner, 1997; Frederiksen & Werner, 2013; Castro Flores et al., 2017; Lund et al., 2018; Eriksson, 2020; Sorknæs et al., 2020; Müller et al., 2020; Averfalk & Werner, 2020).

Although a heat pump can always be used to utilise a low-temperature heat source, this will be both an investment cost and an operation cost, corresponding to the electricity the heat pump uses. Lower heat distribution temperatures can enable direct use of a heat exchanger, resulting in a major cost reduction. However, this chapter does not estimate CRG values for this kind of cost reduction.

This chapter's CRG values allow a first annual cost reduction estimate to be obtained for any district heating system using the same heat supply technology. This cost reduction can be obtained as the product of the estimated CRG value, the proportion of heat delivered to customers and the projected temperature reduction.

2.1 More geothermal heat extracted

To achieve the greatest possible thermal output and, thus, an efficient and cost-effective heat supply from geothermal heat, return temperature must be as low as possible. If available, deep geothermal heat is always a suitable source for district heating systems because it is independent of seasonal variations. For hydrothermal applications, there must be a heat exchanger between the geothermal and district heating cycles. This heat exchanger should be designed with a substanti-

> More heat extracted from geothermal wells because lower temperature geothermal fluid can be returned to the ground.

al thermal length to reduce the temperature gap between the district heating network and the geothermal source as much as possible. Depending on the temperature of the water from the geothermal well, there are several possibilities for connecting the geothermal system to the district heating network. The best-case scenario involves the temperature of the water from the geothermal wells being above the district heating network's supply temperature; this enables direct use of geothermal energy to raise the district heating water from the return to the supply temperature.

Riem Case Study: The importance of a low return temperature is demonstrated by the Riem geothermal plant (Munich, DE), as presented in Figure 9. The system's design was based on a planned district heating network return temperature of 45 °C. With a district heating network supply temperature of 90 °C, the geothermal plant can provide a thermal output of 13.7 MW (≙ 100%). However, the district heating network provides return temperatures of 58 °C. Given the deviation from the target value, the achievable output is reduced to 9.6 MW, a performance reduction of 30%.

This is due to the lower temperature difference between supply and return flow: the temperature spread is 32 °C instead of the prescribed 45 °C (SWM, 2015).

Given benefits are associated with lower return temperature, which enables the supply temperature from the geothermal well to be maintained, a network's annual temperature reduction is equal to the return temperature reduction.

For an investment case featuring unchanged heat output, (Geyer, 2020) estimated an annual cost reduction of 140 000 euro for an annual output of 41.8 GWh with geothermal heat of 90 °C and a return temperature reduction of 5 °C (50 to 45 °C). Assuming 10% heat distribution heat loss, the CRG is 0.74 euro/(MWh·°C). A corresponding CRG estimation by (Averfalk & Werner, 2020) indicated 0.45 euro/(MWh·°C) for a system dominated by 80 °C geothermal heat.

Meanwhile, for one existing case, (Averfalk & Werner, 2020) estimated 0.67 euro/(MWh·°C) for a system dominated by 80 °C geothermal heat; this calculation incorporates the benefit of decreased heat distribution loss. In a similar scenario featuring the same geothermal temperature, Müller et al. (2020) calculated CRG to be 0.68 euro/(MWh·°C). In both of the studies cited, the CRG increased for lower geothermal temperatures and decreased for higher temperatures.

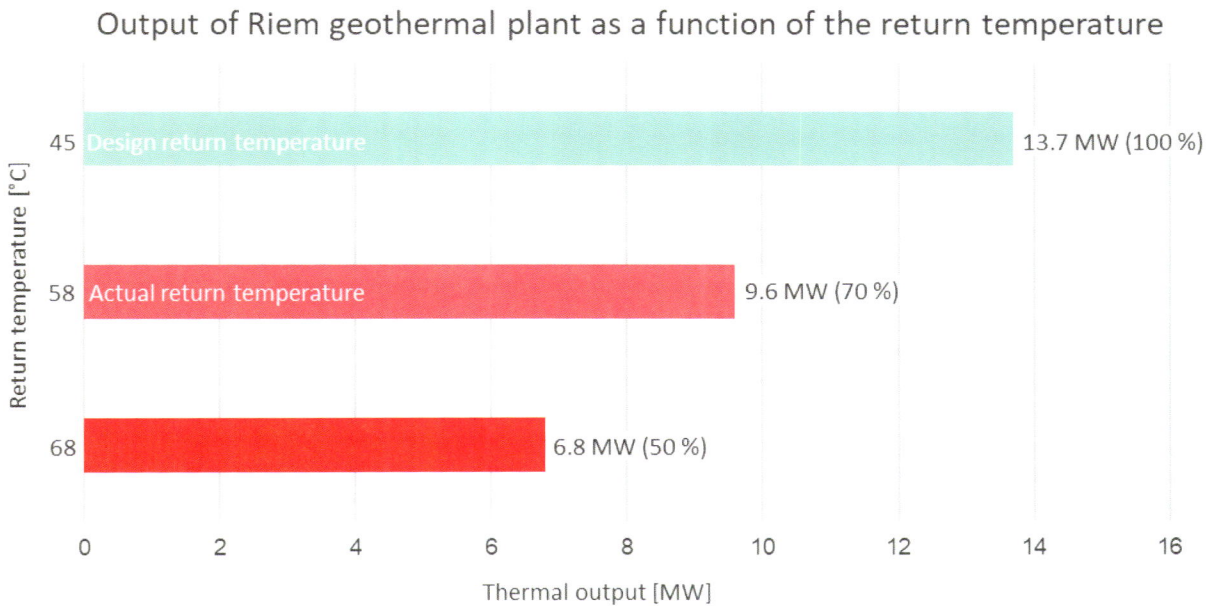

Figure 9. Importance of return temperature demonstrated by the Riem (Munich, DE) geothermal plant (adapted from (SWM, 2015))

Concerning geothermal heat:
Two investment cases have produced CRG estimates of 0.45 and 0.74 euro/(MWh·°C).
Two existing cases have produced CRG estimates of 0.67 and 0.68 euro/(MWh·°C).

2.2 Less electricity used in heat pumps

Heat pumps are already used widely within heat distribution networks and using different heat sources. A survey indicated that large-scale heat pumps featuring thermal output above 1 500 MW are operating in European district heating networks, most commonly using sewage water, ambient water (seas, lakes and rivers) and industrial waste heat as heat sources (David et al., 2017).

The coefficient of performance (COP) can be estimated based on the temperature levels (source and sink) using a general approach. The performance depends strongly on the low and high temperature levels (Equation 1). The theoretically maximum achievable performance coefficient 'COP$_{Carnot}$' of a heat pump is limited by the reciprocal of the Carnot efficiency (Equation 2). Here, the absolute temperature (in kelvins) must be used. The 2nd law efficiency for heat pumps (η_{2nd}) is formulated using the ratio of practical performance (COP) to ideal performance (COP$_{Carnot}$) at the level of temperature used. In practice, heat pumps can achieve 2nd law efficiencies up to 0.64 (Lund et al., 2018; Averfalk & Werner, 2020). Previous research has indicated a range of achievable η_{2nd} between 0.4 and 0.6 (Arpagaus et al., 2018). For COP calculation, an efficiency of approximately 0.5 is a useful rule of thumb because measurement data from different heat pumps on a certification organisation's test bench confirm the value (Grosse et al., 2017). Performance also depends on the heat pump's operational load and temperature levels (source and sink).

$$COP_{Carnot} = \frac{1}{\eta_{Carnot}} = \frac{T_{sink}}{T_{sink} - T_{source}}$$

Equation 1 Calculation of the COPCarnot for a heat pump based on temperature levels (source and sink)

$$\eta_{2nd} = \frac{COP}{COP_{Carnot}} \rightarrow COP = COP_{Carnot} \cdot \eta_{2nd}$$

Equation 2 Calculation of the heat pump's practical achievable COP considering 2nd law efficiency

$$COP = \frac{\dot{Q}_{th}}{P_{el}}$$

Equation 3 Calculation of the COP based on heat flow output and electrical drive input

COP$_{Carnot}$	Maximum achievable COP according to Carnot efficiency law	[-]
η_{Carnot}	Carnot efficiency	[-]
T$_{sink}$	Temperature level of the sink ($\cong$ hot side)	K
T$_{source}$	Temperature level of the source ($\cong$ cold side)	K
η_{2nd}	2nd law efficiency of the heat pump (~ can be assumed to 0.5)	[-]
COP	Practical achievable COP of the heat pump	[-]
$\dot{Q}_{th}$	Heat flow output	kW$_{th}$
P$_{el}$	Electrical drive input	kW$_{el}$

Reduced sink temperatures promote higher heat pump COPs. This means less drive energy (electricity) is required for the same service (heat output), and the same amount of electricity can generate more heat.

Figure 10 shows a heat pump's achievable Carnot efficiency (COP_{Carnot}) with variable temperatures for heat source and heat sink ($\cong$ useful temperature level for the customer). The lower the temperature difference, the higher the achievable COPs. If a high sink temperature is required, using industrial waste heat rather than a conventional heat source – such as ambient heat – can achieve the necessary high COP. An assumed sink temperature of 70 °C and a source temperature of 35 °C produces a COP_{Carnot} of 10. Multiplying the Carnot efficiency by the assumed heat pump efficiency (η_{2nd}) of 0.5 produces a practical achievable COP of 5. This means that to produce 5 units of heat, 1 unit of electricity is required as drive energy (cf. Equation 3). Depending on the temperature difference between the sink and source, practically achieved COPs for compression heat pumps range from 2 to 6 (Geyer et al., 2019).

For unchanged heat output in an investment case, (Averfalk & Werner, 2020) calculated the CRG to be 0.41 euro/(MWh·°C) for a system dominated by heat pumps.

For unchanged heat output in an existing case, (Geyer, 2020) reported a case with an annual cost reduction of 48 400 euro (from lower electricity costs), indicating an estimated CRG of 0.67 euro/(MWh·°C) with an annual heat output of 8 GWh, 10% heat distribution heat loss and 10 °C temperature reduction.

For greater heat output in an existing case, (Averfalk & Werner, 2020) estimated a CRG of 0.63 euro/(MWh·°C) for a system dominated by heat pumps.

2.3 More waste heat extracted

The recovery and use of waste heat increase primary energy efficiency and reduce emissions, as well as potentially contributing to cost savings. The ability to feed recovered waste heat into a district heating network is essential for this approach, with the integration of waste heat from industrial processes having been identified as an important research area for the EU's heating and cooling strategy (European Commission, 2016) and (Schmidt et al., 2020) observing increased efforts to use both conventional and non-conventional waste

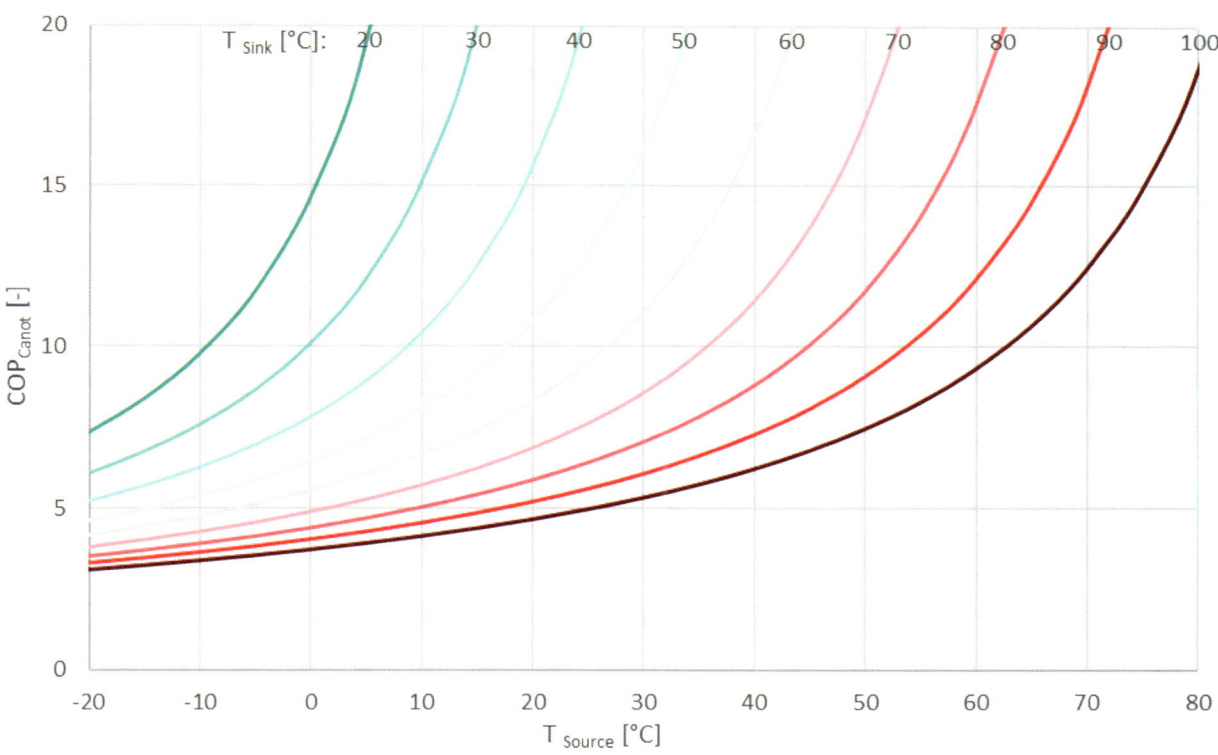

Figure 10. Characteristics of achievable COP_{Carnot} as a function of sink and source temperatures

Concerning heat pumps:
For an investment case, CRG has been estimated as 0.41 euro/(MWh·°C).
Existing cases have produced CRG estimates of 0.63 and 0.67 euro/(MWh·°C).

This section only assesses heat recycling not using heat pumps. Hence, the estimations consider cases where the waste heat sources feature temperatures above and close to the heat distribution temperatures. The economic benefits are greater for low-temperature sources and lower for waste heat sources with higher temperatures (Averfalk & Werner, 2020).

 Concerning waste heat:
The CRG for an investment case has been estimated as 0.65 euro/(MWh·°C).
For existing cases, one CRG estimate is 0.51 euro/(MWh·°C).

heat sources to further decarbonise district heating networks.

For an investment case with unchanged heat output, (Geyer, 2020) estimated an annual cost reduction of 126 000 euro for an annual heat output of 43 GWh with a temperature reduction of 5 °C. Considering 10% heat distribution heat loss, the CRG would be 0.65 euro/(MWh·°C).

For an existing case with higher heat output, (Averfalk & Werner, 2020) estimated a CRG of 0.51 euro/(MWh·°C) for a system dominated by waste heat with a temperature of 80 °C.

2.4 More heat obtained from solar collectors

The efficiency of solar thermal collectors depends considerably on their operating temperatures, with low system temperatures in heat distribution networks positively impacting solar yield. Depending on the temperature requirements, suitable collector types should be selected. A simple way to calculate efficiency is using the parameters (η0, a1, a2) shown on collector data sheets in Equation 4. These parameters are determined according to the European standard EN12975 (CEN, 2018). The optical losses are constant regardless of temperature. To minimise these losses, anti-reflective layers are used. These reduce reflection, meaning more radiation passes through the glass. Usable energy and heat loss are temperature dependent. The lower the average collector temperature, the lower the heat loss and the higher the amount of usable heat.

$$\eta_c = \eta_0 - \frac{a_1 \cdot (T_m - T_a)}{G} - \frac{a_2 \cdot (T_m - T_a)^2}{G}$$

Equation 4 Calculation of collector efficiency

η_c	Collector efficiency	[-]
η_0	Optical efficiency (achievable efficiency without heat loss)	[-]
a_1	1st order heat loss coefficient	W/(m²·K)
a_2	2nd order heat loss coefficient	W/(m²·K²)
G	Total (global) irradiance on the collector surface	W/m²
T_m	Mean collector fluid temperature (Exit–Entry)	°C
T_a	Ambient air temperature	°C

 More heat can be obtained from solar collectors at lower temperatures because heat loss from the solar collectors decreases, increasing conversion efficiency.

Compared to flat plate collectors (FPCs), evacuated tube collectors with compound parabolic concentrators (ETC-CPCs) allow for increased efficiencies at higher collector temperatures. Still, in most solar district heating plants, FPCs are used instead of ETC-CPCs because of the better price-performance ratio, a result of the specific turnkey costs for FPCs being about half the costs for ETC-CPCs (Grosse et al. 2017).

Figure 11 shows the specific solar heat output for three locations at various district heating supply temperatures for an FPC and an ETC-CPC. The specific solar heat output is the thermal energy input in the district heating network per square meter of the gross collector area. This demonstrates a significant difference between the two collector types. The ETC-CPC proved beneficial for higher-temperature applications, with solar heat output decreasing less strongly with increasing set district heating supply temperatures. In contrast, the FPC was more temperature-sensitive. All solar heating systems demonstrated maximum solar heat output at low district heating supply temperatures. At the highest district heating supply temperature of 90 °C, all solar systems showed a significantly lower heat output.

For the FPC, the solar heat output decreased by between about 34% (Bologna) and 40% (Stockholm) compared to a set district heating supply temperature of 55 °C. Meanwhile, the ETC-CPC demonstrated a more stable solar heat output, decreasing by between about 17% (Bologna) and 20% (Stockholm) compared to the set district heating supply temperature of 55 °C. These results highlight the benefits of reducing supply temperature for direct solar heat integration (Jentsch et al., 2020).

For an investment case with unchanged heat output, (Geyer, 2020) estimated annual cost reduction for FPCs to be 12 900 euro, with an annual heat output of 953 MWh and temperature reduction of 20 °C. Considering 10% heat distribution heat loss, the CRG would be 0.75 euro/(MWh·°C) for a Vienna site. Meanwhile, (Averfalk & Werner, 2020) estimated a CRG of 0.35 euro/(MWh·°C) for a Strasbourg location, as well as citing a previous estimation for a Stockholm location of 0.64 euro/(MWh·°C). This indicates higher CRGs for solar collectors at locations with lower solar irradiation and lower outdoor temperatures.

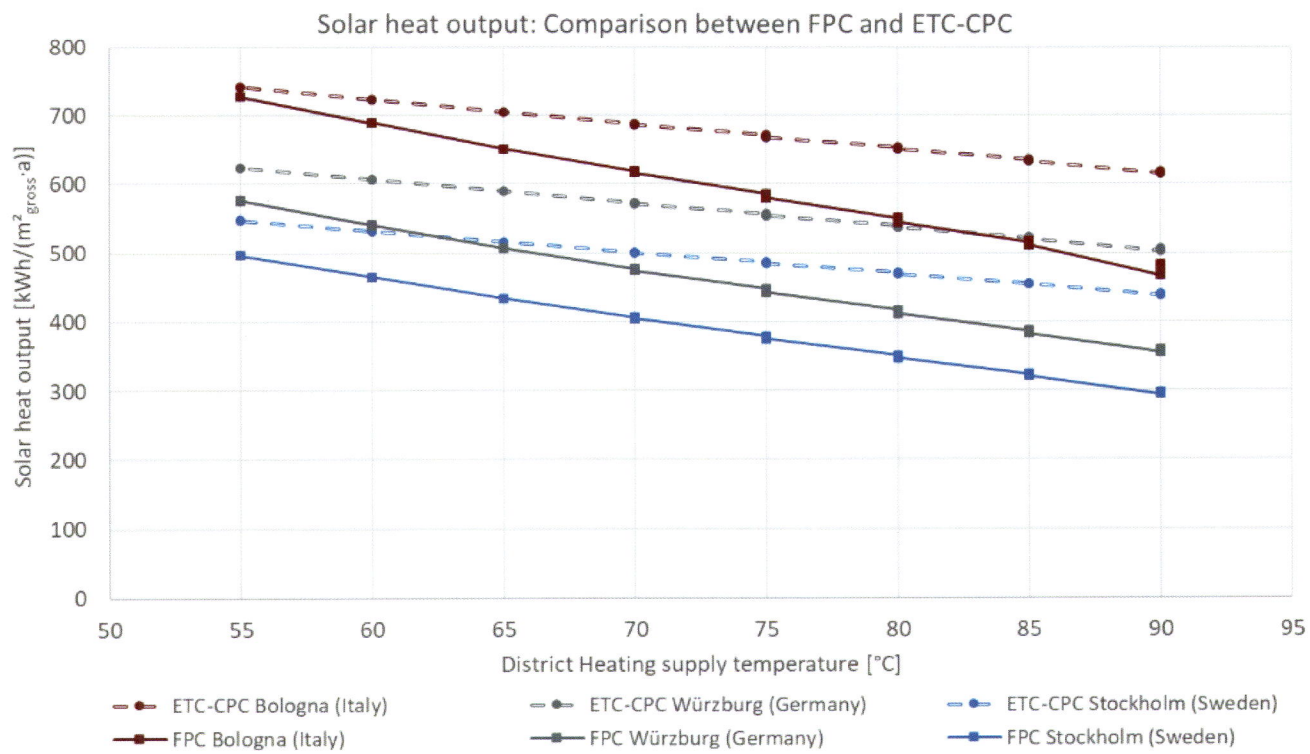

Figure 11. Comparison of specific solar heat output for flat plate collectors (FPCs) and evacuated tube collectors with compound parabolic concentrator (ETC-CPCs) at various district heating supply temperatures (direct feed-in unit, return-to-supply feed-in, with pre-heating loop, without storage) for five different locations

For flat plate collectors, two CRG estimates for investment cases have been observed: 0.35 and 0.75 euro/(MWh·°C).
For evacuated tube collectors, one CRG estimate has been observed: 0.26 euro/(MWh·°C).

For an ETC-CPC investment case, (Geyer, 2020) estimated a CRG of 0.26 euro/(MWh·°C), a product of lower temperature dependency.

No assessments of existing cases have been identified.

2.5 More heat recovered from flue gas condensation

Flue gas condensation units enable the recovery of low-temperature heat from the combustion of fuels with substantially high water content, which significantly increases total furnace efficiency and saves fuel. However, these effects can only be achieved if the return temperature provided by the district heating system is sufficiently low to cool the flue gas below the dew point (which depends on flue gas composition, water content and excess oxygen content). Figure 12 presents increased furnace efficiency according to flue gas temperatures after heat recovery. The return temperatures required to achieve such flue gas temperatures need to be 5–10 °C lower due to the terminal temperature difference in the flue gas condenser.

The loss of sensible heat results from the temperature difference between the exhaust gas and the ambient or combustion air and corresponds to the linear behaviour of Figure 12 which sees increases following increases to the excess air number (λ). The loss of unused condensation heat increases with increasing water content (w) in the exhaust gas. The latent area can be identified by a kink and its non-linear behaviour. The water dew points shown in Figure 12 range from 53 to 65 °C, with the most important parameters influencing the efficiency of the flue gas condensation being

- the temperature of the heat sink (return temperature of the district heating network),
- the water content (w) of the fuel, and
- the excess air during combustion (λ).

For modern continuously operated biomass furnaces, the optimum excess air number usually ranges from 1.4 to 1.8. The water content of solid biogenic fuels usually ranges from 30 to 50% for biomass plants operating on the MW scale. Exhaust gas quenching with water allows

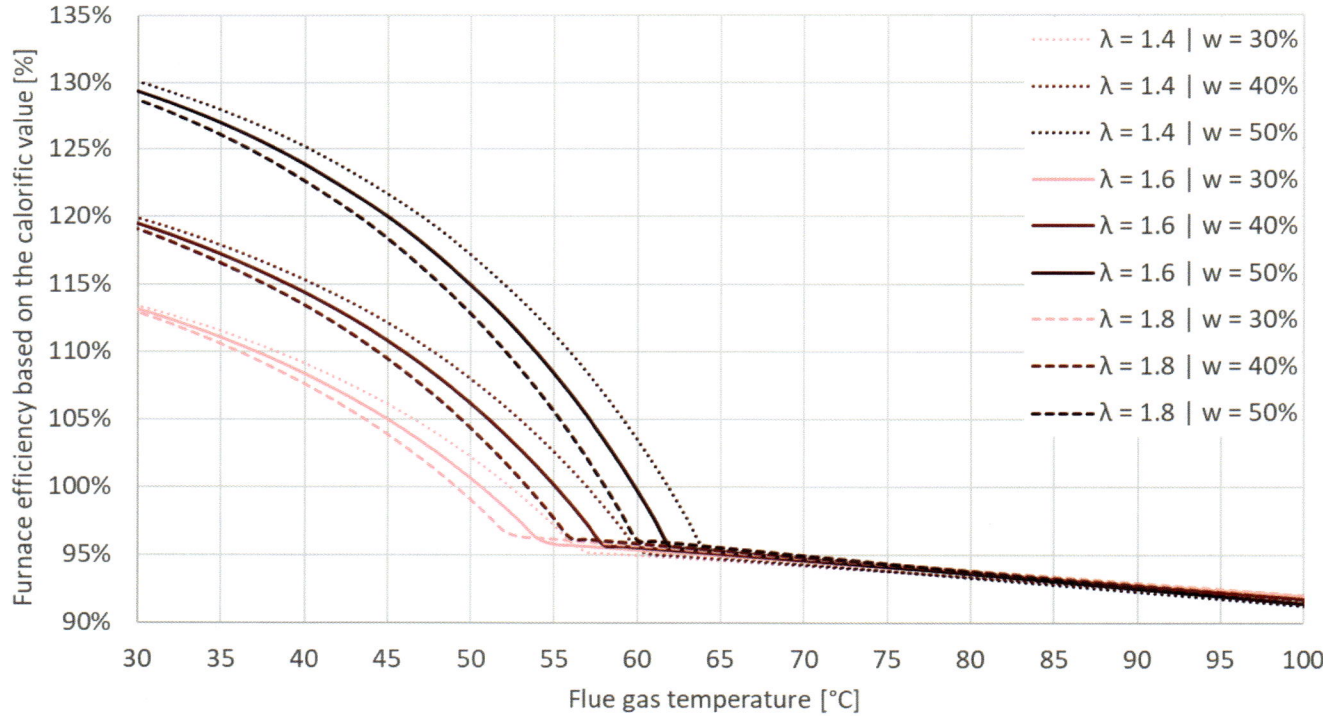

Figure 12. Furnace efficiency of a biomass boiler (according to lower calorific value) as a function of flue gas temperature using wood chips as fuel (w = water content; λ = excess air number).

 Concerning flue gas condensation:
No CRG estimates have been identified for investment cases.
For existing cases, two CRG estimates have been observed: 0.10 and 0.13 euro/(MWh·°C).

for higher dew points and greater heat recoveries (Kaltschmitt et al., 2016).

For an existing case with higher heat output, (Geyer, 2020) estimated annual cost reduction by assuming conditions of λ=1.6 and w = 40%, observing that the efficiency of a biomass boiler could then be increased from 106% to 114% by reducing the flue gas temperature from 50 to 40 °C. Given the plant's higher output, heat generation costs could be reduced to produce a CRG of 0.10 euro/(MWh·°C). Meanwhile, (Averfalk & Werner, 2020) estimated a CRG of 0.13 euro/(MWh·°C) for a system dominated by a biomass boiler using flue gas condensation.

2.6 More electricity generated in combined heat and power plants

The use of combined heat and power generation (CHP) can improve energy utilisation of fuels according to the principle of using (waste) heat generated during electricity generation for heating purposes. This better utilises primary energy compared to generating power separately in a power plant and generating heat with a boiler. An optimally designed and operated CHP plant can save up to a third of the primary energy that would be required for the separate generation of useful electrical and thermal energy (Schaumann & Schmitz, 2010).

Reduced temperature levels for heat extraction positively impact the CHP process. Based on calculations, the effects of reducing a district heating network's supply temperatures, especially in the electricity generation context, can be modelled for the two typical CHP generation configurations depicted using simplified schemes in Figure 13: back-pressure turbine (solid lines of Figure 13) and extraction-condensing turbine (solid and dotted lines of Figure 13).

Figure 14 summarises the effects of changing power-to-heat ratios as a function of steam turbine condenser outlet temperature ($\cong$ supply temperature of a district heating network plus terminal temperature difference of the heat exchanger) on a wood-chip-fired CHP with 40 MWh heat output (the horizontal grey line at 0%). The heat output is set as constant during the parameter variation; to maintain this, fuel heat input and mass flow must be variably adapted. As the figure demonstrates, a district heating network's supply temperature is critical to attainable condenser pressures. This directly affects the achievable turbine output and, thus, the efficiency of the steam power process.

Back-pressure turbines are heavily impacted, with power generation changing 0.71 %/°C. For extraction-condensing turbines, the change is less pronounced

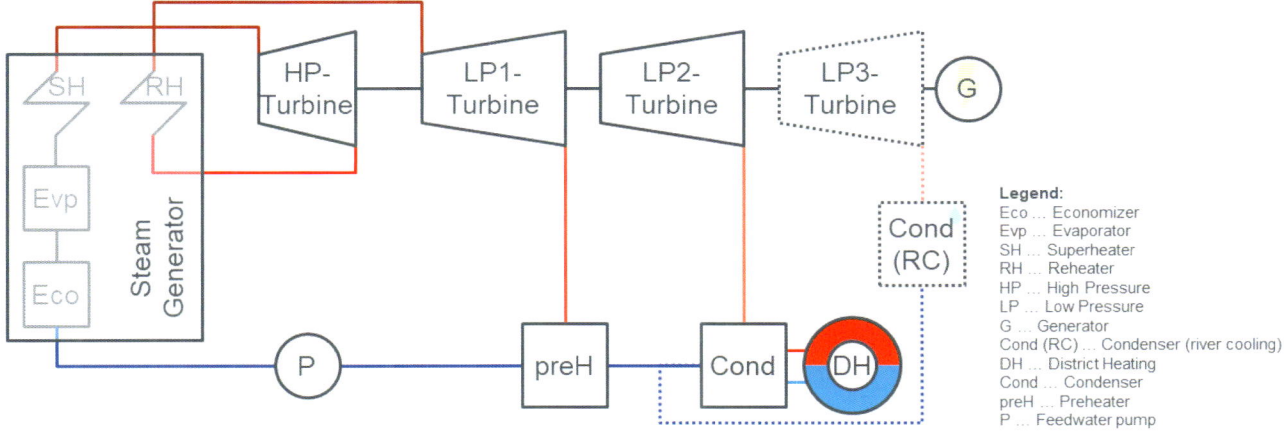

Figure 13. Visualisation of the combined heat and power generation (CHP) model and its main components. The components represented using solid lines represent the back-pressure turbine. The additional components on the right (dotted lines) represent the extraction-condensing turbine.

More electricity is generated per unit of heat in combined heat and power plants because higher power-to-heat ratios are obtained by lower steam pressures in the turbine condensers.

 Note: A CRG estimation can be conducted for part-load conditions and the designing or planning of new plants because the heat output remains constant during the simulation for comparison purposes. In full-load conditions, the steam generation in the boiler is the limiting parameter, requiring increased power generation to be compensated by decreased heat generation, which must be substituted with more expensive heat generation.

(0.30 %/°C) because steam extraction provides a higher degree of freedom. For example, it is assumed that the CHP generates 10 MW of power with the steam turbine condenser at 80 °C. According to Figure 14, reducing the supply temperature in the district heating network by 10 °C leads to power generation of 10.71 MW in a back-pressure turbine and 10.3 MW in an extraction-condensing turbine.

For CHP plants, lower heat distribution temperatures generate more electricity with the same heat output, making annual cost reduction the net benefit, a result of increased electricity revenues and increased fuel cost from increased electricity generation.

In one existing case involving a back-pressure biomass CHP plant with unchanged heat output, (Geyer, 2020) estimated a net annual cost reduction from increased electricity revenues of 274 000 euro for an annual heat output of 320 GWh and a temperature reduction of 10 °C. Considering 10% heat distribution heat loss, the CRG was estimated as 0.10 euro/(MWh·°C). Meanwhile, (Averfalk & Werner, 2020) estimated a CRG of 0.16 euro/(MWh·°C) for a system dominated by a biomass CHP plant using flue gas condensation. A similar case study by (Castro Flores et al., 2017) estimated CRG as 0.12 euro/(MWh·°C).

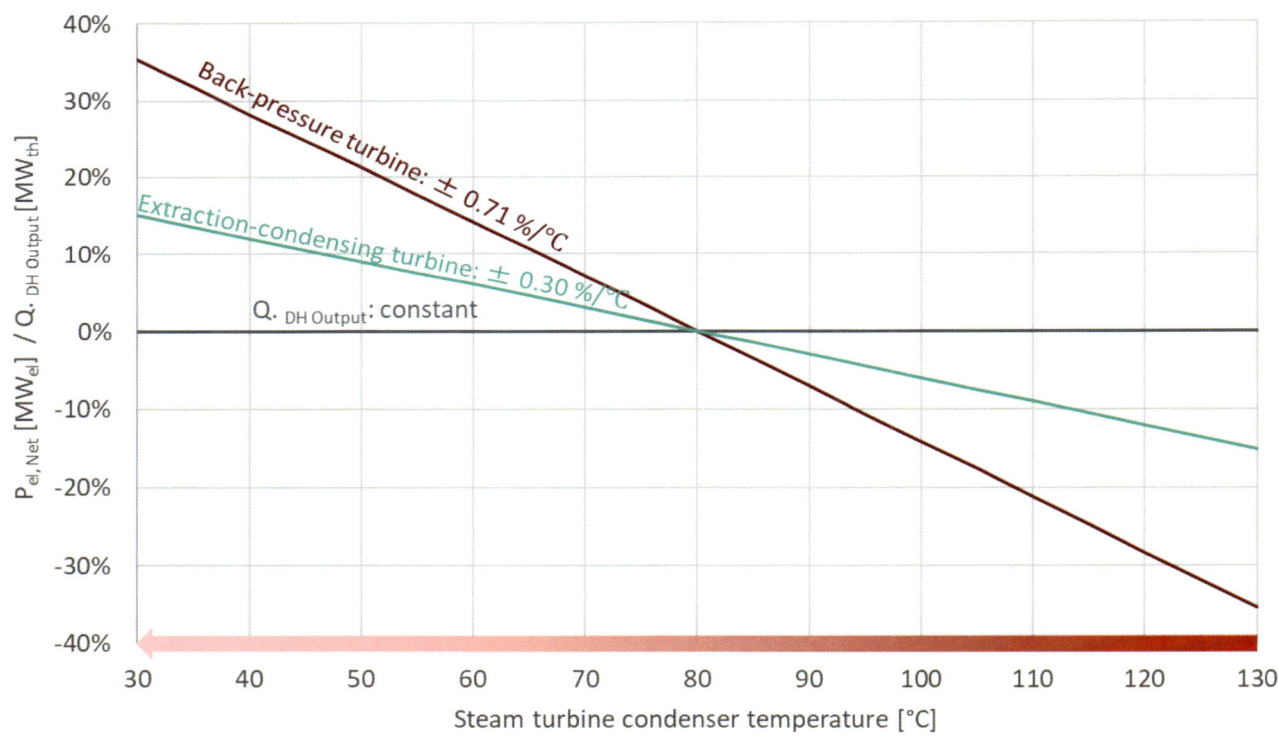

Figure 14. Change in the power output according to heating output (power-to-heat ratio) as a function of steam turbine condenser temperature (≅ supply temperature of a district heating network plus the terminal temperature difference of the heat exchanger). This diagram is only valid for part-load conditions, when further steam can be obtained from the boiler. The values were simulated over a wide temperature range in 10 K steps. The given gradients correspond to average values over the entire temperature range analysed.

For existing cases of back-pressure biomass CHP plants, CRGs have been estimated between 0.10 and 0.16 euro/(MWh·°C).
Concerning an existing extraction CHP plant, a CRG estimate of 0.09 euro/(MWh·°C) has been observed.
Concerning an existing waste CHP plant, a CRG estimate of 0.07 euro/(MWh·°C) has been observed.

In an existing case involving an extraction biomass CHP plant with unchanged heat output, (Geyer, 2020) estimated a net annual cost reduction from increased electricity revenues of 259 000 euro, based on an annual heat output of 320 GWh and a temperature reduction of 10 °C. With 10% heat distribution heat loss, the CRG was estimated as 0.09 euro/(MWh·°C).

In an existing case involving a waste CHP plant with unchanged heat output, (Averfalk & Werner, 2020) estimated a CRG of 0.07 euro/(MWh·°C) when the plant was equipped with flue gas condensation, with the lower estimate likely impacted by the negative variable cost derived from gate fees received for the waste.

2.7 Higher heat storage capacities

Storage is a vital element in a district heating network, creating degrees of freedom by decoupling heat generation from heat demands over time. Reducing return temperatures of district heating networks enables increased storage capacity (direct integration) because the temperature spread is increased to maintain the same supply temperature. Economic benefits are only obtained by maintaining a high charging temperature. This can be accomplished by locating storage next to a high-temperature heat source, with the increased storage capacity being obtained for free at lower district heating temperatures, which allows the capacity gain to be used to store more heat over the course of a year.

For example, a storage operating at a supply temperature of 90 °C and a return temperature of 50 °C features a temperature spread of 40 °C. If the return temperature is reduced by 10 °C, the achievable capacity increase would be 25%. The storage temperature can be higher than the temperature required in the district heating network to achieve greater temperature spreads and thus increased storage capacity. This enables an efficient network operation at the lowest possible temperatures. Reduced operating temperatures also favour lower thermal loss from storage.

The analytical expression of CRG for investment cases is presented as Equation 5, with the corresponding expression for existing cases presented as Equation 6. However, Equation 5 is simplified by assuming that the specific investment cost is the same for the original storage and the smaller storage. However, this is not wholly accurate because heat storage features a high degree of economy-of-size.

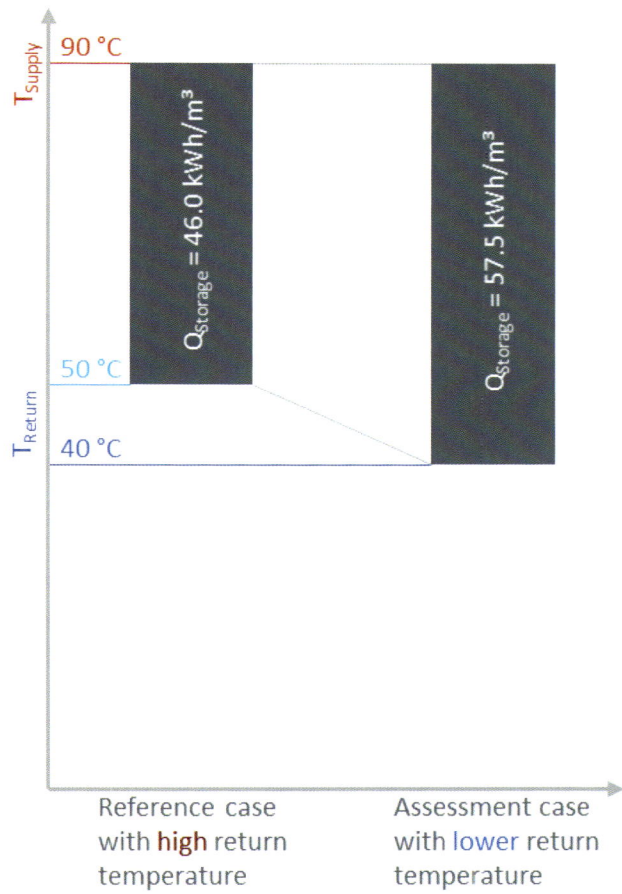

Figure 15. Impact of temperature spread on storage capacity

 Storage capacity depends on temperature spread, with greater temperature spreads increasing the amount of usable storage capacity.

$$CRG = \frac{a \cdot i}{\Delta T_{after}} \cdot \left(\frac{V_{before}}{Q_s}\right)$$

Equation 5 Analytical expression of CRG for investment cases

$$CRG = n \cdot \frac{V}{Q_s} \cdot \rho \cdot c_w \cdot \Delta c$$

Equation 6 Analytical expression of CRG for existing cases

a	Annuity for the investment in heat storage	[-]
i	Specific investment cost for heat storage	[euro/m³]
ΔT_{after}	Temperature difference after the reduction of the return temperature	[°C]
V	Water volume of the heat storage	[m³]
V_{before}	Water volume for the heat storage before the reduction of the return temperature	[m³]
Q_s	Annual amount of heat sold to customers	[MWh]
n	Number of cycles performed using the heat storage	[-]
ρ	Water density	[kg/m³]
c_w	Specific heat for water	[MWh/kg°C]
Δc	Specific cost difference between substituted heat and discharged heat	[euro/MWh]

It is worth considering two example CRG estimates for district heating systems selling 70 GWh heat per year. The first is a daily storage that is used for 100 cycles per year and installed as tank thermal energy storage. The second is a seasonal storage that is used once per year and installed as pit thermal energy storage. According to (Gadd & Werner, 2020), a typical daily storage has a volume of 8 m³ per TJ heat delivered, while a seasonal storage is about 100 times larger. Thus, these two typical storage sizes should be 2000 and 200 000 m³. The corresponding specific investment costs are 370 and 25 euro/m³ (Gadd & Werner, 2020). For investment cases, according to Figure 15, annuity is assumed to be 0.051, and the temperature difference after the return temperature reduction is 50 °C. For existing cases, the specific cost difference between substituted heat, and discharged heat is assumed to be 20 euro/MWh.

For daily storage, the CRG can be estimated as 0.011 euro/(MWh·°C) for the investment case and as 0.067 euro/(MWh·°C) for the existing case.

For seasonal storage, the CRG can be estimated as 0.073 euro/(MWh·°C) for the investment case and as 0.067 euro/(MWh·°C) for the existing case.

In existing cases, the CRG for the daily and seasonal storage is equal because the same amount of heat is

Concerning daily heat storage:
For an investment case, the CRG has been estimated as 0.011 euro/(MWh·°C).
For an existing case, the CRG has been estimated as 0.067 euro/(MWh·°C).
Concerning seasonal heat storage:
For an investment case, the CRG has been estimated as 0.073 euro/(MWh·°C).
For an existing case, the CRG has been estimated as 0.067 euro/(MWh·°C).

stored during the year, with daily storage used 100 times and seasonal storage, being 100 times larger, used only once.

2.8 Lower heat distribution loss

Heat distribution loss impacts the efficiency and economic benefits of district heating networks. Heat loss from district heating networks is determined by the temperature levels (supply and return), the thermal insulation (thickness and condition) of the pipes and the ambient temperature, which depends on the installation type. Target values for annual heat loss in district heating networks are below 10%. However, this is not achieved by all networks, with small and rural district heating networks, in particular, demonstrating heat distribution loss of about 20% and sometimes even higher. In summer, with low heat demand, loss in networks with low heat density can rise above 50%. Depending on the temperature levels, 60–70% of the loss can be dedicated to the supply because the temperature level is higher than the return. The bandwidth depends on the mode of operation between supply and return (average annual temperature difference). (Nussbaumer & Thalmann, 2016; Good et al., 2004)

A reduced system temperature's impact on heat loss was assessed for a small district heating network with annual heat production of 5 GWh. The annual average supply temperature was reduced from 80 to 70 °C, and the average annual return temperature was reduced from 45 to 35 °C, with the heat transfer capacity remaining constant (see Figure 16). The district heating network operated all year (8760 h/a), with total operating hours for customers assumed to be 2000 h/a. The pipe diameter was DN100, and the route length was 4 km, producing a linear heat density of 1.25 MWh/(rm·a).

Heat loss was calculated as 786 MWh/a (15.7%), based on an annual average outdoor temperature of 10.1 °C. Reducing the system temperature by 10 °C, with constant temperature spread (≙ keeping the same transfer capacity), heat loss was calculated as 636 MWh/a (13.1%).

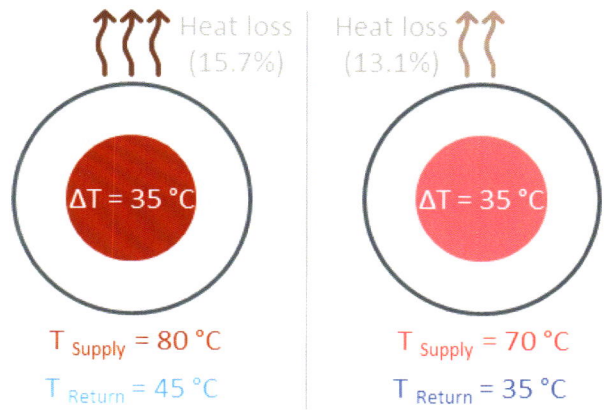

Figure 16. Example of heat loss in a district heating network for design system temperatures (80/45 °C, left) and reduced system temperatures (70/35 °C, right)

For an existing case, this example provides an estimated annual cost reduction of 3000 euro/year, which is associated with an annual loss reduction of 150 MWh and an assumed heat loss cost of 20 euro/MWh. According to (Geyer, 2020), the CRG can then be estimated as 0.07 euro/(MWh·°C), based on annual heat sales of 4214 MWh and a temperature reduction of 10 °C. Meanwhile, (Averfalk & Werner, 2020) estimated CRGs between 0 and 0.13 euro/(MWh·°C), depending on the heat supply technology. The lowest estimation was obtained for the waste CHP plant due to negative operation costs arising from gate fees for received waste.

2.9 Ability to use plastic pipes instead of steel pipes

In LTDH networks, it is possible to use plastic pipes instead of steel pipes, with plastic pipes being unsuitable for use at high temperatures and high pressures (cf. Figure 16). Flexible piping systems are especially worth investigating because they allow for faster installation and feature lifetimes of 50 to 100 years at temperatures below 70 °C. Given the longer pipe length and flexibility, the number of connections is minimised, and most fittings can be mounted prior to installation in the trench. This allows for significantly smaller trenches compared to steel pipes. Additionally, using flexible pipes can reduce installation time by up to 80%

 Lower heat distribution loss result from a decreased average temperature difference between the fluids in the heat distribution pipes and the environment.

 Concerning heat distribution loss:
For existing cases, estimated CRGs have been observed between 0 and 0.13 euro/(MWh·°C).

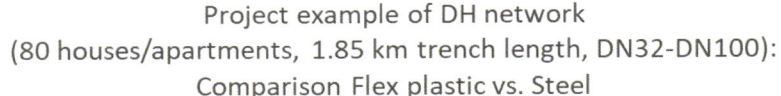

Figure 17. Total costs of a completed example project installed in a smaller district heating network (Engel, 2020)

compared to steel pipes, with the economic benefits of plastic pipe systems – based on practical project experience – ranging from 10–20%. This considers material, installation and civil works. Figure 17 presents an example project featuring a 1.85 km trench, pipe sizes ranging from DN32 to DN100 and connections with 80 houses or apartments. A low amount of civil work was required because the majority (82%) of the trench length was installed in greenfield. Section 7.10 features another example, where plastic pipes are to be used for 2.8 km of a total 6.5 km of trench.

Figure 18 provides initial technical information for typical piping systems in the European market, including temperature and pressure limits. Only the steel system completely covers 2GDH, 3GDH and 4GDH. All of the other systems are flexible, based on plastic medium pipes with a limited application range. However, plastic piping systems are especially preferred for LTDH solutions.

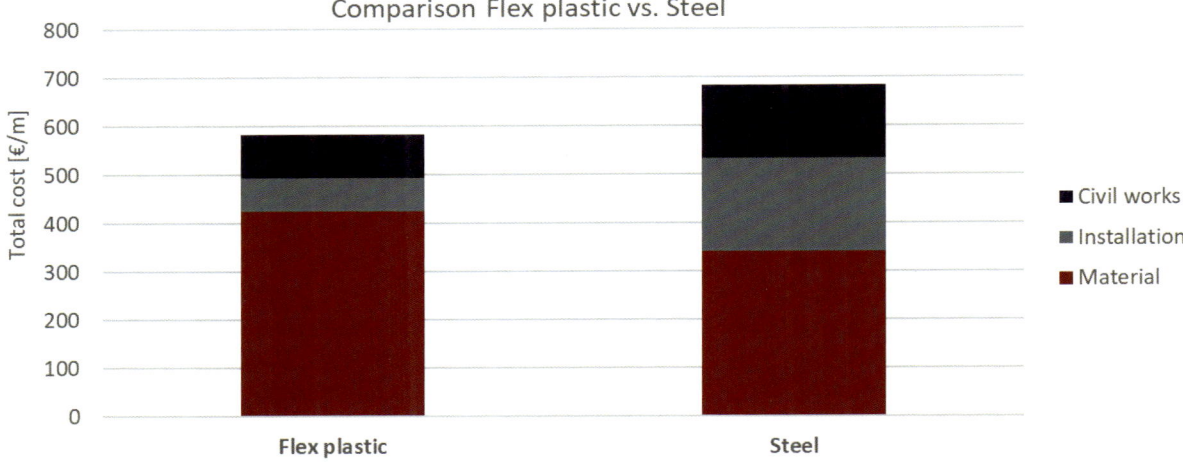

Figure 18. Typical examples of pre-insulated piping systems and their application field in terms of pressure and temperatures (Engel, 2018)

2.10 Non-economic benefits of lower temperatures

Beyond economic benefits, there are other benefits derived from solutions outlined in this section.

1. Flue gas condensation is usually used for biomass boiler outputs from beyond approximately 1 MW. In addition to the efficiency increase, flue gas condensation units promote significant reduction of dust and other pollutants in the flue gas due to a scrubbing effect.
2. Fewer variations in supply temperatures reduce the risk of cycle fatigue of steel pipes. This leads to longer lifetimes for steel pipes in existing networks. This can increase the lifetime of piping systems by up to 100 years, especially in systems using plastic pipes, if network temperatures do not exceed 65 °C, see (CEN, 2020).
3. Less temperature loss in the supply pipe (due to less heat loss) allows for reduced supply temperatures from heat supply plants (this is associated with the previous heat loss benefit).
4. There is a lower risk of scalding during pipe maintenance – high temperatures have caused lethal accidents in existing systems.

2.11 Summary of economic benefits

Without motivating factors driving decreased temperatures from utilities, there are no concrete incentives for building developers, owners or customers to initiate measures to reduce dependence on high supply temperatures or minimise return temperatures. Consequently, high-temperature heating systems will continue to be used in new buildings, establishing the need for sustained use of high-temperature heat sources (a lock-in effect).

To negate this lock-in effect, this chapter has highlighted the numerous benefits that reduced system temperatures promote, including efficiency improvements, capacity increases, decreased distribution loss and easier integration of alternative heat sources.

Importantly, CRG estimations are not always additive. Especially in large, complex systems with many different heat generation units, cancellation effects can appear. In such cases, detailed simulation analyses are necessary. For example, this situation can occur in a CHP plant with flue gas condensation when more heat is recovered by the return pipe in the flue gas condensation, which is located before the turbine condenser. This leads the turbine condenser to then obtain return temperatures similar to those before the temperature reduction, negating the increased power generation effect because heat generation in the turbine condenser is reduced. Nevertheless, the estimated CRG values are useful good indicators allowing an approximate estimation of the economic benefits of reduced temperatures.

Table 1 provides an overview of the economic benefits by listing the CRG for seven district heating networks (four of which are also studied as successful cases in Section 4.4 of this guidebook). In addition to heat delivery specific CRGs, absolute values are listed as LCOH Benefit in euro per MWh to provide an impression of the impact. Additionally, three references are included providing CRG estimates on a national level.

Given the Borås case is over twenty years old, the CRG value is low, a function of low electricity prices in the 1990s providing fewer economic benefits for increased electricity generation.

The Enköping case features a typical Swedish biomass CHP plant with flue gas condensation.

The Middelfart case observes a temperature reduction of 11.8 °C between 2009 and 2015 and an annual heat delivery of 96 GWh. The annual cost reduction (560 000 euro) mainly resulted from very high incentives from the heat supplier to achieve low return temperatures. Hence, this CRG does not represent a true cost reduction.

 Assessments of **cost reduction gradient** (CRG) **facilitate** quantification of the **positive economic impacts** of **lower system temperatures**. These impacts are summarised in Table 1 and include both investment and existing cases. The most significant financial benefits can be achieved by new alternative heat sources, with the **comparison** showing that the examined **CRGs** are in **similar orders of magnitude**. This **strengthens** the **evidence for the economic benefit** of **reduced system temperatures**.

Table 1. Overview of projected economic effects, according to the cost reduction gradient (CRG) in euro/(MWh·°C), of reduced system temperatures.

Chapter section and heat supply technology (either the technology itself or as the dominant component of a system)	Cost reduction gradient (CRG) in euro/(MWh·°C)	
	Investment cases where investment costs are reduced	Existing cases where operation costs are reduced
2.1 Low-temperature geothermal heat	0.45–0.74	0.67–0.68
2.2 Heat pump	0.41	0.63–0.67
2.3 Low-temperature waste heat	0.65	0.51
2.4 Solar thermal – flat plate collectors	0.35–0.75	Not available
2.4 Solar thermal – evacuated tube collectors	0.26	Not available
2.6 Biomass-CHP with back-pressure turbine	Not available	0.10-0.16
2.6 Biomass-CHP with extraction turbine	Not available	0.09
2.6 Waste-CHP with flue gas condensation	Not available	0.07
2.7 Daily storage as tank thermal storage	0.01	0.07
2.7 Seasonal storage as pit thermal storage	0.07	0.07
2.8 Heat distribution loss	Not available	0–0.13

The Viborg case observes a temperature reduction of 11 °C between 2002 and 2016 and an annual heat delivery of 210 GWh. The annual cost reduction was estimated as 670 000 euro, mainly the result of lower heat distribution loss based on a high specific cost of about 45 euro/MWh.

The Göteborg case features a CRG value that is somewhat higher than values obtained for traditional systems because the system features large heat pumps from a sewage treatment plant and heat recoveries from two oil refineries.

The Gleisdorf case observes a temperature reduction of 10 °C between 2016 and 2020, an annual heat delivery of 6.5 GWh and an annual cost reduction of 7 100 euro/a, mainly the result of higher solar thermal yields and lower heat distribution loss. Considering the newly installed heat pump and the continuous growth of the district heating network promoting heat sale increasing to by approximately 9.2 GWh by 2025, the annual cost reduction amounts to 25 700 euro/a.

Finally, the high CRG value for the prospective Aalborg system depends on many new heat sources, including industrial waste heat, heat pumps and solar thermal. The estimation is based on an annual cost reduction of 17 million euro, annual heat delivery of 1330 GWh and a temperature reduction of 22.5 °C.

The Sweden estimation represents the average of CRG values from 27 systems analysed between 1996 and 2010. These CRG estimations vary between 0.04 and 0.38 euro/(MWh·°C). The highest estimations were obtained from systems with high proportions of low-temperature heat sources. For Austria, the actual CRG estimation is based on available statistics (2019 data). The future CRG is estimated on the projected generation mix. As biomass is estimated to be the dominant fuel in 2050, the CRG is lower compared to the Danish estimation. The prospective estimation for Denmark is based on similar assumptions to those used for the prospective Aalborg case.

Several conclusions can be drawn from the technology overview of Table 1 and the systems overview of Table 2. First, traditional combustion processes in CHP plants without flue gas condensation feature CRG values between 0.10 and 0.13 euro/(MWh·°C). Second, corresponding CRG values for low-temperature heat sources – such as geothermal, heat pumps, and waste heat – are found in the interval between 0.5 and 0.7 euro/(MWh·°C). Hence, these new low-temperature heat sources feature CRG values about five times higher than the corresponding values for traditional heat supply systems.

Accordingly, a general forecast can be established for European district heating systems. The future economic benefits of a low-temperature district heating system can be estimated at about 0.5 euro/(MWh·°C). This indicates a total cost reduction potential of 14 billion euro per year, assuming future annual EU heat sales of 950 TWh (see Figure 1 of this guidebook's introduc-

Table 2. *Overview of identified economic benefits in various systems (indicated by cost reduction gradient).*

	Cost Reduction Gradient [euro/(MWh·°C)]	**LCOH Benefit[1] [euro/MWh]**	**Reference**
Local district heating systems			
Borås (SE; 1996)	0.05	1.5	(Dahlberg & Werner, 1997)
Enköping (SE; 2015)	0.12	3.6	(Castro Flores et al., 2017)
Middelfart (DK; 2015)	0.49	15	(Sipilä & Rämä, 2016)
Viborg (DK; 2016)	0.29	8.7	(Diget, 2019)
Göteborg (SE; 2017)	0.20	6	(Eriksson, 2020)
Gleisdorf (AT; 2020)	0.11	3.3	Estimation by AEE INTEC
Gleisdorf (AT; 2025)	0.28	8.4	Estimation by AEE INTEC
Aalborg (DK; 2050)	0.57	17	(Sorknæs et al., 2020)
National estimations			
Sweden (1996–2010)	0.12	3.6	(Frederiksen & Werner, 2013)
Austria (2019)	0.12	3.6	Estimation by AIT
Austria (2050)	0.38	11	Estimation by AIT
Denmark (2050)	0.55	17	(Lund et al., 2018)

[1] temperature reduction is assumed to be 30 °C in average

tion) and a temperature reduction of 30 °C. This cost reduction represents a net present value of more than 200 billion euro.

2.12 Major conclusions concerning economic benefits

Future district heating networks will be dominated by diversified, alternative heat sources. This chapter's assessments indicate that system temperatures will have a significantly stronger influence on profitability compared to the existing generation mix, which is dominated by combustion technologies. As Figure 19 indicates, the CRG will be approximately five times higher for low-temperature (e.g. geothermal) compared to high-temperature (e.g. CHPs or heat-only boilers with flue gas condensation) generation technologies. Continuous improvements and changes to low-temperature systems are ongoing, allowing greater economic benefits compared to high-temperature supply systems. Therefore, measures to reduce system temperatures must be taken sooner rather than later to enable district heating networks to facilitate future advancements in heating systems.

Finally, contrary to broadly held assumptions, the greatest economic benefits are derived from heat generation units rather than decreasing heat distribution loss in the grids.

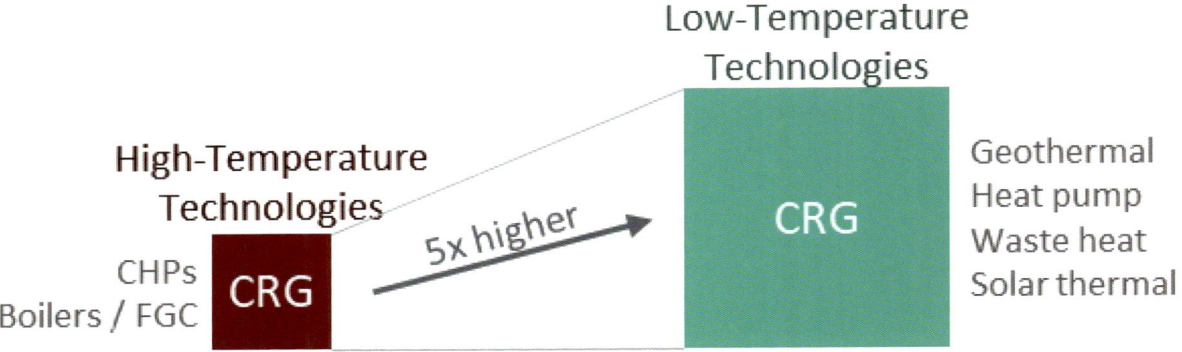

Figure 19. Indicative comparison of cost reduction gradient (CRGs) between HT- and LT-Technologies

2.13 Literature references in Chapter 2

Arpagaus, C., Bless, F., Uhlmann, M., Schiffmann, J., & Bertsch, S. S. (2018). High temperature heat pumps: Market overview, state of the art, research status, refrigerants, and application potentials. Energy 152 (2018) (S. 985-1010). https://doi.org/10.1016/j.energy.2018.03.166: ELSEVIER.

Averfalk, H., & Werner, S. (2020). Economic benefits of fourth generation district heating. In Energy 193 (2020) 116727 (S. 11). https://doi.org/10.1016/j.energy.2019.116727: ELSEVIER.

Castro Flores, J. F., Lacarrière, B., Chiu, J. N.-W., & Martin, V. (2017). Assessing the techno-economic impact of low-temperature subnets in conventional district heating networks. In Energy Procedia 116 (2017) (S. 206-272). 10.1016/j.egypro.2017.05.073: ELSEVIER.

CEN. (2018). EN 12975: Thermal solar systems and components – Solar collectors. Brussels: European Standard Organization.

CEN. (2020). prEN 15632-2: Annex A - Application of Miner's Rule - Calculation of the calculated design life of PB-PE-X and Multilayer piping systems. Brussels: European Standard Organization.

Dahlberg, C., & Werner, S. (1997). Lower Temperature Level in The Borås District Heating System. Borås: EH&P-Unichal 1997.

David, A., Mathiesen, B., Averfalk, H., Werner, S., & Lund, H. (2017). Heat Roadmap Europe: Large-Scale Electric Heat Pumps in District Heating Systems. In Energies, 10, 578 (S. 18). doi:10.3390/en10040578: MDPI AG.

Diget, T. (2019). Motivation tariff - the key to a low temperature district heating network. In HOT|COOL, Journal No. 1/2019 (S. 19-22). Frederiksberg: DANISH BOARD OF DISTRICT HEATING.

Engel, C. (2020). Practical project experiences between 2010-2020 (cross-checked with 4 publications about DH network costs). Gödersdorf: Austroflex Rohr-Isoliersysteme GmbH.

Engel, C. (2018). Installation techniques of District heating. Seminar at CoolHeating.eu. Vienna.

Eriksson, M. (2020). Fourth Generation District Heating - The prospects of Gothenburg. An investigation of the 4GDH concept and the motivations to implement it in Gothenburg. Gothenburg: Chalmers University of Technology.

European Commission. (2016). An EU Strategy on Heating and Cooling. COM(2016) 51 Final. Brussels: European Commission.

Frederiksen, S., & Werner, S. (2013). District Heating and Cooling. Lund: Studentlitteratur AB.

Gadd, H., & Werner, S. (2020). 21 - Thermal energy storage systems for district heating and cooling. In L. F. Cabeza (Ed.), Advances in Thermal Energy Storage Systems, Second edition (pp. 625-638): Woodhead Publishing.

Geyer, R. (2020). Reduzierte Systemtemperaturen in Wärmenetzen: Eine energie-ökonomische Bewertung der Effekte. Pinkafeld: FH Burgenland.

Geyer, R., Hangartner, D., Lindahl, M., & Pedersen, S. (2019). Heat Pumps in District Heating and Cooling Systems. Vienna, Lucerne, Gothenburg, Aarhus: IEA Heat Pumping Technologies Annex 47.

Good, J., Biedermann, F., Bühler, R., Bunk, H., Deines, T., Gabathuler, H., . . . Rakos, C. (2004). Planungshandbuch. www.qmholzheizwerke.at: Arbeitsgemeinschaft QM Holzheizwerke.

Grosse, R., Binder, C., Wöll, S., Geyer, R., & Robbi, S. (2017). Long term (2050) projections of techno-economic performance of large-scale heating and cooling in the EU. Luxembourg: Publications Office of the European Union.

Jentsch, A., Wagner, U., Paulick, S., Rühling, K., Mangold, D., Ott, S., . . . Fröhlich, D. (2020). DELFIN – Decentralized Feed-In: Prognose der Auswirkungen dezentraler Einbindung von Wärme aus erneuerbaren Energien und anderen Wärmeerzeugern in Fernwärmenetze. Frankfurt am Main: AGFW, Technische Universität Dresden, Solites.

Kaltschmitt, M., Hartmann, H., & Hofbauer, H. (2016). Energie aus Biomasse - Grundlagen, Techniken und Verfahren. Heidelberg: Springer-Verlag Berlin Heidelberg.

Lund, H., Østergaard, P. A., Chang, M., Werner, S., Svendsen, S., Sorknæs, P., . . . Möller, B. (2018). The status of 4th generation district heating: Research and results. In Energy 164 (2018) (S. 147-159). https://doi.org/10.1016/j.energy.2018.08.206: ELSEVIER.

Müller, A., Heimrath, R., Leoni, P., & Schrammel, H. (6-7 October 2020). How much to invest? Balancing investment costs and economic benefits of reducing the temperature levels in existing district heating networks. In 6th International Conference on Smart Energy Systems (S. https://smartenergysystems.eu/conference-2020/). Online: #SESAAU2020.

Nussbaumer, T., & Thalmann, S. (2016). Influence of system design on heat distribution costs in district. In Energy 101 (2016) (S. 496-505). ELSEVIER.

Schaumann, G., & Schmitz, K. (2010). Kraft-Wärme-Kopplung. Heidelberg: Springer-Verlag Berlin Heidelberg. doi:10.1007/978-3-642-01425-3

Schmidt, R.-R., Geyer, R., & Lucas, P. (2020). Discussion Paper: The barriers to waste heat recovery and how to overcome them? Brussels: Euroheat & Power aisbl.

Sipilä, K., & Rämä, M. (2016). Low Temperature District Heating for Future Energy Systems. Subtask D: Case studies and demonstrations. Frankfurt am Main: IEA DHC|CHP Annex TS1.

Sorknæs, P., Østergaard, P. A., Thellufsen, J. Z., Lund, H., Nielsen, S., Djørup, S., & Sperling, K. (2020). The benefits of 4th generation district heating in a 100% renewable energy system. In Energy 213 (2020) 119030 (S. 11). https://doi.org/10.1016/j.energy.2020.119030: ELSEVIER.

SWM. (2015). Fernwärme und Rücklauftemperatur in modernen Niedertemperaturnetzen. München: Stadtwerke München.

3 LOWER TEMPERATURES INSIDE BUILDINGS

Authors: Dorte Skaarup Østergaard, Svend Svendsen, Michele Tunzi, Theofanis Benakopoulus, DTU; Oddgeir Gudmundsson, Danfoss.

The main purpose of district heating systems is to deliver heat to buildings to provide comfort for building occupants. District heating temperatures and heat loads are therefore linked to the comfort requirements of occupants and the design of building installations for space heating, domestic hot water, and ventilation. If the heating demands and heating installations of a building require a high supply temperature, this can mean that supply temperatures need to be high in the whole district heating network. Similarly, if the heating installations in buildings are not designed or operated properly, the district heating water will not be cooled sufficiently, and district heating return temperatures will be high. High supply and/or return temperatures will lead to inefficient district heating systems where heat losses are higher than necessary and heat production is more expensive because low-temperature sources, e.g., waste heat, and flue-gas condensation, cannot be utilized properly. Lower district heating temperatures therefore require taking a closer look at the heat demands and heating installations of buildings to ensure that heating systems are designed, operated, and controlled properly, so the lowest possible district heating supply temperatures can be achieved.

This chapter will provide the basic knowledge that is needed to improve the operation of building installations in both existing and new buildings so that supply temperatures in district heating systems can be reduced. The first two sections of the chapter contain a general introduction to buildings and district heating installations in buildings. After this, current knowledge on how to lower the district heating temperatures in buildings are described. Ideas for how to deal with currently high supply temperature requirements in buildings and how to fix simple typical malfunctions in substations are described, and then move on to addressing the more complex issues that need resolving in internal heating installations. Finally, information is provided about how to design new building installations for lower temperatures and what to consider when connecting existing buildings to a district heating system. The chapter is structured in sections as follows:

1. Temperature requirements and heat demands in buildings
2. Building installations and their impact on district heating temperatures
3. Customers with high supply temperature requirements
4. Substation malfunctions and faults that lead to higher temperatures
5. Use of data to identify heating system improvements
6. Actions to lower the temperatures in space heating systems
7. Actions to lower the temperatures in domestic hot water systems
8. Actions to lower the temperatures in ventilation systems
9. How to design new building installations for low temperatures
10. What to consider when existing buildings are connected to district heating
11. Main conclusions on lower temperatures in buildings

3.1 Temperature requirements and heat demands in buildings

Temperature requirements for space heating are often linked to maintaining an indoor air temperature around 22 °C. To deliver this indoor temperature, space heating systems require higher supply temperatures. Although older radiator heating systems were designed for supply temperatures of 80–90 °C, actual operation temperatures are commonly in the range 40–70 °C (Averfalk, Werner, et al., 2017; Jangsten, Kensby, et al., 2017; Østergaard & Svendsen, 2016b). This is because the heating installations were designed for extremely low outdoor temperatures that occur only rarely, which means that design heating system temperatures apply only rarely in practice. The actual heat demands in buildings are further reduced as buildings are renovated during the years, for example through the necessary replacement of old worn out single-pane windows. Since the heat demand in buildings is significantly lower during the normal operations than in the design situations, actual heating system temperatures can be reduced correspondingly. Furthermore, limited system sizes and safety factors applied by designers led to general oversizing of heating system components. In new buildings, the design supply temperatures in the heating systems are often required to be 50–60 °C for radiators and 35–45 °C for floor heating systems (Averfalk, Werner, et al., 2017; Danish Standard, 2013).

Temperature requirements for domestic hot water installations vary between countries (Dalla Rosa, Li, et al., 2014), and therefore national regulations on this subject should always be consulted. For comfort purposes, it should be sufficient to provide a 45 °C domestic hot water temperature at the kitchen tap, while 40 °C hot water is often preferred for showers (Brand, Gudmundsson, et al., 2016; Danish Standards, 2009). Nevertheless, it is common to apply higher domestic hot water temperatures to avoid the growth of *Legionella* bacteria. Growth of *Legionella* bacteria is mainly an issue in installations where large volumes of domestic hot water are stored for long periods (Krøjgaard, 2011; Mathys, Stanke, et al., 2008). This means that domestic hot water tanks and large domestic hot water circulation systems are the main focus areas. For domestic hot water tanks, it is common to maintain domestic hot water temperatures in the range of 55–70 °C (Dalla Rosa, Li, et al., 2014), and temperatures in all parts of the domestic hot water system need to be maintained above a certain temperature threshold of 50–60 °C depending on national legislation and building type (Dalla Rosa, Li, et al., 2014; Danish Transport, Building and Housing Authority, 2020). In ultra-low-temperature district heating systems, where supply temperatures are below this threshold, supplementary electric boosting solutions can be installed to achieve the temperatures required in the domestic hot water installations.

In buildings with balanced mechanical ventilation systems, district heating can provide heat to a heating coil

What is *Legionella*?
Legionella is a bacterium that is present in fresh water. When the concentration of *Legionella* bacteria is high and aerosols are created, there is an increased risk of inhaling the bacteria which can lead to illnesses such as Pontiac Fever or Legionnaires disease, which can be lethal. *Legionella* growth takes place at water temperatures in the range of 20–45 °C, while temperatures above this level will cause bacterial decay (Frederiksen & Werner, 2013). Long periods with lukewarm stagnant water provide perfect growth conditions for *Legionella* bacteria. Furthermore, biofilms and dirty installations increase the risk of bacterial growth and can make it possible for *Legionella* bacteria to survive at higher temperatures during disinfection cycles (Bagh, Albrechtsen, et al., 2004; Krøjgaard, 2011).

How to reduce the risk of *Legionella* growth
Lower district heating temperatures do not constitute an increased risk of *Legionella* bacteria growth, but the district heating provider should still be aware of providing temperatures that allow in-house installations to be operated in accordance with national temperature recommendations for reducing the risk of *Legionella*. Research shows that the risk of *Legionella* in domestic hot water systems is reduced when temperatures are above 50 °C and even more so for temperatures above 55 °C (Barna, Kádár, et al., 2016; Kruse, Wehner, et al., 2016; Dalla Rosa, Li, et al., 2014; Mathys, Stanke, et al., 2008). International standards recommend keeping the domestic hot water supply temperatures above 55 °C, but depending on national guidelines it may not be advisable to use supply temperatures above 60 °C, because this will cause increased lime precipitation in installations using calcium-rich cold-water supply and also increase the risk of scalding.

Legionella contaminations often occur locally, which means that only some parts of a domestic hot water system are affected (Kruse, Wehner, et al., 2016). Local contaminations can be caused by insufficient temperature control in domestic hot water circulation systems, infrequent domestic hot water tapping, insufficient cleaning of shower hoses, etc. (Krøjgaard, Krogfelt, et al., 2011; Nielsen & Aggerholm, 2019). It is therefore important to ensure proper circulation in the domestic hot water system to avoid dead pipe ends and pipes with temperatures below 50 °C (Buhl, 2012). Recent research has evaluated whether the growth of *Legionella* could be limited by designing installations with small domestic hot water volumes, short pipe lengths, and high water-flow rates (Buhl, 2018; Dalla Rosa, Li, et al., 2014; Rühling & Rothmann, 2020). However, more research is needed in these alternative protection measures.

that can pre-heat the fresh air intake to ensure that the temperature of the inlet ventilation air does not become too low and compromise occupant comfort. The comfort temperature required depends on the type of ventilation, the location of the air inlet, and the air velocity. In practice, it is recommended that the design inlet air temperature is at least 16 °C (Brand, Lauenburg, et al., 2012), in which case the remaining heat necessary to ensure a comfortable indoor air temperature will be provided by the space heating system and through heat gains. The temperatures required from the district heating system in order to deliver this temperature depend largely on the design of the heating coil. Newer ventilation heating coils are typically designed for supply/return temperatures of around 60/30 °C (Swedenergy AB, 2016). In modern ventilation systems, a heat recovery system will typically provide sufficient heating of the inlet air for most of the year, and additional heating is usually only necessary to maintain comfortable air temperatures when outdoor temperatures are sub-zero and the heat recovery system may be turned off due to the risk of ice formation.

Historically, most of the heat demand inside buildings in cold climates has been for space heating. In Denmark and Sweden, space heating typically accounts for around 60–70% of the heat demand in existing homes (Bøhm, Schrøder, et al., 2009; Jangsten, 2016). In office buildings and schools, where domestic hot water demands are often small, space heating can make up 90% of total heat demands (Bøhm, Schrøder, et al., 2009). In new buildings, the demand for space heating is reduced due to new requirements for higher levels of insulation, lower U-values, and ventilation with heat recovery, while the demand for domestic hot water is not similarly reduced (Bøhm, Schrøder, et al., 2009; Lidberg, Olofsson, et al., 2019). The share of district heating utilized for preparation of domestic hot water in a new or energy-renovated multi-family building is therefore likely to be in the range of 45–50% (Best, Braas, et al., 2020; Bøhm, Schrøder, et al., 2009; Lidberg, Olofsson, et al., 2019). The importance of these findings is that, while domestic hot water installations typically have a smaller impact on the district heating supply and return temperatures in existing buildings and especially offices, in new and renovated apartment buildings domestic hot water installations can be expected to determine both supply and return temperatures for most of the year (Lidberg, Olofsson, et al., 2019). Figure 20 summarizes the typical shares of heat for space heating and domestic hot water in three different building types in Denmark.

The heating consumption in a building includes both the actual heat demand to cover space heating and domestic hot water and the heat losses from distributing heat and domestic hot water throughout the building. The heat losses from the pipes in the domestic hot water circulation system may even make up the largest part of the heat consumption for domestic hot water. Estimations made on the basis of measurements in existing Danish buildings suggest that the actual use of domestic hot water makes up only 15% of the total heat used in domestic hot water installations in office buildings, 33% in apartment buildings, and 48% in sin-

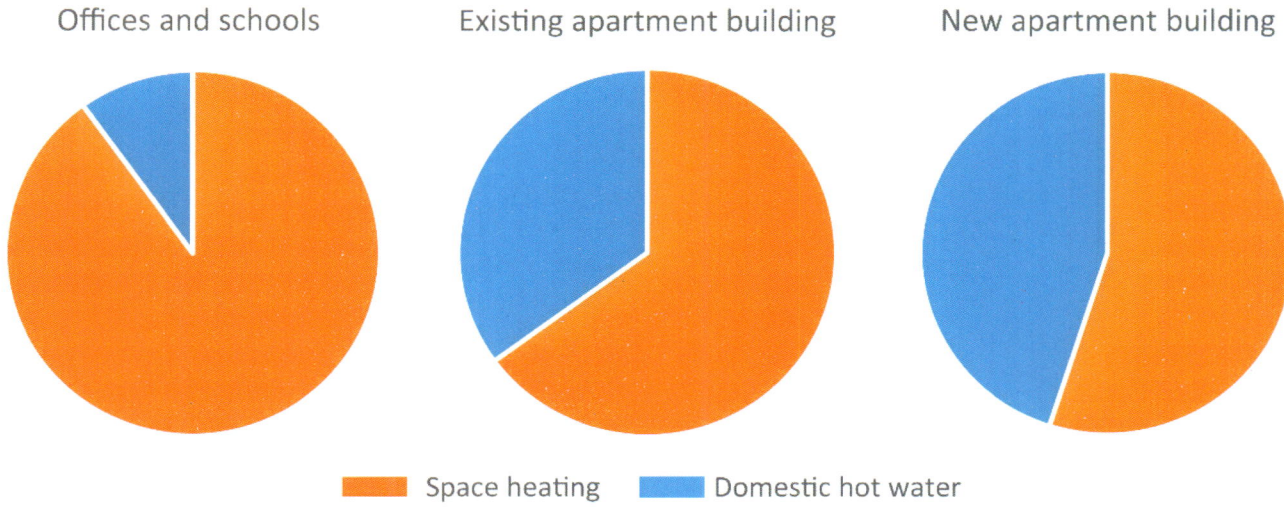

Figure 20. Shares of heat for domestic hot water and space heating in three different building types in Denmark (based on Bøhm, Schrøder, et al., 2009)

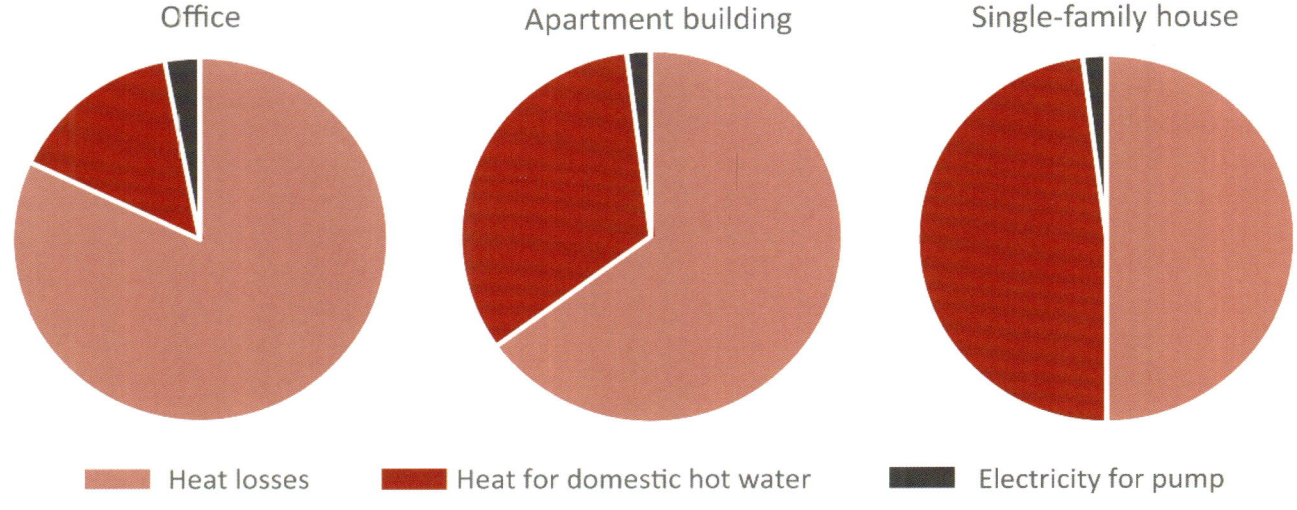

Figure 21. Energy use for typical Danish domestic hot water installations (based on Bøhm, Schrøder, et al., 2009)

gle-family houses (Bøhm, Schrøder, et al., 2009). The remaining heat is wasted mainly due to heat loss from domestic hot water circulation pipes and, to a smaller extent, also from domestic hot water tanks, as shown in Figure 21. Overall, the system efficiency of domestic hot water installations in Danish apartment buildings was found to vary between 30% and 77% (Bøhm, 2013). Similar system efficiency trends are found for domestic hot water systems in Chinese buildings (29–56%) and for Polish buildings (30–43%) (An, Yan, et al., 2016; Cholewa, Siuta-Olcha, et al., 2019).

New installations for domestic hot water in buildings supplied by low-temperature district heating are improved to avoid circulation of domestic hot water and reduce domestic hot water heat losses.

3.2 Building installations and their impact on district heating temperatures

Building heating demands gradually decrease with increasing energy-saving awareness and building code requirements, but there is no clear and obvious relationship between the age of a building and its ability to cool the district heating water (Kristensen & Petersen, 2020). This is because it is not the insulation level of the building that determines the ability to cool the district heating water, but the design and operation of the building's heating installations. Improving the operation of these installations may therefore make it possible to reduce the current district heating supply and return temperatures and limit the gap between the high temperatures supplied in many district heating systems today and the comfort temperatures required inside the buildings. The following paragraphs provide background knowledge on how typical heating installations are expected to operate, which gives a baseline for understanding why installations do not always work as expected and how the issue can be corrected. Recommendations on this topic are also given by Euroheat & Power (Euroheat & Power, 2008), and Danfoss provides an in-depth explanation of control solutions for district heating substations (Danfoss, 2012).

A typical space heating substation consists of either a heat exchanger whereby heat is transferred between the district heating network and the internal space heating system, or a direct connection where the district heating enters directly into the pipes of the space heating system in the building. Figure 22 illustrates typical layouts. The space heating substation should preferably be equipped with a controller (A) that makes it possible to reduce the space heating supply temperature automatically when the outdoor temperature increases (Euroheat & Power, 2008). The system typically uses measurements from temperature sensors (B) on the space heating supply pipe and in the outdoor air to regulate the supply temperature in the space heating system in accordance with the chosen supply temperature curve. If the space heating supply temperature is above the chosen set-point, an actuator closes the control valve (C) located on the district heating pipe, which reduces the heat transfer to the space heating system. If the space heating supply temperature is below the set-point, the actuator opens the valve to increase the heat transfer. A differential pressure controller (D) can be installed across the control valve to ensure that the

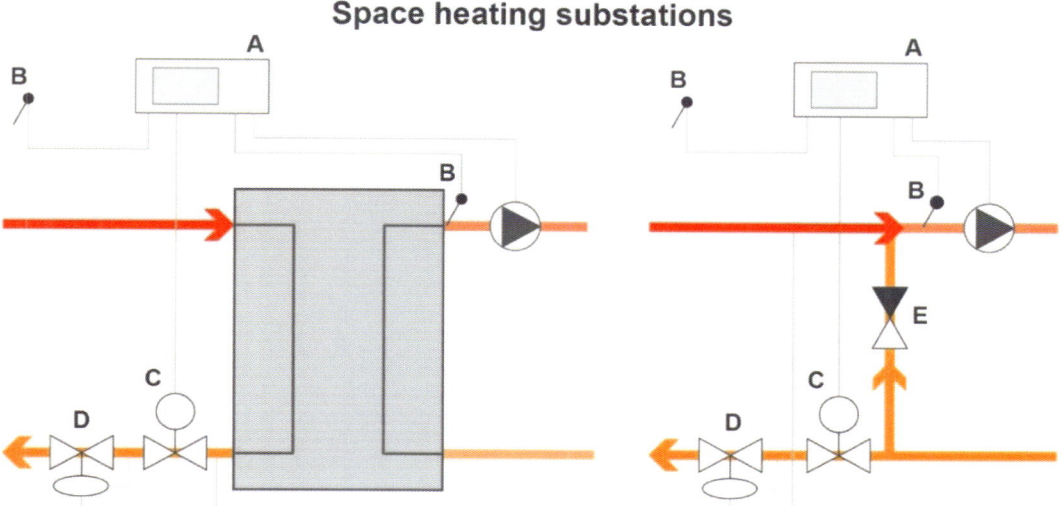

Figure 22. Example of the basic layout for an indirect (left) and a direct (right) space heating substation. (A) Controller, (B) Temperature sensors, (C) Actuator and control valve, (D) Differential pressure controller, and (E) Non-return valve.

valve has the appropriate operating conditions. In the case of a direct connection, the supply temperature can be regulated in accordance with a chosen set-point by mixing the space heating return water into the supply through a non-return valve (E).

Domestic hot water is typically heated in a heat exchanger or a tank. In both cases, the fresh water in the domestic hot water system is kept physically separated from the district heating water.

Instantaneous heat exchangers are often preferred for the heating of domestic hot water in district heating systems. In this installation, the domestic hot water is heated instantly at the time when the hot water tapping occurs. This reduces the volume of stored domestic hot water, which reduces the risk of *Legionella* bacterial growth. On the other hand, it requires that the peak heat demand for domestic hot water can be supplied instantaneously. This means service pipes and heat exchangers in single-family houses, including terraced houses, have to be wider to deliver the high-peak heat demand (Thorsen, Christiansen, et al., 2011). In apartment buildings, these needs are reduced dramatically because the peak domestic hot water demands of the various occupants do not occur simultaneously and the domestic hot water demand is spread throughout the day.

The local storage of domestic hot water in a tank requires precautions to be taken to avoid the growth of *Legionella* bacteria. Tank solutions are mainly recommended in for example sports facilities or similar, where large hot water demands occur frequently, but for short periods of time. A domestic hot water tank is also worth considering in low-energy buildings with a limited heating season or in buildings with long service pipes, because in these cases an instantaneous heat exchanger requires frequent bypass flows (Euroheat & Power, 2008). Local water storages can help reduce the peak heat demands in district heating service pipes and local networks, if the tank is operated with a suitably low charging rate. On the other hand, the generally low simultaneity in the use of domestic hot water means that such tanks do not reduce central district heating peak demands. Furthermore, they increase the heat loss from the domestic hot water installations inside the building, leading to higher heat demands for domestic hot water purposes.

Figure 23 shows layouts for typical domestic hot water substations. The substations are connected to the controller (A), which ensures that the actuator regulates the district heating flow through the control valve (C) to achieve a certain domestic hot water temperature. The actuator is controlled by a sensor (B) located at the outlet of the domestic hot water heat exchanger or in the domestic hot water tank. The regulation has to be fast due to the rapid changes in demand for domestic hot water, and the actuator regulating the control valve has to be activated only when there is a need for heating in the domestic hot water circuit or tank. A differential pressure controller (D) can be installed to ensure stable operation conditions for the control valve. The dome-

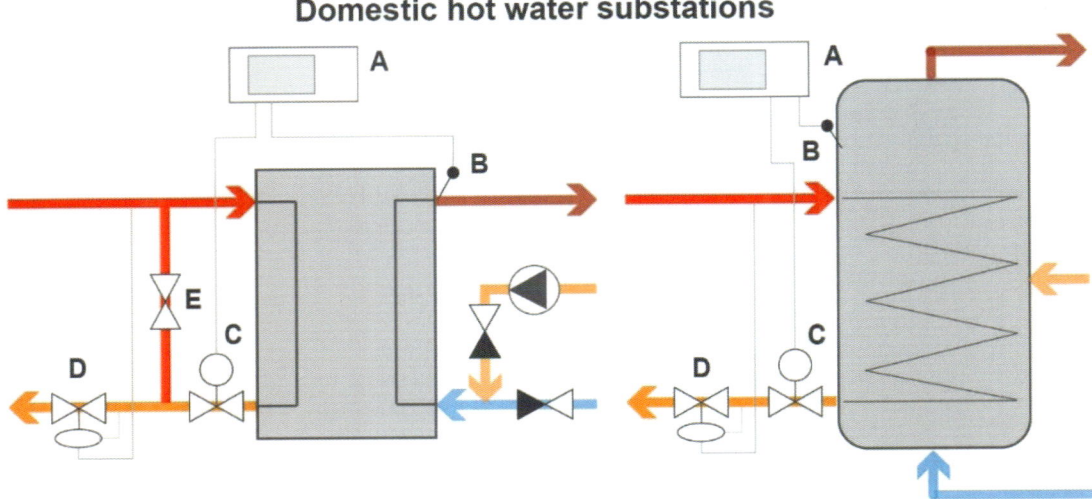

Figure 23. Example of the layout of a domestic hot water heat exchanger (left) and a domestic hot water storage tank (right). (A) Controller, (B) Temperature sensor, (C) Actuator and control valve, (D) Differential pressure controller, and (E) Bypass valve (may be located internally in the unit).

stic hot water heat exchanger is often equipped with a small bypass (E) to ensure that the district heating service pipe is kept warm so that domestic hot water can be available with very little delay. This is especially relevant for summer periods when there is no space heating demand. The bypass can be regulated by a temperature-controlled valve, which opens when the district heating supply temperature at the heat exchanger drops below a certain set-point, typically around 35–40 °C. For a domestic hot water tank, the connection should be in the middle of the tank to ensure the best possible cooling of the district heating water.

In a mechanical balance ventilation system, Figure 24 shows a typical layout for a heating coil in a situation where heat recovery is bypassed due to cold outdoor temperatures. The heating coil is supplied with hot water at a constant water flow, and the controller (A) is used to vary the supply temperature to the heating coil so as to deliver the heat necessary to achieve the specified air temperature as measured by a temperature sensor in the inlet air (B). The supply temperature to the heating coil is regulated by a control valve (C) in the mixing loop. If the air temperature is too high, the mixing valve is closed to recirculate more return water into the supply. If it is too low, the valve is opened to increase the flow of supply water to the coil. The heating coil is

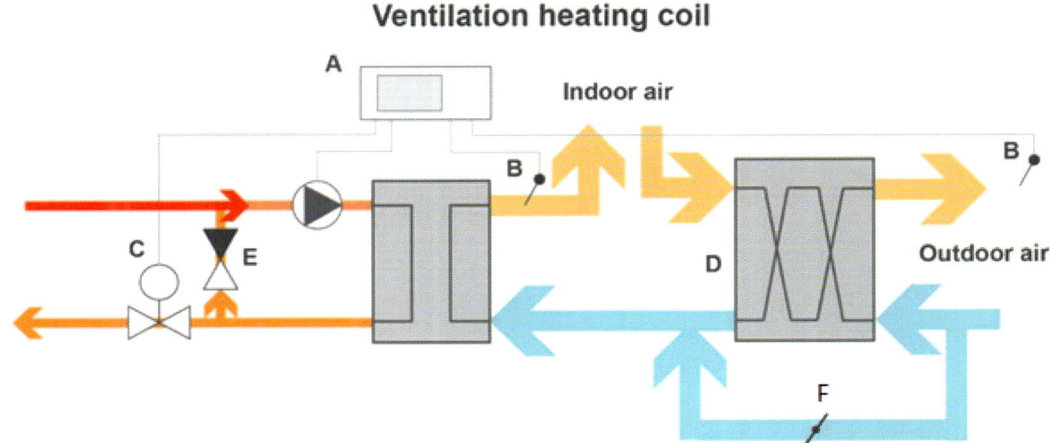

Figure 24. Example of the layout of a ventilation heating coil in a ventilation system where heat recovery is bypassed due to cold outdoor temperature. (A) Controller, (B) Temperature sensor, (C) Control valve, (D) Heat recovery unit, (E) District heating bypass and (F) Bypass damper.

located after the heat recovery unit (D) and is therefore only active if additional heating of the air is necessary. This is mainly the case when outdoor air is too cold to pass through the heat recovery unit and is instead passed through the bypass damper (F). In some cases, the control of the heating coil can be carried out with a constant supply temperature, and with the water flow being varied instead to ensure the correct transfer of heat. In this case, it may be possible to achieve a greater temperature difference in the heating coil, but there would be a risk of freezing at low flows or if the ventilation system malfunctions, unless the controller is set to ensure that the water temperature never decreases below a specified set-point.

3.3 Customers with high supply temperatures requirements

Reductions in supply temperature are the main focus area for most district heating companies, because lower supply temperatures are of major importance for economic and energy-efficient utilization of many future heat sources, such as heat pumps, waste heat, etc. A natural first step for many district heating companies therefore is to focus on customers who have higher supply temperature requirements than the average district heating customer.

Before taking action, it is important to investigate in detail whether the old supply temperature requirements are still valid today. For example, modern hospitals often do not require the same high temperatures as they used to. In one case, a hospital in one district heating area had always been supplied with temperatures around 100–110 °C, but these temperatures were now being reduced to 70–80 °C in a local shunt before the heat was supplied to the hospital. Another case demonstrated similar tendencies, where the local hospital boiler plant could be eliminated, because the hospital no longer required the same temperatures and could be heated by the local biomass CHP plant. Instead of designing new installations similar to existing substations or delivering high temperatures based on old contract specifications, it is therefore well worth thoroughly investigating the actual supply temperature requirements and putting extra effort into customer communication on this topic.

If high supply temperatures are required for only a few industries or special customers, one solution could be to install a local temperature booster in these cases. Such a booster might be a high-temperature heat pump, which can provide an environmentally sustainable solution (Elmegaard, Zühlsdorf, et al., 2017; Khan, Zevenhoven, et al., 2020; Olvondo Technology, 2020). Examples include using high-temperature heat pumps to raise district heating temperatures of 80–85 °C to 184 °C to produce steam for sterilization processes in dairy production (Tveit, 2017). Heat pumps are available for many different purposes and temperatures, and they can offer favourable solutions with payback times

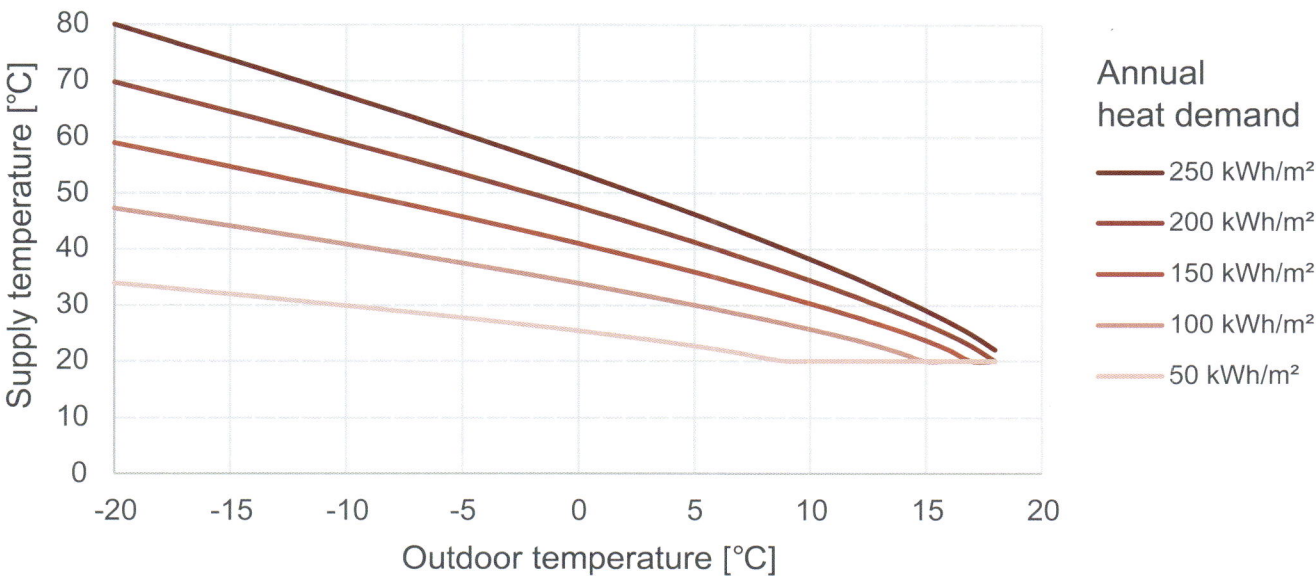

Figure 25. Example of supply temperatures required in a radiator system when various levels of energy renovation are carried out in a building.

of 2–7 years, depending on electricity prices, natural gas prices, and the specific industrial services required (Elmegaard, Zühlsdorf, et al., 2017).

For buildings where supply temperatures are kept high due to the high temperature demands of their heating installations, it is worth keeping in mind that energy renovations carried out in the buildings in years to come will have a positive effect on the possibility of lowering the district heating temperatures. Energy renovations will mean that both radiators and space heating heat exchangers will be over-sized compared to the new heat demands of the building. The operation of the heating system should therefore be optimized so as to operate the components with correspondingly lower temperatures (Oevelen, Vanhoudt, et al., 2018). Figure 25 provides a generic overview of how the supply temperature curve in a heating system changes if the existing radiators are kept while the heat demand of the building is reduced due to various levels of energy renovation. In a renovation example in Borlänge, Sweden, new windows, some additional insulation, and a ventilation system with heat recovery reduced the total heat demand in a 1970s apartment building from 149 kWh/m² to 95 kWh/m². The renovation therefore made it possible to reduce the supply and return temperatures in the existing radiator system to 48 °C/36 °C.

3.4 Substation malfunctions and faults giving higher temperatures

A natural second step for many district heating companies is to take a closer look at customers with high return temperatures. Lowering high return temperatures from relevant customer substations may make it possible to increase the capacity of the district heating network and reduce supply temperatures without compromising the pressure limits of the network. Many buildings that return high district heating temperatures have been found to have some malfunction or fault in the district heating substations (Gadd & Werner, 2015; Månsson, Johansson Kallioniemi, et al., 2019; Zinko, Lee, et al., 2005). These have traditionally been identified by manual data processing, but will soon be automatically identified using software tools (Fester, Bryder, et al., 2019; Jensen, 2020), which can also provide possible explanations for the issues identified. Many substation malfunctions and faults can be identified by conducting on-site visits to buildings with high return temperatures. Some problems are both easy and cheap to fix, which means it is often possible to identify and eliminate the worst issues and achieve a reduction in district heating temperatures quite quickly (Zinko, Lee, et al., 2005). The most common malfunctions and faults identified in building installations for district heating can be summarized as follows:

- *Malfunctions related to the secondary side of the heating installation, in particular poor balancing of space heating systems and shunts or bypasses*
- *Too high supply temperature set-points for both space heating and domestic hot water*
- *Oversized control valves and broken valves or actuators*
- *Broken controllers and misplaced temperature sensors*
- *Faults in relation to heat exchangers*

The frequency of the faults is illustrated in Figure 26, which shows an overview of 520 specific faults identified in the years 1992–2002 in 246 Swedish district heating substations with high return temperatures.

In addition to errors and malfunctions in the district heating substation there can be malfunctions and errors in the building heating installations that impact the operation of the district heating system. *Malfunctions in the secondary side components* make up around 30% of the malfunctions experienced by Swedish district heating companies (Frederiksen & Werner, 2013; Månsson, Johansson Kallioniemi, et al., 2019), and many of these faults are related to poor balancing of space heating systems or missing or broken thermostatic radiator valves (Månsson, Johansson Kallioniemi, et al., 2019). Other common issues on the secondary side are over-sized pumps or pump flows that are too large. The solutions for these types of issue are described in detail in Section 3.6 dealing with improvements in internal building installations.

Other typical malfunctions on the secondary side of the heating installation include inappropriate use of shunts, mixing valves, or bypasses in both space heating systems and domestic hot water systems (Frederiksen & Werner, 2013; Zinko, Lee, et al., 2005). These shunts, mixing valves, and bypasses are often inappropriate leftovers from when the buildings were heated by fossil-fuel boilers that required a high flow and a high return temperature on the secondary side. When changing from individual high temperature building boilers to district heating, it is important to remove all parts of

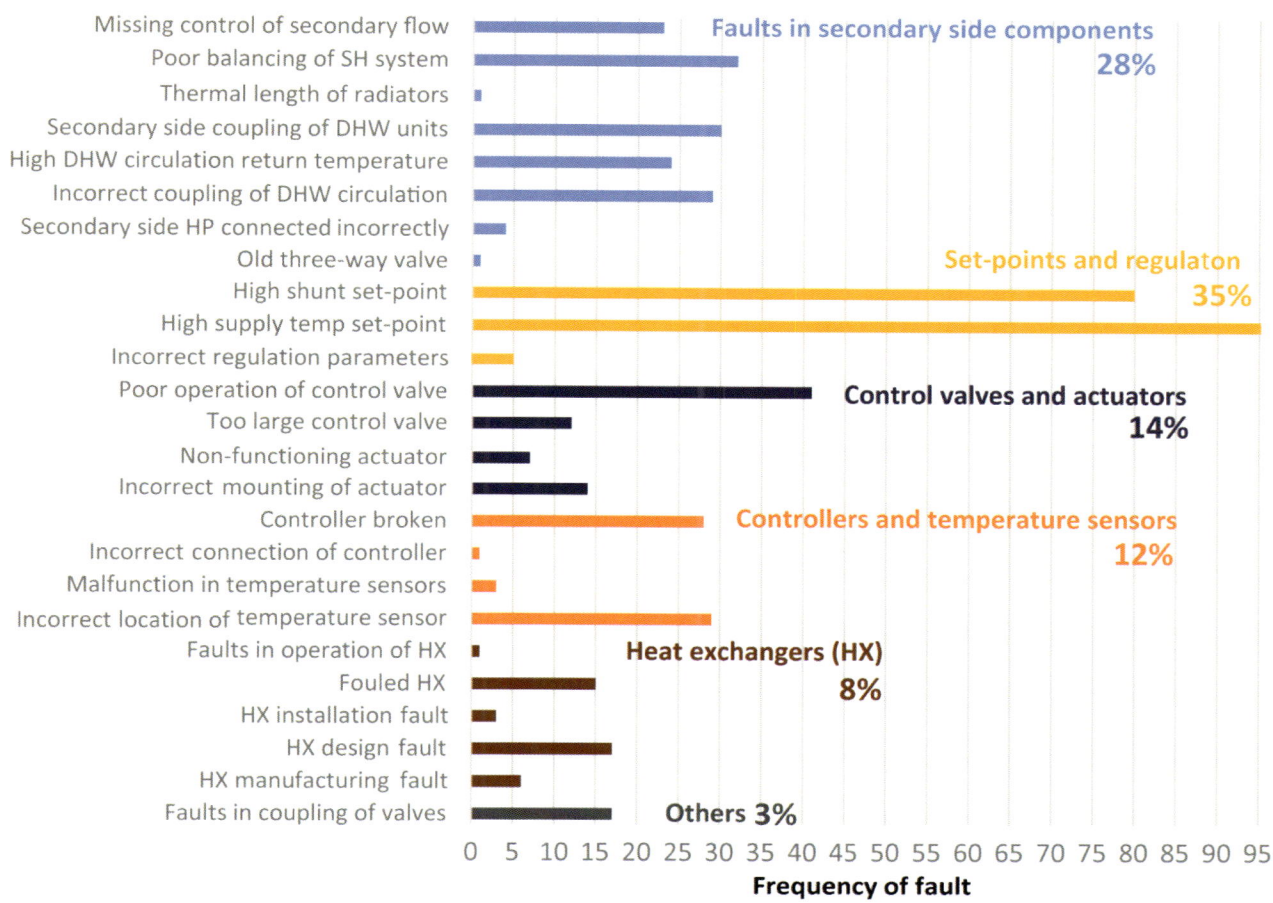

Figure 26. Categorization of 520 faults identified in 246 Swedish substations with high return temperatures. The figure is based on Figure 10.21 in Frederiksen & Werner, 2013, which has been updated based on personal communication.

the old installation that are not required for operating the new district heating based heat supply, this includes all shunts, mixing valves, and bypasses in the boiler room. The modern standardized district heating substation have all the necessary flow controls and logic to supply the building heating installation.

Bypasses can also cause issues in the ventilation system heating coil in buildings equipped with balanced mechanical ventilation. Problems can occur when the heating medium is circulated through a bypass in a heating coil to avoid freezing in the pipes even though the heating coil is no longer active or is currently not in use. In this case, the flow will return with high temperatures because no heat is transferred to the ventilation air. Inactive heating loops should be removed, and temperature control should be added to active ventilation circuits to ensure a minimum flow that will prevent freezing risks.

High temperature set-points are a common problem reported in both Swedish and Danish district heating systems (Frederiksen & Werner, 2013; HOFOR, 2019; Zinko, Lee, et al., 2005). Set-point temperatures should never exceed the supply temperature available in the district heating network. Space heating supply temperatures should be set in accordance with the heat demands of the building (see Section 3.6), and the heat load variation should fit the given building type (Gadd & Werner, 2015). For example, night setbacks are generally not recommended for buildings heated with LTDH. Temperature set-points in domestic hot water installations should comply with national regulations. A suitable supply temperature set-point is often 55 °C, while circulation return temperatures should be at least 50 °C to ensure efficient operation, minimize lime precipitation, and reduce growth of *Legionella* (Euroheat & Power, 2008; HOFOR, 2019). High air temperature set-points in ventilation heating coils can cause the ventilation system to serve as a major source of space heating. This is not only inefficient, but causes high district heating return temperatures because heating coils in the systems are not designed for this.

Control valves and actuators regulate the flow of the district heating through the primary heat exchangers and hot water tanks (see Figure 22 and Figure 23 in Section 3.2), and several guidelines stress the importance of using the right valve sizes (Euroheat&Power, 2008; Olsen, Christiansen, et al., 2014). Over-sized control valves lead to poor control of the flow on the primary side of the heat exchanger and therefore often lead to increased district heating return temperatures. Over-sized control valves can also cause the actuator to break down prematurely due to increased wear (Månsson, Johansson Kallioniemi, et al., 2019). Over-sizing may be due to the valves being sized for extreme peak heat demands that do not occur in practice. The oversizing of the control valves at locations with higher differential pressure levels than the minimum specified by the district heating utility can be avoided by utilizing pressure independent control valves.

Other common problems are that the control valve cannot close entirely or that it is stuck in closed position. The first issue can occur due to erosion of the valve if it

Why are malfunctions often present in building installations?
Malfunctions are often due to wear of components, manual changes in installation settings, or the replacement or modification of installation components. Studies have indicated that around 5% of district heating substations experience a temperature fault every year (Gadd&Werner, 2014). Customers are also more likely to react to issues that impair their comfort than to those that lead to inefficient operation without impacting comfort. However, many issues that occur in the heating installations have a negative impact on the district heating temperatures without affecting occupant comfort.

What are the benefits of fixing the malfunctions?
In addition to the economic benefits related to lower operating temperatures as described in Chapter 2, the effort increases customer satisfaction by reducing customer costs and through the convenience that district heating can offer compared to individual heating solutions. Close customer relations along with access and a mandate to improve customer substations are some of the most important means used by Swedish district heating companies to reduce district heating return temperatures (Månsson, Johansson Kallioniemi, et al., 2019).

How to plan the work of fixing typical malfunctions in building installations
A detailed description for how district heating companies can plan the work to identify and rectify substation malfunctions is available (Zinko, Lee, et al., 2005). The process can be taken on in two steps:

1. Identification of substations with problems
Carry out regular, for example monthly, checks of billing data, quality indexing of substation performance, analysis of return temperatures and flows, or use the over-flow method (Månsson, Johansson Kallioniemi, et al., 2019; Zinko, Lee, et al., 2005). Another method is suggested in Chapter 4 of this handbook. Automatic fault detection is a rapidly developing area of research and automated software for the identification of problems in substations can be expected to become a standard solution in years to come.

2. Service check of customer substations
When a substation with problems has been identified, carry out a physical on-site visit to look for typical issues and malfunctions that lead to high temperatures. The service checks can be performed by an employee of the district heating company or the job can be done by subcontractors (Månsson, Johansson Kallioniemi, et al., 2019). Some researchers give detailed lists of possible faults in the district heating installations of buildings (Geyer, 2018; Geyer&Schmidt, 2017; Zinko, Lee, et al., 2005), and some also provide possible working routines and flow charts for how to identify faults (Euroheat&Power, 2008; Zinko, Lee, et al., 2005). In the near future, automatic fault detection software will also be available to provide a diagnosis of the malfunction that has occurred. Regular service checks of installations can also be planned to be carried out in combination with other tasks, such as regular servicing of energy meters or simply whenever staff are present at the building site.

is operated near to a closed position; the second arise after the summer period when there is no heat demand and the space heating system was turned completely off for the summer (Månsson, Johansson Kallioniemi, et al., 2019). Furthermore, general leaking or broken actuators and control valves are also common malfunctions that lead to poor operation of the district heating installations (Månsson, Johansson Kallioniemi, et al., 2019; Zinko, Lee, et al., 2005). The issues experienced with control valves and actuators can sometimes be resolved by exercising the valve or tightening the screws that hold the valve or actuator together to improve operation. Broken components on the other hand require replacement. Issues due to over-sized control valves can be resolved by replacing them or installing a differential pressure controller to adjust the differential pressure across the control valve (HOFOR, 2019).

Broken controllers or broken or misplaced temperature sensors are reported as a less common but still important issue encountered in buildings with poor utilization of district heating (HOFOR, 2019; Månsson, Johansson Kallioniemi, et al., 2019). Temperature sensors for the regulation of supply temperatures for both domestic hot water and space heating should be located close to the exit of the secondary side supply pipe in the heat exchangers. This will ensure the best regulation of the control valve, whereas other locations can mean that the temperature sensor will not respond properly to increased flow on the primary side of the heat exchanger, which will cause a short circuit on the primary side. This is also a common issue if the heating system is turned off during the summer and the flow on the primary side is not closed, as it should be (HOFOR, 2019). In ventilation systems, sensor faults or improper tuning have also been found to be common problems that lead to unnecessary bypass of heat recovery (La Fleur, Moshfegh, Rohdin, 2017; Visier and Buswell, 2010). Such issues can also lead to increased heat demands and temperature requirements in ventilation heating coils.

Heat exchangers have generally not been found to be a common source of problems (Frederiksen & Werner, 2013; Månsson, Johansson Kallioniemi, et al., 2019), and modern plate heat exchangers generally work well and provide proper cooling of the district heating water. Fouling of heat exchangers and heat exchanger design errors, however, do occur in some cases (Frederiksen & Werner, 2013; HOFOR, 2019; Månsson, Johansson Kallioniemi, et al., 2019). Fouling of heat exchangers can often be identified by a temperature difference between the primary and secondary return temperatures in the heat exchangers of more than 3–5 °C when operating close to the design conditions (Euroheat & Power, 2008; HOFOR, 2019). In some cases, the issue may be resolved simply by cleaning the heat exchanger. Over-sized heat exchangers are generally not a problem in district heating substations, because they can provide improved cooling of the district heating supply (Oevelen, Vanhoudt, et al., 2018).

Other faults that can cause high district heating return temperatures include issues with expansion vessels, leaking installations, and incorrect coupling of valves. These issues should be resolved through general maintenance, with leaking components and broken expansion vessels being replaced, and ensuring the correct pre-pressure in the expansion vessel (HOFOR, 2019).

3.5 Use of data to identify heating system improvements

One major aspect of improving heating installations is to identify not only faults but also improvement potentials. This process can be made more efficient by the use of high-resolution data and automated software solutions. Digital solutions developed with focus on improvements in the operation of internal building installations have been developed, for instance by Neogrid Technologies, Leanheat (Danfoss), NorthQ, and Egain. These solutions make use of measurements of indoor air in critical building zones, solar radiation, and wind and weather forecasting to optimize operation patterns and provide the desired comfort with the least energy consumption. However, even if detailed software solutions are not installed in a building, it is still possible to use data to make simple analyses that provide a basis for estimating possible improvement potentials. Such analyses can be carried out based on data sources that are available to most building operators, such as heat meters, electronic controllers, and heat cost allocators, provided they have been implemented on the basis of current national regulations (Castellazzi, 2017). These sources can be supported by supplementary measurements of return temperatures on the primary side of the substations, from the space heating system, and from the domestic hot water circulation, if these are not available through any of the above devices.

The following key performance indicators for the operation of each part of the heating installation (the space heating system, the domestic hot water system, and the ventilation system) can typically be calculated based on the data available:

1. Share of heat used out of total heat supplied.
2. Average supply and return temperatures
3. Share of heat losses out of heat supplied.

These key performance indicators provide an overview of where in the heating installations the largest improvement potentials can be found. They can also be used as a simple benchmark to compare different buildings and get an understanding of how far the operation of a building installation is from what might be expected. The heating system operation can also be analysed to identify inefficient operation patterns. This can be done using two simple methods:

4. Graphical analysis for visualisation of operation patterns
5. Linear regression analysis for the identification of typical operation modes.

The use of the key performance indicators and these simple analysis methods is illustrated in the case of TU Darmstadt's Campus Lichtwiese in Chapter 5, and the methods are described in detail below.

1. The share of heat used for space heating, domestic hot water, and heating of ventilation air can be used to identify which systems have the highest heat demands and therefore the greatest effect on heat demands and district heating temperatures. The key performance indicator is calculated by comparing the heating consumption in each individual system to the total heating consumption in the building.

2. Average supply and return temperatures can be used to provide an overview and identify whether high temperatures are mostly an issue in some parts of the heating installation. Average temperatures can be calculated either as a time-based average or a mass-based average. While the time-based average is an average of the temperatures measured, e.g. every hour, the mass-based average takes into account the fact that the water mass may be greater in some hours than others, so that the temperatures of those hours weigh more. Although the latter is more accurate, mass measurements are not always available, so time averages are often used. In this case, you should consider exclu-

ding from the analysis periods with very low heat demands in order to calculate the average temperature for the periods where the heating installation is in use. Time and mass averaged supply and return temperatures can be calculated using Equation 7 and Equation 8 respectively:

$$\bar{T}_{S/R} = \frac{1}{n_{ts}} \sum_{i=1}^{n_{ts}} T_{S/R,i} \qquad \text{Equation 7}$$

$$\bar{T}_{S/R} = \frac{\sum_{i=1}^{n_{ts}}(\dot{M}_i \cdot T_{S/R,i})}{\sum_{i=1}^{n_{ts}}(\dot{M}_i)} \qquad \text{Equation 8}$$

where $\bar{T}_S$ and $\bar{T}_R$ are the average supply and return temperatures during a period of n_{ts} time steps (e.g. hours), T_{S_i} and T_{R_i} denote the temperatures measured in time step i out of n, and $\dot{M}_i$ denotes the mass flow during time step i.

3. The share of heat losses in the systems compared to the actual supply of heat can be calculated for both the space heating system and the domestic hot water system. This can show whether the heat losses or heat demands have the largest impact on system operation. The share of heat losses in the space heating system can be estimated based on the total heating consumption in the space heating system and the estimated heat emitted from radiators and other heating elements, where this data is available from the heat cost allocator company.

The share of circulation heat losses in the domestic hot water system can be shown by the system efficiency, which can be estimated based on the total heating consumption for domestic hot water including losses and the heat used specifically to heat the domestic hot water used in the building ($Q_{DHW,use}$), as shown in Equation 9 (Bøhm, Schrøder, et al., 2009). If the latter is not available directly, it can be calculated based on the volume of domestic cold water used in the domestic hot water installation (V_{DCW}), and the temperatures of the domestic hot water and cold water respectively, as shown in Equation 10. In Denmark, hot and cold water temperatures are typically assumed to be around 55 °C and 10 °C, respectively, but the temperatures depend on the operation of the individual installations, the temperature of cold water delivered by the utility, and the temperature in the substation room (Bøhm, 2013).

$$DHW\ system\ efficiency\ [\%] = 100 \cdot \frac{Q_{DHW,use}}{Q_{DHW,total}} \qquad \text{Equation 9}$$

where $Q_{DHW,total}$ is the total heating consumption for the domestic hot water including circulation losses, and $Q_{DHW,use}$ is the heat used specifically to heat the domestic hot water used in the building.

$$Q_{DHW,use} = V_{DCW} \cdot \rho \cdot c_w \cdot (T_{S,DHW,Build} - T_{DCW}) \qquad \text{Equation 10}$$

where V_{DCW} is the volume of domestic cold water used in the domestic hot water installation, ρ is the density of water, c_w is the specific heat capacity of water, and $T_{S,DHW,Build}$ and T_{DCW} are the domestic hot water supply temperature and the domestic cold-water temperature respectively.

It should be kept in mind that some of the losses from both the space heating and the domestic hot water system will be utilized for space heating inside the buildings and will therefore not be totally lost. In one specific Danish case, it was estimated that approximately 30–40% of heat loss from the domestic hot water system was utilized for heating (Bøhm, Schrøder, et al., 2009). However, in buildings with cooling demands, heat losses from domestic hot water systems may end up with an additional negative impact in the form of higher cooling requirements.

4. Visualization of data in graphs can be a useful tool to illustrate changes in temperature over a period of time and identify daily or seasonal patterns. The tool represents a simple way of getting an overview of the heating system operation and identifying faults or unexpected operation trends. Time series graphs can be produced for the space heating system, the domestic hot water system, and the ventilation system individually, and it may be relevant to include supply and return temperatures and, if possible, also the mass flows in the systems. The analyses can be used to illustrate whether peaks with high temperatures occur and whether for example night setback is applied in the heating system control, as shown from the examples in the case of Campus Lichtwiese at TU Darmstadt in Chapter 5.

5. Linear regression analysis can be used to investigate whether a heating installation is operated in an efficient operation mode. For instance, the analysis can be used to investigate the relationship between return temperatures in the heating installation and the measured heat demands. In a well-functioning heating installation, return temperatures generally increase with increasing heat demands. The relationship between the two variables can be investigated, for example in Excel, by plotting the measured return temperatures at given heat loads into a graph. Then a linear trend line can be added to the graph based on linear regression, which aims at fitting a linear equation to the data presented. The linear relation is given by Equation 11.

$$\dot{Q} = \beta_0 + \beta_1 \cdot T_R \qquad \text{Equation 11}$$

where $\dot{Q}$ is the heat flow demand and T_R is the return temperature from the heating system.

If return temperatures increase with decreasing heat flow demands, the linear trend line will be declining and the relationship between return temperature and heat flow demand will be negative (negative β_1). This can indicate a malfunction in the operation of the given heating installation. If the heating installation is operated with different operation patterns in different periods, the correlation coefficient (R^2) will be low, which indicates that the estimated linear equation does not fit well with all the given data points. This may indicate inefficient periodic or seasonal operation. Examples of this are shown for the ventilation system in the case of Campus Lichtwiese at TU Darmstadt in Chapter 5.

3.6 Actions to lower the temperatures in space heating systems

A well-functioning space heating system can typically be supplied with temperatures that vary over the year in the range of 40–70 °C. Return temperatures can optimally be as low as 25–35 °C. The lower range of return temperatures is demonstrated for example in well-functioning heating systems in single-family homes with a direct district heating connection. Well-functioning heating systems in multi-family homes should be able to provide annual average district heating return temperatures in the range of 30–35 °C, provided that malfunctions in the district heating installations are corrected, the space heating system is equipped with heat exchangers with the correct thermal length, and hydronic balancing is carried out correctly (Petersson & Werner, 2003; Rämä, Lindroos, et al., 2019). If time series analysis of data and calculated average temperatures indicate that the space heating temperatures are far above these values, this indicates a large potential for improvement.

Examples of how data can provide information about possible heating system improvements:

1. **Where is the main issue?**
   By comparing the share of heat used in space heating, domestic hot water, and ventilation systems it is possible to evaluate whether the use of heat corresponds to expectations (see Section 3.1), and show which installation to focus on.
2. **Are the heating system temperatures reasonable?**
   Average supply and return temperatures can be compared to benchmark values. Examples of benchmark values are given in Sections 3.2, 3.5, and 3.6 to illustrate improvement potentials, but may also consist of comparison with other similar buildings.
3. **What is shown by average heating system temperatures?**
   High average return temperatures in space heating systems could indicate incorrect control curve settings, a lack of hydronic balancing, or small heating elements. Average temperatures from the domestic hot water system can show whether the system is operating correctly to avoid *Legionella* and ensure energy efficiency.
4. **How much heat is wasted as heat loss in building installations?**
   If the shares of heat loss in space heating and domestic hot water systems are large, it may be worth considering adding additional pipe insulation to the systems (especially in unheated rooms) or reducing temperatures as much as possible. The share of heat loss in the domestic hot water circulation system has a large impact on the return temperatures that can be achieved, as shown in Section 3.7.
5. **Is the heating system operated without heating system setbacks?**
   Time series analysis can be used for the simple identification of night setbacks that are generally undesirable in LTDH systems.
6. **Are there bypasses in the heating system?**
   Time series analysis can show whether supply and return temperatures in the internal building installations are similar and follow the same pattern irrespective of the load situation. This can indicate that there is a bypass in the installation.
7. **Does the heating system operate as expected?**
   For a well-operated space heating system, the supply and return temperatures are generally expected to decrease with increasing outdoor temperatures.

Causes for space heating return temperatures to be above the desired level of 25–35 °C can be anything from simple malfunctions and inefficient weather compensation curves to heating elements or heat exchangers being too small. A common excuse for needing high return temperatures is that current heating elements are too small to allow low-temperature heating. However, this is seldom the case. Studies have shown that average radiator sizes in up to 80% of Danish buildings are large enough to accommodate the transition to LTDH (Østergaard, 2018). This is supported by measurements from a number of existing heating systems in Sweden and Switzerland showing that typical mean supply temperatures in a number of existing radiator systems are as low as 50–60 °C at an outdoor temperature of -5 °C (Averfalk, Werner, et al., 2017; Jangsten, Kensby, et al., 2017).

Rather than small radiator sizes, common reasons for high space heating return temperatures are simple malfunctions or inefficient operation of the internal space heating system. Poor control of the heating system can cause water mass flows through some heating elements to be so large that the combined water from all the elements is not cooled down properly. Large water mass flows with high return temperatures from even one heating element can easily overrule the effect of low return temperatures from many well-functioning radiators that have small water mass flows (Benakopoulos, Salenbien, et al., 2019). Such a situation can occur when:

- There are unintentional direct bypasses in internal space heating installations, such as where a radiator is removed and the space heating supply and return pipes are connected instead of the correct plugging of the pipes (Sørensen, 2017).

- A few heating elements are not equipped with proper thermostats to regulate the water mass flow, and the valves are large and do not have pre-settings to reduce the valve opening.
- Supplementary hydraulic floor heating circuits in bathrooms or hydraulic towel rails are not equipped with return temperature thermostats.
- If some occupants operate their heating system by only turning on a few of the radiators, causing some radiators to heat a much larger area than intended. Examples have shown that it is not uncommon that only 1/3 of radiators are typically in use.
- Thermostats are operated with an on/off schedule of thermostat set-points, which causes large water mass flows during re-heating periods (Liao, Swainson, et al., 2005).

The problems will be worsened if:

- The heating system is not balanced with string balancing valves on the risers. A lack of balancing commonly leads the heating system operator to increase pump pressure and supply temperature to ensure sufficient water flow and heat reach the far end of the installation. This in turn results in large water flows and high return temperatures in cases where the heating elements close to the pump are not controlled properly.
- The supply temperature curve is not set correctly in accordance with the heating system design and the actual heat demand in the building. Often supply temperatures are increased to ensure the thermal comfort of all occupants, which results in an escalation of the issues with high return temperatures, for example from unintentional bypasses.

It is therefore worth investigating how much the space heating temperatures can be improved by improving the control of the space heating system. A first step in this regard is to remove unintentional bypasses by ensuring that all radiators are equipped with thermostatic radiator valves, that there is suitable control of hydraulic bathroom floor heating circuits and towel rails, and that there are no unintentional direct connections between supply and return in the space heating system. This first step is essential for further improvements to have the intended impact. Moreover, the installation of thermostatic radiator valves is often a financially beneficial measure to apply, because thermostats can provide large heat savings and short pay-back times (Cholewa, Siuta-Olcha, et al., 2017). Thermostats are currently only installed in 24% and 43% of collective housing in France and the UK respectively (Averfalk, Werner, et al., 2017).

Secondly, it can be worth investigating if the installed pumps fit with the internal installations and if they are operated correct. Over-sized pumps are not uncommon, and it is also not uncommon that pumps are operated at a wrong setting with too high speed. As for high supply temperatures, these settings are often chosen as an easy fix to ensure occupant comfort and avoid complaints. However, over-sized pumps and wrong operation can lead to high flows in the heating system that provide poor operation conditions for valves in the system and often cause high return temperatures. It is therefore essential as a minimum to check if the pump speed should be reduced.

The next step is to consider whether to improve the hydronic balancing of the space heating system through the installation of string-balancing valves and radiator valves with proper pre-settings on all radiators (if these are not already installed). Proper hydronic balancing can make it easier to ensure that the correct amounts of heat are delivered to all parts of the building, and thereby reduce heating consumption, improve the thermal comfort of occupants, and help regulate and limit the water mass flows in the heating system (Cholewa, Siuta-Olcha, et al., 2017; Euroheat & Power, 2008). Proper hydronic balancing in a space heating system equipped with thermostats can provide heat savings in the range 9–15% and the investment can be paid back in 1–2 years (Cholewa, Siuta-Olcha, et al., 2017; Trüschel, 2005). The savings achieved by installing thermostats and implementing improved hydronic balancing depend on the situation before the measures are introduced. In addition to energy savings, hydronic balancing can also lead to a reduction in occupant complaints or a reduction in heating system temperatures respectively in cases where poor occupant comfort or poor cooling in the district heating installations is the main issue (Trüschel, 2005).

Whether hydronic balancing is carried out in detail or not, it is good practice to evaluate whether the space heating supply temperature curve is achieving the lowest possible heating system temperatures. Supply temperatures at unnecessarily high levels can be chosen as an easy fix to ensure occupant comfort and avoid complaints, however this may lead to unnecessarily high district heating supply temperatures and higher return temperatures. The potential for lowering the heating system temperatures in a building can be evaluated through detailed calculations based on the design temperatures of the heating elements in the building and the design heat demand of the building, as described in text books and research on hydronic balancing (Kärkkäinen, 2010). As a more practical alternative, the potential can also be demonstrated through a number of simple calculations based on the actual heat demand and the design heat output of the heating elements in the building.

This simple stepwise approach to minimise the supply return temperatures can be defined as follow. First, the actual measured annual heating consumption for space heating in the building (from energy meters) is divided by the annual number of heating degree days at the building location and multiplied by 24 to get a value per hour. This provides a global heat demand coefficient for the building in kW/°C. Second, this coefficient is used to estimate the heat demand at each outdoor temperature, simply by multiplying the coefficient by the number of heating degree days at the given outdoor temperature. Third, the part load operation at each outdoor temperature can be calculated from knowing the heat demand at various outdoor temperatures (step 2) and the design heat output of the heating elements installed in the building. The information about the design heat output of the installed heating elements can be obtained from the heat cost allocator company or from blueprints of the heating system. This will make it possible to calculate the logarithmic mean temperature difference that is necessary for the heating elements to cover the total heating demand at each part-load and outdoor temperature. Then, the minimised supply temperature curve is estimated assuming that the heating systems are optimally controlled and operated and that the design flow will be circulated in the system in all operating conditions.

Figure 27 shows an example of the potential minimised supply and return temperatures calculated by using this method for a Danish apartment building from the 1990s that was designed for supply and return temperatures of 70/40 °C. Since the method described assumes a perfect distribution of the heat demand throughout the building and that all heating elements are used, it provides a reference for the potential temperatures that could ideally be achieved in the building. The method can be used as a baseline for a stepwise reduction of current space heating supply temperatures in order to find an applicable level without using detailed calculations but only using the energy measurements from the energy meters, space heating design capacity and the heat degree days distribution for the specific year.

Heating systems can be operated with either a small temperature difference and high flow, or a larger temperature difference and less flow. In practice, the supply strategies are often somewhere in between (Jangsten, 2016). If the heating system is well-balanced and equipped with both string balancing valves and thermostatic radiator valves with pre-settings, the supply temperature curve can be chosen to support a relatively stable water mass flow in the heating system at various outdoor temperatures (Jangsten, 2016; Kärkkäinen, 2010). Weather compensation will then allow valves and thermostats to be operated in the intended operating conditions, leading to better control and a longer lifetime of components. Alternatively, a strategy with a higher flow and a lower temperature difference can be chosen, which will make it possible to apply the lowest possible supply temperature in the heating system and thereby reduce the heat loss from the space heating distribution system. This strategy is especially relevant if the billing is based on heated square meters or if the heating system is not equipped with control and balancing equipment, because it is more robust in the face of faults and malfunctions in the heating system (Benakopoulos, Salenbien, et al., 2019; Trüschel, 2002). If the strategy is applied in buildings with balancing valves, it may require an update of critical balancing valves, but then the strategy encourages tenants to use all radiators to heat their apartment, which is recommended.

Obtaining proper hydronic balancing and ensuring necessary flow limitations can be a challenging manual process. In the study in case 13, a new electronic radiator valve with return temperature limiting functionality was therefore developed, with the aim of providing an easy way to ensure sufficient cooling across each heating element and provide a degree of automated hydronic balancing in the space heating system. The valves were tested in two apartment buildings in Denmark, and it was demonstrated that the valve was successfully able to limit the return temperature from the radiators to a given set-point. The tests indicated that the valve could

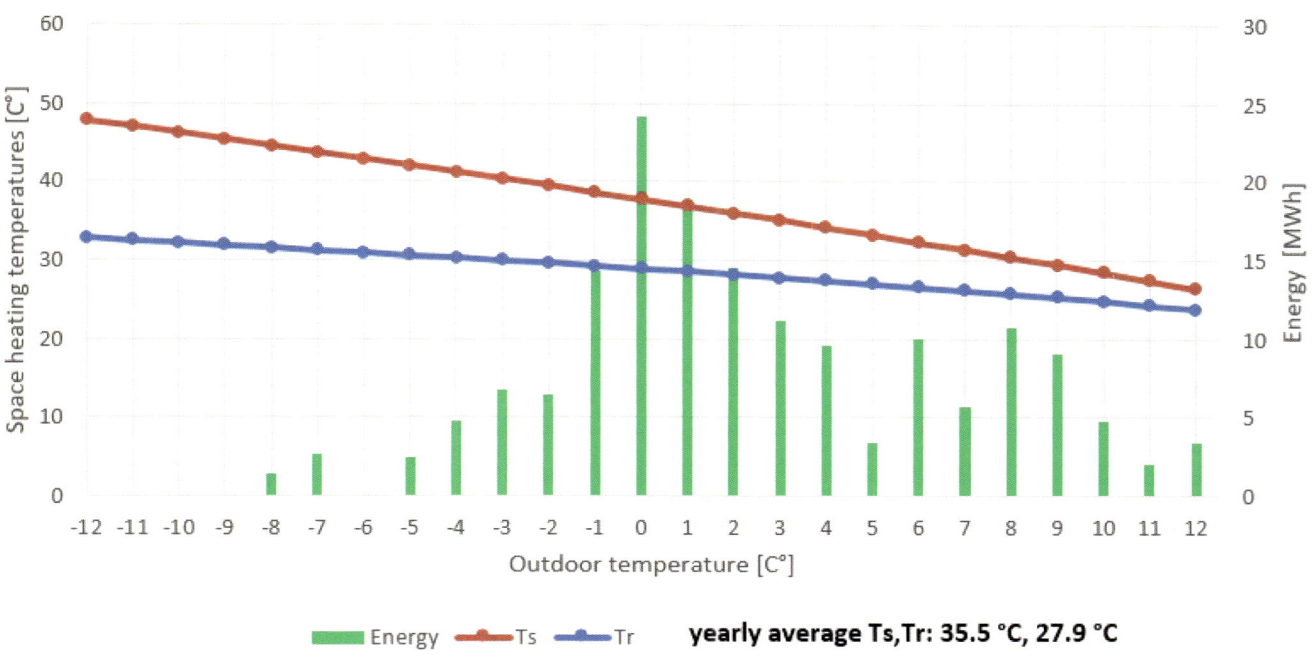

Figure 27. Example of a typical heating system supply temperature curve (and expected return temperatures) in a radiator system supplied with similar water mass flows irrespective of the time of the year.

help ensuring a larger cooling in the heating system, and in one case it was seen that return temperatures were reduced by 5 °C during the period where the valve was installed on all radiators. As for other valves, the control is still sensible to faults and malfunctions in the heating system, and return temperatures where therefore higher than expected in cases where the pump was operated at too high speed and in some cases where the occupants dismantled some of the valves. Nevertheless, if the space heating system is operated properly, the valve may provide a good alternative to manually pre-set radiator valves, and in the given case studies the valve was seen to help reduce space heating return temperatures to around 35 °C.

When large renovations or replacements of heating system components are carried out, for example because old installations break down or can no longer deliver the comfort occupants expect, it may be worth considering a general re-design of the existing installations. For district heating substations and ventilation systems, old installations that should be replaced include shell-and-tube heat exchangers, which are both less robust and less energy-efficient than the new plate heat exchangers (Gummerus, 2016). New heat exchangers and heating coils should be designed for low-temperature operation, as illustrated in Figure 28, where the design temperatures of the old space heating heat exchangers are compared to the temperatures applied for heat exchangers in a LTDH heating network. Even though the temperature difference on the primary side is reduced, it should not be necessary to increase the flow, because heat demands are also expected to be reduced in the future.

Other relevant improvements of the space heating system include replacing radiators in for example living rooms or bathrooms, where occupants often prefer high indoor comfort temperatures, especially if the existing radiators are small and pose a limit to reducing the overall heating system temperatures (Østergaard & Svendsen, 2016a). Another option is to replace old single-pipe space heating systems, because these cannot deliver as low return temperatures as a well-balanced two-pipe system. Although buildings with one-pipe heating systems can also be connected to district heating systems, a two-pipe heating system is recommended and more common (Euroheat & Power, 2008).

Figure 29 summarizes the proposed actions in space heating systems to lower the primary district heating temperatures.

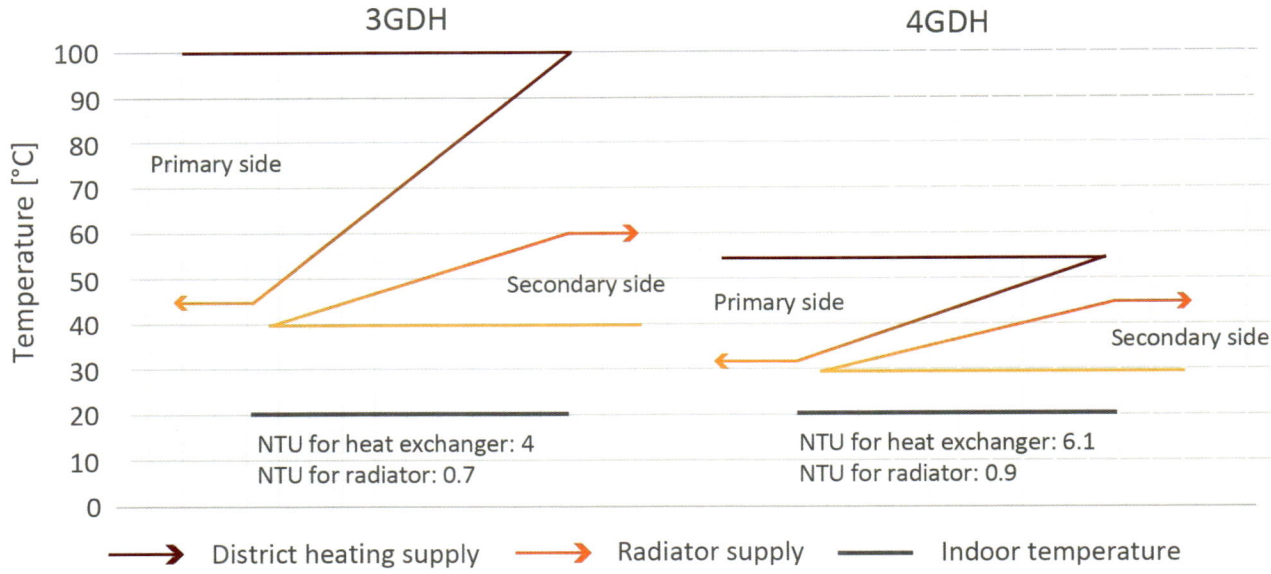

Figure 28. Example showing the difference between the design of a heat exchanger for space heating in an old district heating system and the design of a heat exchanger intended for 4GDH.

3.7 Actions to lower the temperatures in domestic hot water systems

The temperatures in a well-functioning domestic hot water system depend on the specific type of installation. As described in Section 3.1, minimum supply temperatures can range from 50 °C to 65/70 °C depending on national requirements in respect of *Legionella*. Optimal return temperatures from domestic hot water installations can potentially be as low as 25 °C for installations with instantaneous heat exchangers with no circulation of domestic hot water and with short service pipes, while well-functioning installations with a domestic hot water tank and circulation of domestic hot water will be able to deliver return temperatures of about 35 °C (Rämä, Lindroos, et al., 2019).

District heating return temperatures are often as high as 40–45 °C in domestic hot water installations. This is

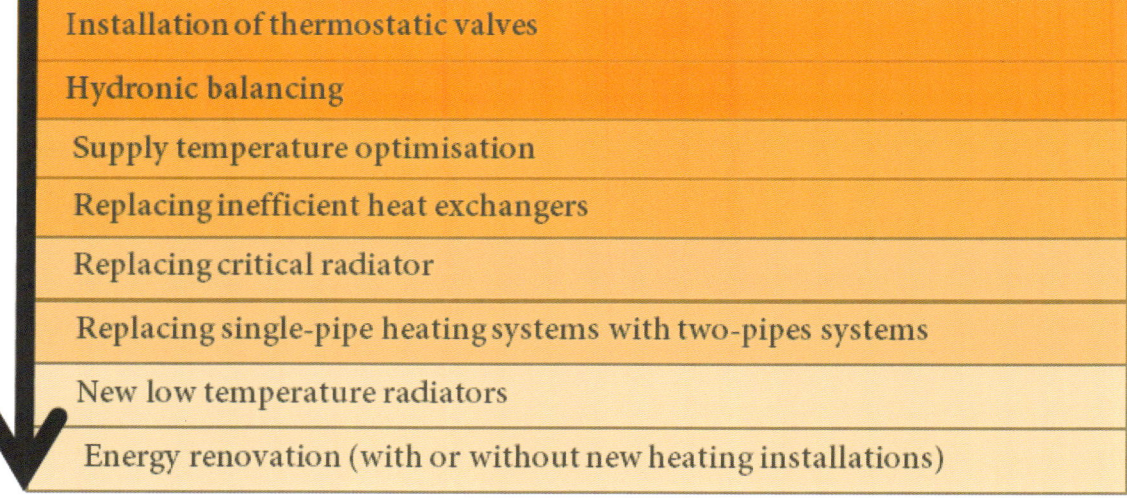

Figure 29. Actions to lower the temperature in space heating systems

because domestic hot water circulation temperatures are kept at a minimum of 50 °C to avoid the growth of *Legionella* bacteria (Bøhm, Schrøder, et al., 2009). The circulation system has considerable influence on the return temperature in cases where circulation heat losses make up most of the heat demand for domestic hot water, which is not an uncommon situation (see Section 3.2) (Best, Braas, et al., 2020; Bøhm, Schrøder, et al., 2009; Lidberg, Olofsson, et al., 2019). In such cases, it is relevant to reduce circulation heat losses, which should initially be done by adding insulation to the domestic hot water pipes. Further improvements could include the replacement of old domestic hot water tanks, removing unnecessary circulation pipes in the attic or basement, and installation of pipe-in-pipe circulation for reduced heat losses. A study combining all these measures was found to reduce the annual heat losses from the domestic hot water system by approximately 25%, even though pipe insulation was limited due to lack of space (Bøhm, 2013).

High return temperatures from installations with a domestic hot water tank may be a result of poor control, which causes district heating mass flow rates to be higher than necessary. Limiting the primary side flow-rate that charges the domestic hot water tank can slow the reheating of the tank and thus help solve the issue. The limitation can be implemented by installing a smaller control valve, though this can lead to customer dissatisfaction if the charging becomes too slow to deliver the heat necessary for domestic hot water. Instead, an improved solution is currently being developed, whereby the central electronic controller regulates the charging flow rate to the tank in accordance with the average domestic hot water demand, while allowing greater charging flows in periods with extraordinarily large domestic hot water demands (Huang, Yang, et al., 2019)The concept was demonstrated in a new building in Copenhagen, where peak load of the domestic hot water tank was successfully reduced from 70-80 kW to 25 kW and the average return temperature from the tank was reduced by 10 °C (Yang & Honoré, 2019).

The return temperature reduction potential of this type of solution depends on the available district heating supply temperature and the share of circulation heat losses. Figure 30 shows that high share of circulation heat losses and high supply temperature would lead to high return temperatures. It is important to highlight that, based on the DHW system efficiency defined in Equation 9, the curve where circulation heat losses are three times bigger than the DHW use (Qcir/Quse = 300%) corresponds to a DHW efficiency of 25%; whereas, this is increased to 80% for a DHW system where the circulation heat losses are only one quarter of the DHW consumption (Qcir/Quse=25%). It should be noted that the graph is based on the domestic hot water tank being of optimal size, design and functionality illustrating the potentials for the return temperatures in different conditions.

Due to the impact on the return temperature in DHW installations with storage tanks, an alternative solution was developed by integrating a micro booster heat pump to compensate for the circulation heat losses. The demonstration in an old multi-family building in Copenhagen, characterised by domestic hot water circulation losses equal to 70% of the total DHW consumption, shows that the return temperature was successfully reduced from 47 °C to 22 °C (Thorsen, Skov, et al., 2019).

In many cases, district heating companies provide supply temperatures that are higher than what is necessary to comply with the temperature requirements to avoid the growth of *Legionella* in building installations. The potential to reduce the supply temperatures without compromising supply temperature requirements at each customer, can be demonstrated through the use of frequent data measurements from smart energy meters in the buildings. Supply temperatures should always be high enough to allow the building operators to deliver sufficient temperatures in domestic hot water installations to avoid the growth of *Legionella*. However, they may be lowered further from the current values, if the specific building installations for domestic hot water are modified to allow this.

The installation of disinfection equipment in the domestic hot water system can be one way of making it possible to reduce district heating supply and return temperatures in the future, if it is acceptable for the occupants and fulfils the regulation requirements. If *Legionella* bacteria are eliminated from the domestic hot water system through disinfection, domestic hot water temperatures can be reduced to the comfort requirement of having 45 °C warm water available at the hot water taps. This means that the domestic hot water installation can in principle be operated with lower supply and return temperatures, which will also reduce

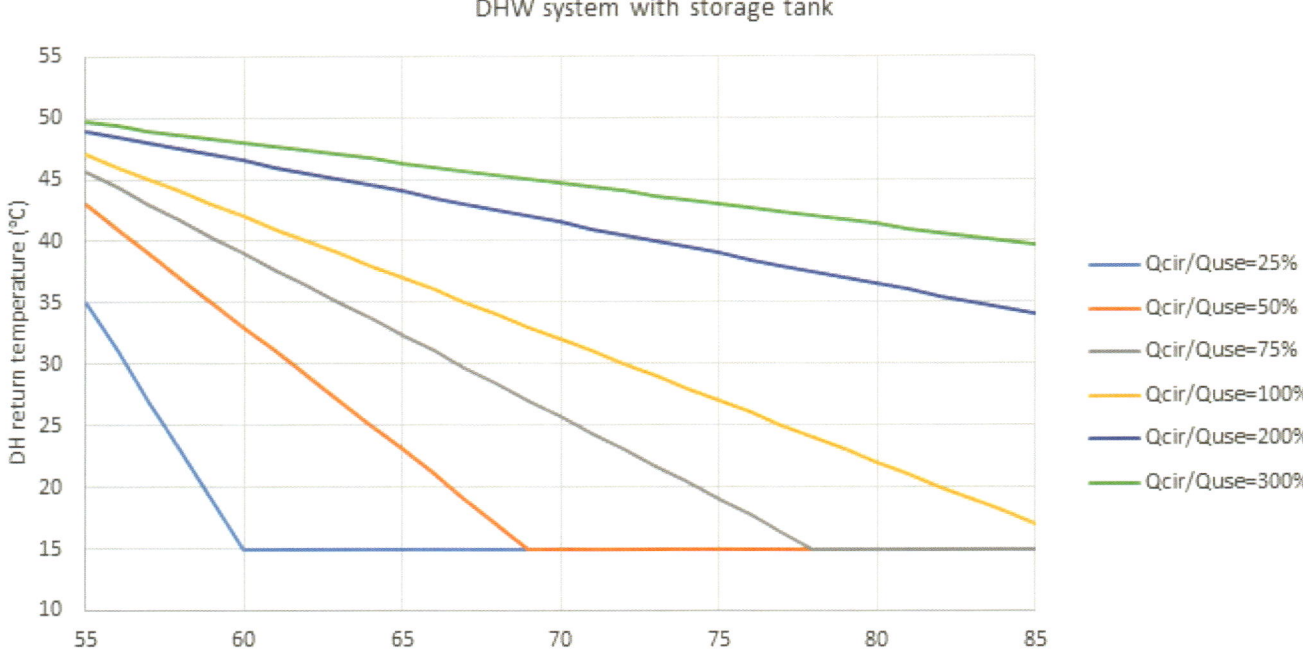

Figure 30. Impact of circulation heat losses and supply temperature on return temperature for a domestic hot water system with storage tank

heat losses. Disinfection also reduces the risk of *Legionella* related illnesses as *Legionella* bacteria are efficiently removed from the domestic hot water. In a case reported in Chapter 7, a disinfection method based on electrolysis of saltwater was found to enable a reduction in domestic hot water circulation temperatures from 55 °C/51 °C to approximately 47 °C/44 °C (Danish Clean Water, 2020). With such reductions in domestic hot water circulation temperatures and optimization of the circulation strategy, the heat loss from the domestic hot water circulation system could be reduced by 38% or more (Benakopoulos, Tunzi, et al., 2021). Alternative disinfection methods, for example using chlorine dioxide, have also been successfully applied (Dalla Rosa, Li, et al., 2014) .

In some cases, it may be relevant to install electric solutions for domestic hot water installations in order to reduce heat demands or district heating supply and return temperatures. In office buildings where domestic hot water taps may be spread out in discrete locations and hot water demand is very small, local electric heaters can be used for the preparation of domestic hot water instead of central domestic hot water preparation and circulation systems. In other cases, it may also be relevant to install a domestic hot water circulation heat pump unit to reduce typically high return temperatures from domestic hot water circulation systems. The heat pump unit will use district heating as a medium to heat the domestic hot water circulation loop while cooling the district heating. A recent test demonstrated that this solution could reduce district heating return temperatures from approx. 45 °C to 20 °C and that electricity would cover 15–20% of the heating needed in the domestic hot water circulation system (Thorsen, Ommen, et al., 2019; Thorsen, Svendsen, et al., 2020).

In cases where district heating supply temperatures are reduced to below 50 °C, often referred to as ultra-low temperature district heating, individual domestic hot water heat pumps or in-line electric heaters can be installed to boost the domestic hot water temperature in the building to suit the required comfort level. Electric boosting solutions in combination with district heating supply temperatures around 40–45 °C, in some cases with low heat densities, have been found to be an energy-efficient alternative to high-temperature district heating (Best, Orozaliev, et al., 2018; Brand, Gudmundsson, et al., 2016; Danish Energy Agency, 2015; Yang & Svendsen, 2018). However, electric boosting solutions are currently in most cases not as economical as LTDH due to the additional cost of electricity and components (Best, Orozaliev, et al., 2018; Danish Energy Agency, 2015; Gudmundsson, Thorsen, et al., 2020;

Lund, Østergaard, et al., 2017; Meesenburg, Ommen, et al., 2020; Yang & Svendsen, 2018). For district heating supply temperatures around 45 °C, recent studies have shown that a simple in-line electric heater combined with a micro-hot-water tank will often be more favourable than a heat pump solution (Yang, 2016; Yang, Li, et al., 2016b). It should however be ensured that the electrical fuses in the house and in the local electricity grid can support the additional peak electricity consumption (Brand, Gudmundsson, et al., 2016).

In the case of multi-family buildings that require large-scale renovation or where the existing domestic hot water system causes excessive heating consumption, poor occupant comfort, or a continuous risk of *Legionella* contamination, it may be worth replacing the domestic hot water system with new decentralised systems based on the flat station concept. These can provide extra safety with regard to *Legionella*, and they will simplify billing for heating and enable a higher degree of occupant choice about heating system operation in individual apartments. Studies have estimated that these benefits mean a flat station system is not necessarily much more expensive than the installation of a new standard domestic hot water system (Thorsen, Christensen, et al., 2017). A renovation where flat stations were installed and a one-pipe heating system was replaced with a two-pipe system was found to provide annual energy savings of about 30% and a reduction in the return temperature from 65 °C to 35 °C (Thorsen, Christensen, et al., 2017). An estimated 8% energy savings were derived from the reduced length of distribution pipes, 10% were from eliminated central substation heat losses, and 12% were from increased occupant awareness and individual optimization of heating supply. A Danish case study showed that heat losses could be reduced by as much as 30% if the flat station concept was adopted instead of the traditional pipe layout for domestic hot water and space heating (Yang, Li, et al., 2016a).

The proposed actions in domestic hot water systems to lower the primary district heating temperatures are summarised in Figure 31.

3.8 Actions to lower the temperatures in ventilation systems

As shown in the study in Chapter 5, heating coils in ventilation systems can be a major cause for high district heating return temperatures. However, the problem can often be fixed at a rather low cost, and when the heating coil is properly sized and controlled, it should be possible to achieve supply temperatures below 60 °C and return temperatures of 20–30 °C. If the problem persists even when issues with bypasses and temperature set-points have been corrected, further improvements of the system may need to be considered.

The first measure that should be considered in existing ventilation systems is to install an efficient heat recovery unit, if one is not already in place. In Danish residences, as much as 90% of the ventilation heat loss (typically 35–40 kWh/m^2 per year) can be saved if heat recovery is included in the ventilation system. This means heat recovery is essential in buildings with balanced mechanical ventilation, and it also means it is worth installing modern ventilation units in buildings that do not currently have balanced mechanical ventilation. The installation of heat recovery may also make it possible to reduce district heating supply temperatures, because the space heating system can be operated at lower water temperatures when heat demands are lower. Researchers in Estonia modelled the full renovation of a multi-family building, including mechanical ventilation with heat recovery, and the resulting weighted-average return temperature to the district heating network was 22 °C (27.4 °C for space heating; 17.2 °C for domestic hot water). When they simulated a renovation package that instead recovered heat using an air-to-water heat pump with the evaporator in the exhaust air, the district heating return temperature was 10–15 °C higher (Thalfeldt, Kurnitski, et al., 2018).

The most efficient ventilation system is one that achieves the proper balance of supply and exhaust air flows in combination with improvements that help increase the air tightness of the building envelope. These aspects are therefore well worth investigating to identify improvements in cases where ventilation with heat recovery has already been installed. If ventilation airflows are not properly configured during commissioning, unintended imbalances between supply and exhaust airflows may increase the heat loss due to airflows through the building envelope. Poor airflow balancing may derive from measurement errors during the balancing or from air leakage through ducts. According to a 2018 report from the US National Renewable Energy Laboratory, air-duct leakage accounted for 22.7% of the annual energy impact from the top 20 faults in building services in US commercial buildings, and excessive airflow

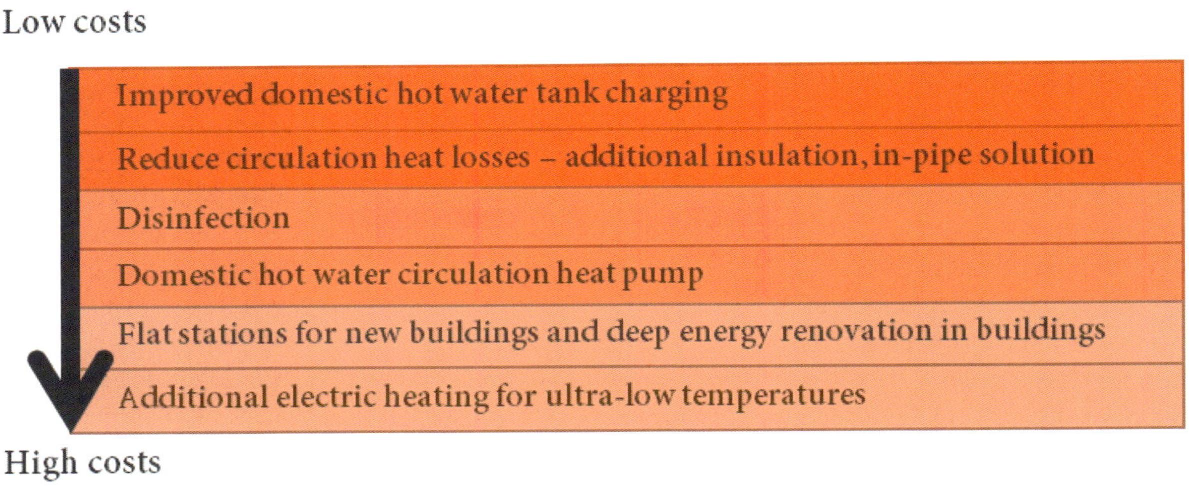

Low costs

Improved domestic hot water tank charging
Reduce circulation heat losses – additional insulation, in-pipe solution
Disinfection
Domestic hot water circulation heat pump
Flat stations for new buildings and deep energy renovation in buildings
Additional electric heating for ultra-low temperatures

High costs

Figure 31. Actions to lower the primary district heating temperatures for domestic hot water systems.

through the building envelope accounted for 26.1%. Together, these two faults accounted for 25.8 TWh in energy losses (Kim, Cai, et al., 2017). To fix these problems and minimise heat demand and heating system temperatures, functional performance testing at regular intervals is recommended, if not continuously via on-line methods, to ensure the proper operation of ventilation systems (Jradi, Liu, et al., 2020; Mattera, Jradi, et al., 2020). Such testing can narrow the performance gap between expected and realised heat demand, thereby allowing lower supply and return temperatures in the district heating system that provides heat to the ventilation heating coil.

3.9 How to design new building installations for lower temperatures

New building installations should be designed with future low temperatures in mind, whether they are installations in new buildings or new installations for the renovation of existing buildings. In this field, most current guidebooks still need updating to comply with future low-temperature requirements, because they are still targeted on old fossil fuel technology. Only a few national standards have been updated for the future requirements of low-temperature heating. A good example is Switzerland, where heating systems have been designed for supply temperatures of 50 °C or below for radiators and 35 °C for floor heating since 2009 (Averfalk, Werner, et al., 2017; SIA Zürich, 2009). Other national standards should be updated correspondingly, and subjects relevant for new low-temperature heating systems should be addressed. These include long thermal lengths in heat exchangers and heating elements, well-functioning hydronic balancing of space heating systems, and designs of domestic hot water systems with no risk of *Legionella* (Averfalk & Werner, 2017).

New heat exchangers and space heating elements should be designed for low-temperature operation. Heat exchangers for space heating systems should have design temperatures of 55/25 °C (Olsen, Christiansen, et al., 2014) or lower. Space heating elements include both floor heating and radiators. Floor-heating systems are well suited for low-temperature supply because they provide a large heat-emitting surface area, which means that they can provide the heat necessary with a low surface temperature. Floor heating design supply temperatures should be maximum 35–40 °C, and the individual supply temperature should be minimized using either local or central mixing shunts to ensure stable operation and low return temperatures (Danfoss, 2019). New radiators can also be designed for LTDH. Suitable design temperatures might be 55/25 °C, 50 °C/30 °C, or 45 °C/35 °C, which would ensure that the radiators can deliver sufficient thermal comfort with low heating system supply temperatures (Brand, Lauenburg, et al., 2012; Danish Energy Agency & EUDP 2010-II, 2014). In low-energy buildings, this will result in radiator sizes similar to earlier decades, due to the reductions in peak space heating demand (Olsen, Christiansen, et al., 2014). In low-energy buildings, it may be relevant to consider using design temperatures with a smaller difference between supply and return to ensure that the flow in the radiators is large enough to ensure proper operation of the thermostatic valves. The benefits of using radiators compared to floor heating are that they are easy to replace, they provide a warm air convection that counterbalances the cold drafts from windows if placed underneath them, they have a fast response time so that heat output is quickly adjusted when the room becomes either too warm or too cold, and lastly radiators are designed to provide a lot of cooling of the heating system water.

It is necessary to pay special attention when convector type radiators are installed in heating systems operated with LTDH. Convectors are generally operated with a small temperature drop, which can make them less suited for low-temperature heating. When a convector is operated with low heating system temperatures and a lot of cooling, the heat output is often reduced heavily (Danish Standard, 2013). In fact, the heat output from a convector operated at low temperatures can be less than 50% of the heat estimated when standard radiator equations are used (Paulsen & Rosenberg, 2012). So, where convectors are included in the heating system, it is necessary to make sure that they can emit sufficient amounts of heat with low heating system temperatures by acquiring specific manufacturer test results from low-temperature operation or through specialised calculations.

Solutions for automatic hydronic balancing in space heating systems are not currently available, and therefore hydronic balancing is still an important and difficult issue in new buildings. The newly developed electronic return temperature radiator valve that is demonstrated in case 13, is the closest one can get to automated balancing. This type of valve could provide a good component for this in the future. Current guidelines recommend installing balancing valves and pre-set radiator valves, having separate heating system loops if different zones of a building require different supply temperature set-points, and commissioning new space heating installations before they are taken into use (Danish Standard, 2013). In this regard, it is important to choose a control strategy that ensures adequate operating conditions, weather-compensated supply temperatures, and optimized pump control. It is still common that heating systems are not properly commissioned, so it is important to ensure that string balancing valves and pre-settings are not only installed, but also adjusted to appropriate levels before the new heating installation begins operation.

In new buildings with large thermal inertia where electronic thermostatic radiator valves are installed, occupants should be urged to avoid inefficient on/off operation or scheduling of thermostat set-points, which is common but makes it difficult to achieve low peak heat demands and low district heating return temperatures (Hu, 2020; Huebner, Cooper, et al., 2013; Liao, Swainson, et al., 2005). Heating system setbacks are generally not recommended for buildings supplied by LTDH. They are especially not suited for heavy floor heating systems where thermal inertia may give the heating element such a long response time that the setback will not be effective within the expected time period.

The *domestic hot water system* in buildings connected to LTDH should be designed to minimise bypasses and circulation heat losses, and it should be able to deliver domestic hot water without risk of *Legionella* at district heating supply temperatures of 55 °C. One solution that satisfies these criteria is to use decentral heat exchangers for the local preparation of domestic hot water in each flat. If domestic hot water is prepared in instantaneous heat exchangers close to the domestic hot water taps, this reduces the need for domestic hot water circulation and the risk of *Legionella*. Furthermore, flat stations provide an easy solution for energy metering, which makes it cheap to distribute energy costs between apartments. Where flat stations are installed, it is important that measurements of individual heating consumption and temperatures are available for the technicians who provide maintenance and fault identification, and that the flat stations are accessible from the stairwell for service and maintenance purposes.

Heat exchangers for domestic hot water should be designed to deliver the design peak load at low design temperatures, such as 55/25 °C (Olsen, Christiansen, et al., 2014). Design temperatures could be even lower for single-family units with small hot water volumes, where in some cases it is possible to reduce domestic hot water supply temperatures to 50 °C. New substations should include quality components, control valves and actuators, and a properly controlled bypass (Crane, 2016; Olsen, Christiansen, et al., 2014). Small deviations in the regulation of the control valve and high stand-by losses or bypass flows can easily have a large negative impact on the district heating return temperature and may vary between substations (Crane, 2016). The DHW peak demand should be carefully chosen and take into account the simultaneity of domestic hot water consumption and realistic peak heat demands for domestic hot water in single-family units. Recommendations are available from Euroheat (Euroheat & Power, 2008), and peak demand values in the range of 25–40 kW have been found to be sufficient for single-family units (Crane, 2016; Thorsen, Christiansen, et al., 2011).

New ventilation systems should be equipped with heat recovery and ventilation heating coils designed for low temperatures. The ventilation systems can be designed either as a centralized solution or with decentralised ventilation units serving individual apartments, floors or zones. Decentralisation limits the size and pressure requirements of ducts. In either case, ventilation heat recovery efficiencies of above 80% can be achieved, which means that the heating coil is only needed when the system partially bypasses heat recovery to keep exhaust air temperatures above freezing to avoid frost accumulation. In temperate climates, it may be worth using an electric-resistance heating coil for these relatively few instances, or no heating coil at all. In colder continental climates with extreme temperatures, a water-based heating coil might be preferred due to its lower cost and primary energy factor with district heating. In any case, in ventilation systems with heat recovery, the heating coil should only be used to ensure comfort in situations with freezing outdoor temperatures, and measures should be in place to monitor its energy consumption.

3.10 What to consider when existing buildings are connected to district heating

When existing buildings are connected to a district heating network, it is vital to ensure that the operation of internal building installations complies with the needs of a modern district heating system. Many building installations were designed and operated for heat supply from a fossil-fuel boiler, in which case it may be necessary to make small adjustments to the control and layout of the installations. It is therefore essential to make a general check of the installations in the substation room while staff are on site to connect the new substation. The check should include typical faults as described in Section 3.4 and the following additional aspects could be taken into account.

In building installations designed and operated with an oil boiler, it is common to apply high-temperature and high flow to avoid low return temperatures that would lead to risk of corrosion. Furthermore, it is common to apply building heating system setbacks, where the space heating system is turned off during the night to save energy. These control patterns are not suited for modern LTDH systems, where continuous operation of the space heating system with the lowest possible temperatures should be applied instead. Apart from modifying the current control settings in internal building installations, it may be necessary to identify and remove bypasses and shunts, which have often been added to ensure the high flow on the secondary side. Most of these have traditionally been installed in the central installation in the substation room, but they can also be present further out in the heating installations, and there may even be a bypass at the furthest end of the space heating loop. Where such bypasses and shunts in building installations have not been identified and removed, this will often show from the energy meter measurements from the building, which will make it possible to correct the issue quickly after the new installation is taken into operation.

It may also be worth taking a look at the internal space heating system when an existing building is connected to a district heating system. Small improvements in the space heating system are often cheaper and easier to include as part of a large heating system installation, whereas it is not easy to convince building owners to spend extra money on building installation improvements only few years after they have invested in a new district heating substation. Typical improvements in the heating system might include installation of thermostatic radiator valves, string balancing valves, or new radiator valves with pre-settings suited for low-temperature heating system operation. In some countries, multi-family buildings do not usually have thermostats installed (Averfalk, Werner, et al., 2017), but the huge benefits of installing them mean that they are commonly required when new heating systems or buildings are designed. In some cases, the installation of thermostats may not add much to the cost of a new district heating substation, but they can ensure higher efficiency in the operation of the district heating system, as well as provide the occupants with improved comfort and therefore a good experience of being connected to the district heating system. When new thermostats are installed, it is worth considering state-of-the-art thermostats with return temperature limitation functionality.

How to ensure that new building connections are designed properly for low temperatures

When buildings are connected to the district heating system, it is important to ensure that they are designed properly for future LTDH. Most installations designed today have high investment costs and will be in service for many decades to come, so if they are not designed and operated properly it will lead to the extra costs of excessive temperature operation for many years. Here are some suggestions for how the district heating company can help ensure that new connections are designed properly:

- Set requirements for the minimum technical standards of all new installations. These can include sketches of recommended layouts and requirements for the installation of specific components.
- Provide guidance for new customers. This could include a list of pre-fabricated substations and heating system components that have been quality-tested to comply with your specific temperature requirements.
- Consider whether to offer the opportunity to lease district heating substations with quality controls to customers. Such leasing can bring increased cash flows, monitoring and fault prevention, and happy customers.
- Consider whether to perform a quality check of new installations before they are taken into use. This can help eliminate malfunctions that cause excessive temperatures or complaints.

3.11 Major conclusions on lower temperatures in buildings

The main take-away from this chapter is that in most cases it is possible to use lower district heating temperatures than those that are typically applied today in both new and existing buildings. In most cases, district heating temperatures are much higher than needed to comply with national temperature requirements for *Legionella* and typical comfort requirements for space heating.

The main obstacle to lower district heating temperatures is the occurrence of simple malfunctions and faults in the district heating substations, and the main task if wanting to lower district heating temperatures is therefore to eliminate these. And if wanting to keep temperatures low, increased maintenance and automatic fault detection in building substations are the main tools that should be applied in district heating systems.

Treatment of *Legionella* is a major focus area when low-temperature operation is discussed. The main risk of *Legionella* growth is often due to the poor design or operation of internal building installations. Alternatives to thermal treatment of *Legionella* are available, but in most existing buildings, district heating companies must rely on frequent readings of energy meters to make sure that the district heating supply temperature on entry to each individual customer is high enough to meet national requirements for the thermal treatment of *Legionella* in domestic hot water installations.

In a longer perspective, energy renovations will enable lower district heating temperatures by reducing the heating demands in existing buildings and therefore reducing the space heating temperatures needed in commonly over-sized existing heating systems. Clever design of new heating installations (in both existing and new buildings) will make it possible to use lower temperatures in the future. The low-temperature design of new installations will make sure that components installed today will be future-proof.

Longer thermal lengths in heat exchangers, *Legionella*-safe supply of domestic hot water, and automatic balancing of space heating systems are the main focus areas for low-temperature heating. Currently best-available technologies in this regard include externally accessible flat stations for domestic hot water supply and smart return temperature thermostats with automatic balancing functionalities.

For the design of new space heating systems and domestic hot water installations, there is a need for new standards that address the use of low-temperature heating based on renewable heat sources in the future. There is also a need for research focused on developing more solutions for how to provide *Legionella*-safe and comfortable domestic hot water without needing high heating consumption for the circulation of domestic hot water at high temperatures.

In summary, the following conclusions can be drawn:
1. District heating temperatures can often be lowered if typical faults and malfunctions in substations are corrected.
2. The process of correcting faults and malfunctions should be supported by digital tools and increased substation servicing.
3. Frequent readings from smart energy meters should help make sure that district heating supply temperatures at individual customer locations are sufficient to meet national temperature requirements for *Legionella*-free domestic hot water systems.
4. Oversized existing heating systems and the energy renovation of buildings will make it possible to apply lower space heating temperatures.
5. Longer thermal lengths in heat exchangers, externally accessible flat station heat exchangers for domestic hot water, minimised supply temperature for space heating systems, and smart return temperature thermostats are currently best-practice solutions for building installations for LTDH.
6. Updated standards on how to design building installations for low-temperature operation should be developed and further research on alternatives to the thermal treatment of *Legionella* should be carried out in the future.

3.12 Literature references in Chapter 3

An, J., Yan, D., Deng, G., & Yu, R. (2016). Survey and performance analysis of centralized domestic hot water system in China. Energy and Buildings, 133, 321–334. https://doi.org/10.1016/j.enbuild.2016.09.043

Averfalk, H., & Werner, S. (2017). Essential improvements in future district heating systems. Energy Procedia, 116, 217–225. https://doi.org/10.1016/j.egypro.2017.05.069

Averfalk, H., Werner, S., Felsmann, C., Rühling, K., Wiltshire, R., Svendsen, S., ... Quiquerez, L. (2017). Annex XI final report: Transformation Roadmap from High to Low Temperature District Heating Systems.

Bagh, L. K., Albrechtsen, H. J., Arvin, E., & Ovesen, K. (2004). Distribution of bacteria in a domestic hot water system in a Danish apartment building. Water Research, 38(1), 225–235. https://doi.org/10.1016/j.watres.2003.08.026

Benakopoulos, T., Salenbien, R., Vanhoudt, D., & Svendsen, S. (2019). Improved Control of Radiator Heating Systems with Thermostatic Radiator Valves without Pre-Setting Function. 1–24. https://doi.org/10.3390/en12173215

Benakopoulos, T., Salenbien, R., Vanhoudt, D., Tunzi, M., & Svendsen, S. (2021). Low return temperature from domestic hot-water system based on instantaneous heat exchanger with chemical-based disinfection solution. Energy, 215, 119211. https://doi.org/10.1016/j.energy.2020.119211

Best, I., Braas, H., Orozaliev, J., Jordan, U., & Vajen, K. (2020). Systematic investigation of building energy efficiency standard and hot water preparation systems' influence on the heat load profile of districts. Energy, 197, 1–12. https://doi.org/10.1016/j.energy.2020.117169

Best, I., Orozaliev, J., & Vajen, K. (2018). Economic comparison of low-temperature and ultra-low-temperature district heating for new building developments with low heat demand densities in Germany. International Journal of Sustainable Energy Planning and Management, 16, 45–60. https://doi.org/10.5278/ijsepm.2018.16.4

Bøhm, B. (2012). Production and distribution of domestic hot water in selected Danish apartment buildings and institutions. Analysis of consumption, energy efficiency and the significance for energy design requirements od buildings. Energy Conversion & Management, 67, 152–159.

Bøhm, Benny, Schrøder, F., & Bergsøe, N. C. (2009). Varmt Brugsvand - måling af forbrug og varmetab fra cirkulationsledninger [Domestic hot water - measurements of consumption and heat loss from circulation pipes] - SBi 2009:10.

Brand, M., Gudmundsson, O., & Thorsen, J. E. (2016). District heating substation with electrical heater supplied by 40 °C ultra-low district heating.

Brand, M., Lauenburg, P., Wollerstrand, J., & Zboril, V. (2012). Optimal Space Heating System for Low-Energy Single-Family House Supplied by Low-Temperature District Heating. Papers Presented at the Conference Passivhusnorden 2012.

Castellazzi, L. (2017). Analysis of Member States' rules for allocating heating, cooling and hot water costs in multi-apartment / purpose buildings supplied from collective systems Implementation of. In Implementation of EED Article 9(3). https://doi.org/10.2760/40665

Cholewa, T., Siuta-Olcha, A., & Anasiewicz, R. (2019). On the possibilities to increase energy efficiency of domestic hot water preparation systems in existing buildings – Long term field research. Journal of Cleaner Production, 217, 194–203. https://doi.org/10.1016/j.jclepro.2019.01.138

Cholewa, T., Siuta-Olcha, A., & Balaras, C. A. (2017). Actual energy savings from the use of thermostatic radiator valves in residential buildings – Long term field evaluation. Energy and Buildings, 151, 487–493. https://doi.org/10.1016/j.enbuild.2017.06.070

Crane, M. (2016). Individual apartment substation testing - Development of a test and initial results. The 15th International Symposium of District Heating and Cooling.

Danfoss. (2014). District heating application handbook - Making applications future proof - all our knowledge is now yours.

Danfoss. (2019). Application guide - Designing hydronic floor heating.

Danish Clean Water. (2020). Clean and safe solutions to control and prevent bacteria.

Danish Energy Agency. (2015). Demonstrationsprojekter om varmepumper eller andre VE- baserede opvarmningsformer [Demonstration projects on heat pumps and other RE-based heating solutions]. (2), 1–50. Retrieved from https://docplayer.dk/19072248-Demonstrationsprojekter-om-varmepumper-eller-andre-vebaserede.html

Danish Energy Agency, & EUDP 2010-II. (2014). Demonstration i Lystrup – Fuldskalademonstration af lavtemperaturfjernvarme i eksisterende bebyggelser - Journalnr. 64010-0479 [Demonstration in Lystrup – Full scale demonstration of low-temperature district heating in existing building areas]. Retrieved from https://docplayer.dk/3044458-Delrapport-demonstration-i-lystrup-energistyrelsen-eudp-2010-ii-fuldskala-demonstration-af-lavtemperatur-fjernvarme-i-eksisterende-bebyggelser.html

Danish Standard. (2013). DS 469 - Heating and cooling systems in buildings.

Danish Standards Assosciation. (2009). DS 439:2009 "Norm for vandinstallationer (Code of Practice for domestic water supply)."

DTU. (2017). Book of presentations of the International Workshop on High Temperature Heat Pumps.

Euroheat & Power. (2008). Guidelines for District Heating Substations.

Fester, J., Bryder, K., Østergaard, P. F., & Nielsen, J. (2019). Smart fjernvarme - Effektivisering af fjernvarmeforsyning ved anvendelse af smart meter data og data fra supplerende sensorer [Smart District Heating - Higher efficiency in district heating by use of smart meter data and data from supplementary sensors].

Frederiksen, S., & Werner, S. (2013). District Heating and Cooling. Sweden: Studentlitteratur, pag. 408.

Gadd, H., & Werner, S. (2014). Achieving low return temperatures from district heating substations. Applied Energy, 136, 59–67. https://doi.org/10.1016/j.apenergy.2014.09.022

Gadd, H., & Werner, S. (2015). Fault detection in district heating substations. Applied Energy, 157, 51–59. https://doi.org/10.1016/j.apenergy.2015.07.061

Geyer, R. (2018). Projekt heat_portfolio (FFG-Nr. 848849) Deliverable D6.1 & D6.2.

Geyer, R., & Schmidt, R.-R. (2017). Vorsicht heiß - Ursachen hoher Rücklauftemperaturen in Fernwärmenetzen. EuroHeat&Power, pp. 50–53.

Gudmundsson, O., Thorsen, J. E., Dyrelund, A., & Schmidt, R. (2020). Case comparison: Low vs Lower Temperature District Heating. Presentation - 6th International Conference on Smart Energy Systems, 1–14.

Gummerus, P. (2016). 10 – New developments in substations for district heating. In Advanced District Heating and Cooling (DHC) Systems. https://doi.org/10.1016/B978-1-78242-374-4.00010-0

HOFOR. (2019). Typiske fejl i varmekælderen [Common malfunctions in the district heating substation].

Hu, N. (2020). Optimisation of operation temperatures in existing heating systems. Technical University of Denmark.

Huang, T., Yang, X., & Svendsen, S. (2019). Multi-mode control method for the existing domestic hot water storage tanks with district heating supply. Energy, (xxxx), 116517. https://doi.org/10.1016/J.ENERGY.2019.116517

Huebner, G. M., Cooper, J., & Jones, K. (2013). Domestic energy consumption—What role do comfort, habit, and knowledge about the heating system play? Energy and Buildings, 66(30), 626–636. https://doi.org/10.1016/j.enbuild.2013.07.043

Jangsten, M., Kensby, J., Dalenbäck, J. O., & Trüschel, A. (2017). Survey of radiator temperatures in buildings supplied by district heating. Energy, 137, 292–301. https://doi.org/10.1016/j.energy.2017.07.017

Jensen, S. S. (2020). EU puts smarter district heating on the agenda-copenhagen leads the way. Euroheat and Power (English Edition), 2020-Janua(1), 47–50.

Jradi, M., Liu, N., Arendt, K., & Mattera, C. G. (2020). An automated framework for buildings continuous commissioning and performance testing – A university building case study. Journal of Building Engineering, 31(January), 101464. https://doi.org/10.1016/j.jobe.2020.101464

Kärkkäinen, A. (2010). Gasfri Påfyllning Av Värme- Och Kylsystem Samt Injustering Av Radiatorsystem.

Khan, U., Zevenhoven, R., & Tveit, T. M. (2020). Evaluation of the environmental sustainability of a stirling cycle-based heat pump using LCA. Energies, 13(17). https://doi.org/10.3390/en13174469

Kim, J., Cai, J., & Braun, J. E. (2017). Common Faults and Their Prioritization in Small Commercial Buildings. Golden, CO.

Kristensen, M. H., & Petersen, S. (2020). District heating energy efficiency of Danish building typologies: Datasets and supplementary materials. Energy & Buildings, nn(nn). https://doi.org/10.17632/v8mwvy7p6r.1

Krøjgaard, L. H. (2011). Legionella in habitations.

Li, H., Svendsen, S., Werner, S., Persson, U., Ruehling, K., Felsmann, C., … Bevilacqua, C. (2014). Annex X Final report Toward 4 th Generation District Heating : Experience and Potential of Low-Temperature District Heating. In IEA Annex X.

Liao, Z., Swainson, M., & Dexter, a. L. (2005). On the control of heating systems in the UK. Building and Environment, 40(3), 343–351. https://doi.org/10.1016/j.buildenv.2004.05.014

Lidberg, T., Olofsson, T., & Ödlund, L. (2019). Impact of Domestic Hot Water Systems on District Heating Temperatures. Energies, 12(4694). https://doi.org/10.3390/en12244694

Lund, R., Østergaard, D. S., Yang, X., & Mathiesen, B. V. (2017). Comparison of Low-temperature District Heating Concepts in a Long-Term Energy System Perspective. International Journal of Sustainable Energy Planning and Management, 12(0), 5–18. https://doi.org/10.5278/ijsepm.2017.17.x

Månsson, S., Johansson Kallioniemi, P. O., Thern, M., Van Oevelen, T., & Sernhed, K. (2019). Faults in district heating customer installations and ways to approach them: Experiences from Swedish utilities. Energy, 180, 163–174. https://doi.org/10.1016/j.energy.2019.04.220

Mathys, W., Stanke, J., Harmuth, M., & Junge-Mathys, E. (2008). Occurrence of Legionella in hot water systems of single-family residences in suburbs of two German cities with special reference to solar and district heating. International Journal of Hygiene and Environmental Health, 211(1–2), 179–185. https://doi.org/10.1016/j.ijheh.2007.02.004

Mattera, C. G., Jradi, M., Skydt, M. R., Engelsgaard, S. S., & Shaker, H. R. (2020). Fault detection in ventilation units using dynamic energy performance models. Journal of Building Engineering, 32(August), 101635. https://doi.org/10.1016/j.jobe.2020.101635

Meesenburg, W., Ommen, T., Thorsen, J. E., & Elmegaard, B. (2020). Economic feasibility of ultra-low temperature district heating systems in newly built areas supplied by renewable energy. Energy, 191(xxxx), 116496. https://doi.org/10.1016/j.energy.2019.116496

Oevelen, T. Van, Vanhoudt, D., & Salenbien, R. (2018). Evaluation of the return temperature reduction potential of optimized substation control. Energy Procedia, 149(September), 206–215. https://doi.org/10.1016/j.egypro.2018.08.185

Olsen, P. K., Christiansen, C. H., Hofmeister, M., Svendsen, S., & Thorsen, J.-E. (2014). Guidelines for Low-Temperature District Heating.

Olvondo Technology. (2020). Highlift by Olvondo Technology.

Østergaard, D. S. (2018). Heating of existing buildings by low-temperature district heating. Technical University of Denmark.

Østergaard, D. S., & Svendsen, S. (2016a). Replacing critical radiators to increase the potential to use low-temperature district heating - A case study of 4 Danish single-family houses from the 1930s. Energy, 110, 75–84. https://doi.org/10.1016/j.energy.2016.03.140

Østergaard, D. S., & Svendsen, S. (2016b). Theoretical overview of heating power and necessary heating supply temperatures in typical Danish single-family houses from the 1900s. Energy and Buildings, 126, 375–383. https://doi.org/10.1016/j.enbuild.2016.05.034

Paulsen, O., & Rosenberg, F. (2012). Dimensionering af radiatorer til lavtemperatursystemer [Dimensioning of radiators for low temperature systems]. HVAC, 48(1), 46–53.

Petersson, S., & Werner, S. (2003). Långtidsegenskaper hos lågflödesinjusterade radiatorsystem.

Rämä, M., Lindroos, T. J., Svendsen, S., Østergaard, D. S., Tunzi, M., Sandvall, A., & Holmgren, K. (2019). IEA DHC Annex XII - Stepwise transition strategy and impact assessment for future district heating systems - final report.

SIA Zürich. (2009). SIA 384/1:2009 Bauwesen - Heizungsanlagen in Gebäuden - Grundlagen und Anforderungen.

Sørensen, I. B. (2017). Implementation of low-temperature district heating in existing buildings in city area Master thesis. DTU.

Swedenergy AB. (2016). District heating substations, Design and Installation, Technical regulations F:101. In Technical regulations F:101.

Thalfeldt, M., Kurnitski, J., & Latõšov, E. (2018). Exhaust air heat pump connection schemes and balanced heat recovery ventilation effect on district heat energy use and return temperature. Applied Thermal Engineering, 128, 402–414. https://doi.org/10.1016/j.applthermaleng.2017.09.033

Thorsen, J. E., Christensen, H., & Boysen, H. (2017). Trend for heating system renovation in multi - family buildings.Thorsen, J. E., Christiansen, C. H., Brand, M., Olesen, P. K., & Larsen, C. T. (2011). Experiences on low-temperature district heating in Lystrup - Denmark. Proceedings of International Conference on District Energy.

Thorsen, J. E., Ommen, T., Meesenburg, W., Smith, K., & Oliver, E. M. (2019). Energy Lab Nordhavn - Deliverable no .: D10.1c - 1 : Heat Booster Substation for Domestic Hot Water and 2: Circulation Booster for Domestic Hot Water. 1–33.

Thorsen, J. E., Skov, M., & Honoré, K. (2019). Installation and demonstration of domestic hot water circulation booster for reducing district heating return temperature. 1–21. Retrieved from http://www.energylabnordhavn.com/uploads/3/9/5/5/39555879/d5.4b_dhw_circulation_booster.pdf

Thorsen, J. E., Svendsen, S., Smith, K. M., Ommen, T., & Skov, M. (2020). Experience with booster for DHW circulation in multi appartment building. Presentation - 6th International Conference on Smart Energy Systems, (6-7 October).

Trüschel, A. (2002). Hydronic heating systems - the effect of design on system sensitivity (Vol. 12). https://doi.org/10.1088/0950-7671/12/9/307

Trüschel, A. (2005). Värdet av injustering [The value of balancing]. Svensk Fjärrvärme [Swedish District Heating].

Tveit, T.-M. (2017). Application of an industrial heat pump for steam generation using district heating as a heat source. 12th IEA Heat Pump Conference 2017, (1395).

Werner, S. (2020). Types of errors in district heating substations.

Yang, X. (2016). Supply of domestic hot Water at comfortable temperatures by low-temperature district heating without risk of Legionella. Technical University of Denmark.

Yang, X., & Honoré, K. (2019). Installation and demonstration of capacity limitation functionality for domestic hot water tank in a new building to reduce peak load , lower return temperature and utilize thermal flexibility in domestic hot water tanks. (i), 1–21. Retrieved from http://www.energylabnordhavn.com/uploads/3/9/5/5/39555879/d5.4c_i__optimum_dhw_new_building.pdf

Yang, X., Li, H., & Svendsen, S. (2016a). Decentralized substations for low-temperature district heating with no Legionella risk, and low return temperatures. Energy, 110, 65–74. https://doi.org/10.1016/j.energy.2015.12.073

Yang, X., Li, H., & Svendsen, S. (2016b). Evaluations of different domestic hot water preparing methods with ultra-low-temperature district heating. Energy, 109, 248–259. https://doi.org/10.1016/j.energy.2016.04.109

Yang, X., & Svendsen, S. (2018). Ultra-low temperature district heating system with central heat pump and local boosters for low-heat-density area: Analyses on a real case in Denmark. Energy. https://doi.org/10.1016/j.energy.2018.06.068

Zinko, H., Lee, H., Kim, B.-K., Kim, Y.-H., Lindkvist, H., Loewen, A., … Wigbels, M. (2005). Improvement of operational temperature differences in district heating systems.

4 LOWER TEMPERATURES IN HEAT DISTRIBUTION NETWORKS

Authors: Roman Geyer, AIT; Christian Engel, Austroflex; Isabelle Best, Uni Kassel; Harald Schrammel, AEE INTEC, Karl Ponweiser, TU Wien; Natasa Nord, NTNU; Johannes Oltmanns, TU Darmstadt

This chapter is dedicated to measures that can be taken on the system/distribution side to keep system temperature levels as low as possible and reduce them further. The chapter is divided into three parts (see Figure 32):

- A collection of advice and considerations for existing systems, including measures that are easy to implement or lower cost;
- Description of successful cases providing insight into previous approaches to reducing system temperatures; and
- Discussion of measures requiring further implementation efforts, and which could be better suited to new installations or systems.

Given customers have a substantial influence on a system's return temperature, a simple method is described for identifying crucial substations/customers from an operator's point of view (Section 4.1). In this context, identifying and fixing unintentional circulation flows is very important (Section 4.2). Then, bottlenecks (Section 4.3) are addressed, along with seven solutions for overcoming them.

Section 4.4 describes four successful cases, including transformation measures for low temperature networks. This section includes diagrams illustrating temperature reductions over several years. Furthermore, the economic benefits are discussed, including evidence for the economic viability of reduced system temperatures. Chapter 7 provides further insight through detailed case descriptions of the given references.

Solutions utilising subnetworks (Section 4.5) and cascading (Section 4.6) address the beginning of the transformation at a local level and according to network subsections. Accompanying this, Sections 4.7 and 4.8 address the integration of alternative and decentralised supply sources. Digitalisation (Section 4.9) enables optimisation potentials through automatic fault detection using artificial intelligence and big data. Following consideration of proposed measures, Section 4.10 presents aspects to consider in the design of new systems. Here, innovative supply and distribution concepts (Section 4.11) are potential solutions enabling overcoming the limitations of the business and supply logic of conventional district heating networks.

The chapter closes with a discussion of lessons learned regarding obtaining lower system temperatures (Section 4.12), including approaches to beginning the transition, before providing the chapter's major conclusions.

4.1 Tracking malfunctioning substations with high return temperatures

Many substations in existing district heating networks do not achieve optimal efficiencies, which often corresponds to high return temperatures. By evaluating heat

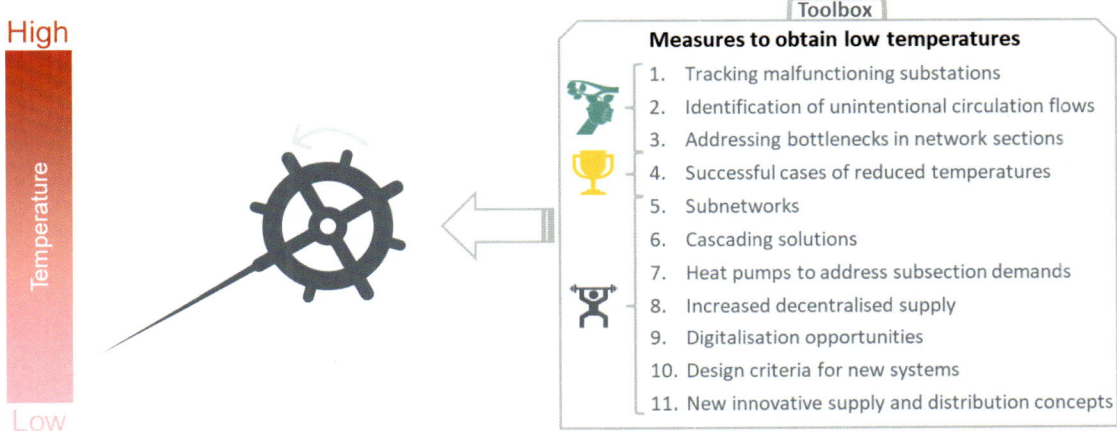

Figure 32. Measures for obtaining lower temperatures addressed in this chapter

meters, utilities providers can identify customer systems causing high return temperatures. However, heat meters must be able to record both temperature levels and mass flow rates. This evaluation can be conducted continuously during regular meter changes or remotely in the case of intelligent meters (so-called Smart Heat Meters). The following pages describe a simple method for doing so. Chapter 5 describes advanced methods – based on the Darmstadt case – requiring a higher level of data.

Prioritisation based on excess flow analysis

From the operator's point of view, it is worth considering the operating data of all components and assessing their performance and impact on the district heating system. To control substations, data such as primary return temperature, secondary supply temperature, supply temperatures of individual heating circuits, hot water tank temperature and ambient temperature are usually recorded. Systems with small temperature differences between supply and return operate at high mass flow rates, resulting in an above-average influence on the return temperature. In small networks, a small number of malfunctioning substations can be decisive for the whole network return temperature. In contrast, improvement measures taken in buildings with the worst performance can substantially impact the system. To benefit from this circumstance, identifying and applying measures at critical substations has the highest impact with minimal effort (Kreisel, 2014).

A simple method is the excess flow analysis (Zinko et al., 2005), through which substations are ranked according to their current impact on the district heating network's return temperature. This allows the operator to identify customers contributing the most to a high return temperature. Customers with high heat demand, large mass flows and low cooling present greater savings potentials and should be highly prioritised (Wirths, 2008). Optimising a few crucial substations can achieve a significant return temperature reduction for the whole district heating network. Visualisation of the data points helps to prioritise the customer plants according to their optimisation potential. The mean annual temperature spread is calculated according to Equation 12.

Figure 33 shows that the current temperature spread at each substation is compared to a set temperature spread (dT_Set) using the example of 40 K. The red dots are the substations causing lower temperature spreads – thus, higher return temperatures – while the green ones are already showing a temperature spread of 40 K. The resulting curve shows substations with the highest impact on return temperature first. The ranking order is equal to the caused excess flow of the substation. The black curve shows how much the return temperature is reduced if measures are taken in the substation ranking order. In this example, this means taking measures in the first 200 substations results in a return temperature of 51 °C, a reduction of 4.5 K.

 Using methods such as excess flow analysis allows the operator to identify critical customers in the system. Taking actions on malfunctioning substations with high return temperatures and the highest impact on the system should be prioritised.

$$\Delta T = \frac{Q}{V \cdot \rho \cdot c_w}$$

Equation 12 Calculation of the mean annual temperature spread

ΔT	Mean annual temperature spread	K
Q	Thermal energy (1 GWh/a ≙ 3 600 000 000 kJ/a)	kJ/a
V	Annual water quantity recorded	m³/a
ρ	Water density (992.2 at 40 °C)	kg/m³
c_w	Specific heat capacity of water (4.18 at 40 °C)	kJ/(kgK)

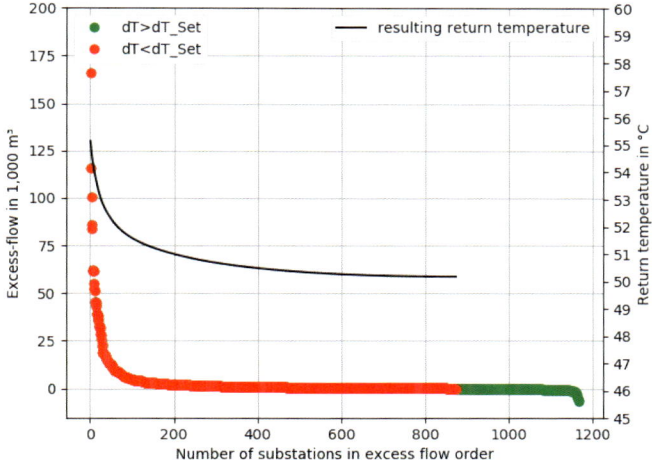
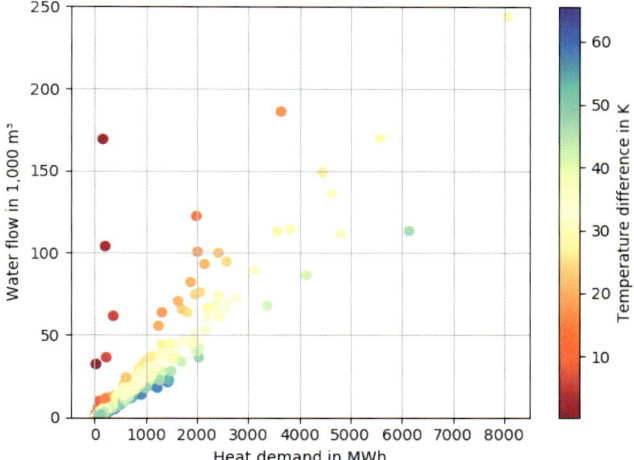

Figure 33. The excess flow analysis approach allows analysis and visualisation of the reduction potential of return temperature in existing district heating networks (left: ranking of substations according to their current impact on the return temperature of the district heating network; right: water flow linked with heat supply).

Figure 33 also includes a scatter plot of the substations water flow as a function of the heat demand (on the right side) from Kassel. The temperature difference at the substation is presented as a colour continuum covering a range between 0 and 65 K. The lower the ratio of annual total water flow (m³/a) to heat demand (MWh/a), the lower the return temperatures. A ratio of 50 (100 000 m³ to 2000 MWh) indicates a temperature difference of 20 K (orange dots). Meanwhile, a ratio of 18 (36000 m³ to 2000 MWh) at the same substation indicates a temperature difference of 45 K (light green dots). The difference between the first and second water flow constitutes the excess flow. For example, an excess flow of 64 000 m³ would result from this substation showing a temperature difference of 20 K rather than 45 K (Bergsträßer et al., 2020). Although this method is a good starting point for systems with low-data availability, it provides only limited information. Chapter 5 describes methods enabling more detailed analyses providing more data and data of higher resolution.

For district heating network operators, the most critical customers – in terms of return temperatures – are not necessarily those with the highest temperatures but those that have the biggest impact on the overall network return temperature. The following discussion refers to this impact as return temperature reduction potential.

The return temperature reduction potential $\overline{\Delta T}_{R,j}$ (see Equation 13) describes the potential reduction of the average network return temperature $\overline{T}_{R,DHN}$ through a reduction of the average return temperature $\overline{T}_{R,j}$ from substation j to the target temperature T_{target}. To calculate it, the measured return temperature $T_{R,i,j}$ for each time step i is compared to T_{target} and weighted with the share of mass flow $\dot{M}_{i,j}$ from substation j in the mass flow of the district heating network $\sum_{j=1}^{n_{sub}} \dot{M}_{i,j}$.

The resulting return temperature reduction potential is valid under the condition that the return temperature from all substations can be lowered to the target temperature at the same time. If the return temperatures from the substations are lowered in succession, the temperature reduction achievable by decreasing the temperature in one substation is smaller because the return temperature reduction promotes decreased mass flow, reducing the individual substation's share of the network's total mass flow. If measures are implemented in only some substations, the consequent reduction per substation will lie between the two cases. (Oltmanns, 2020)

To find out where to most efficiently apply comprehensive reduction measures, the heat demand at substation as a proportion of total heat demand (see Equation 14) is compared to the specific heat demand of each customer (see Equation 15). From a system point of view, the most important customers – in terms of implementing measures – are those with both high absolute and high specific heat demands. These customers represent the highest potential for reduction of

$$\overline{\Delta T}_{R,j} = \frac{1}{n_{ts}} \sum_{i=1}^{n_{ts}} \frac{(T_{R,i,j} - T_{target}) \cdot \dot{M}_{i,j}}{\sum_{j=1}^{n_{sub}} M_{i,j}}$$

Equation 13 Calculation of the return temperature reduction potential

$$f_j = \frac{Q_j}{Q_{DHN}}$$

Equation 14 Share of total heat demand at substation

$$q_j = \frac{Q_j}{A_j}$$

Equation 15 Specific heat demand of each customer

$\overline{\Delta T}_{R,j}$	Time average return temperature reduction potential	°C
n_{ts}	Number of time steps	[-]
$\overline{T}_{R,j}$	Reduction of the average return temperature	°C
T_{target}	Target temperature	°C
$\dot{M}_{i,j}$	Mass flow (per time step and substation)	kg/s
n_{sub}	Number of substations	[-]
f_j	Heat demand at substation j as a share of the total heat demand of district heating network	%
Q_j	Heat demand at substation j	MWh
Q_{DHN}	Total head demand of the district heating network	MWh
q_j	Specific heat demand of the customer	kWh/(m²a)
A_j	Heated area of the customer	m²

both heat demand and temperature level. A high specific heat demand can be due to not only a low-quality thermal envelope (i.e. walls, roofs, windows) but also problems with the ventilation system, especially malfunctioning or missing heat recovery and high air infiltration through the building fabric. Further insights and instructions on applying methods are described in the applied study 'Campus Lichtwiese at TU Darmstadt' (Sections 5.4 and 5.5).

Plant evaluation and customer advice
Access to customer systems is necessary because the probability of detecting faults that lead to high return temperatures increases during operational activities (inspection, maintenance, repair and troubleshooting). After identifying and prioritising the corresponding systems, they must be checked for possible causes of excess return flow temperature. First, the proper functioning of the system should be checked; checklists enable this to be conducted systematically. Checklists should be kept up-to-date by the utilities services using a well-maintained document management system. The possible causes of high return temperatures listed in Section 3.4 provide a point of reference for the creation of checklists.

To achieve low supply temperatures in buildings without using return temperature limiters, utilities should be consolidated for advisory support. This includes consolidating planning and execution (of new constructions and refurbishments), commissioning and additional services offerings (correct application for optimum plant operation). This is because heating systems can only function as intended if the technologies and user behaviour are coordinated. To achieve the desired low-energy consumption and energy savings, additional information must be provided to users. Users who are not familiar with the basic rules for energy-saving heating and ventilation can have a counterproductive effect on the efficient operation of technical systems through their (unconscious) misconduct. Possibilities for educating users include (Overhage, 2014):

- Providing customers with information on, for example, energy-saving and correct heating and ventilation and handling of technical components such as thermostatic valves;
- Instructing users and operating personnel on how to use the system technology;
- Providing lectures and training to operating personnel;
- Conducting plant inspections (involving utilities, plant and heating engineers and users); and
- Energy consultations for customer and plant operators.

In addition to these activities, utilities services could offer financial incentives to their customers, such as a new tariff system offering incentives to customer or 'pay-per-use' tariffs. Customer efforts to comply with the agreed return temperatures would be immediately 'noticeable', which would, in turn, increase their motivation (see Chapter 6 for some information on such initiatives). Identification of unintentional circulation flows

4.2 Identification of unintentional circulation flows

For certain operational purposes, circulation flows are necessary. These are managed by either hot water circulation systems on the customer side or by a valve in a bypass. A bypass (or short circuit) is defined as a connection between the supply and return pipes (see Figure 34). For operational purposes, there are only two cases for circulation flows in district heating systems:

- during summer to enable sufficient flow for direct DHW preparation, and
- during winter in cold countries to avoid freezing (maintaining 8 °C in the pipes is sufficient).

Although bypasses might be necessary during the construction and expansion of district heating networks, eliminating these bypasses is often forgotten. Therefore, each system bypass should be assessed to determine its necessity (e.g. from a control engineering perspective). If it is not necessary, bypasses should be eliminated because they can cause unintentional circulation flows.

Figure 34. Visualisation of controlled (left) and uncontrolled (right) bypasses and indicative impacts on return temperatures

The main disadvantage of circulation flow rates is that, when not used, the hot supply water is injected into the return pipe, preventing the return pipe from cooling down. Consequently, the average temperature increases, inducing extra heat losses in the return pipe that are detrimental to the network's heat generation efficiency. To achieve the lowest possible return temperatures, unregulated circulation flows between the supply and return pipes must be removed or avoided. The annual proportion of circulation flows has been estimated between about 10% (Vojens Fjernvarme, 2020) and 25% (Oltmanns, 2020). Based on 20 simulation cases, proportions of circulation flows have been calculated between 10% and 43% (Averfalk & Werner, 2018). Higher circulation flow proportions have been observed for buildings with improved thermal performance, a result of the decreased annual water volume of delivery flows.

Uncontrolled bypasses can be identified by either manually switching off individual strings or using specific software. In (Kamstrup, 2020a), an uncontrolled bypass was identified in a heat network using software. By eliminating the uncontrolled bypass, the return temperature for the zone was reduced by approximately 1.5 °C.

Furthermore, in most district heating customer substations, circulation flows are installed to retain a certain flow in the pipes to maintain circulation systems for DHW preparation, for moments of limited heat demand or when the network is in stand-by mode. This flow prevents the supply pipe from cooling down too much, enabling it to quickly provide the customer with hot water. This circulation flow rate is substation-specific and can vary substantially (Crane, 2016). Certain innovative concepts have been developed using an additional pipe

If bypasses are necessary, they should be controlled. If not, errors can lead to increased return temperatures, necessitating the removal of the bypass. district heating operators should complete their own analyses using operational data to check the proportions between delivery and circulation flow.

in the network (for more information, see Sections 4.9 and 10.1), with circulation flow from the substation injected into the additional pipe, returning the flow directly to the heat source (Li et al., 2010). Although the water in this pipe will cool slightly, the temperature of the water in the regular return pipe will be unaffected, guaranteeing a significantly lower return temperature.

4.3 Addressing bottlenecks in network sections

Pipes in district heating networks are dimensioned for a certain flow. If the flow increases, the consequent higher velocities promote greater pressure losses. A too-great flow can result in insufficient heat delivery to customers in affected areas (due to the pressure loss being too substantial). This can increase heat demand or reduce supply temperature without a correspondingly lowered return temperature. Notably, district heating networks that are expanded, densified or converted to lower supply temperatures are at a particularly high risk of high specific pressure losses through the pipelines. The resulting low differential pressures constitute a problem known as a bottleneck. The most common short term approaches to overcoming bottlenecks are increased supply temperatures and increased pumping power. However, there are many potential long term solutions to bottlenecks.

Brange et al. (2017) reviewed the literature on different measures for eliminating bottlenecks, as well as surveying Swedish district heating companies with regard to bottlenecks. Of 131 companies, 89 responded, with 75% perceiving having or having had bottleneck problems. The seven measures for eliminating bottlenecks are summarised in Figure 35. According to the survey, the most used methods for eliminating them were introducing wider pipe diameters, increasing supply temperatures and increasing pumping power. Increasing the supply temperature should only be a short term solution; if not, a lock-in effect occurs because whole-network heat losses increase. Additionally, increasing pumping power is a limited measure because the pressure drop increases with the square of the flow.

Brange et al. (2017) concluded that it is important to plan for future developments when planning or extending a district heating network to avoid bottleneck problems, recognising that there are better and cheaper possible solutions to bottleneck problems than those conventionally used. They suggest that these may become more common when more data and better technology become available.

Economic viability is another important factor in the choice of a bottleneck prevention measure. However,

 Bottlenecks are obstacles to lowering supply temperatures in district heating networks. Besides increasing supply temperature, there are several solutions to bottlenecks.

Methods to eliminate bottlenecks:
- Higher supply temperature (short-term solution)
- More pumping power (short-term solution)
- Wider pipe diameters (high costs)
- Local heat supply (see also section 4.7 and 4.8)
- Storages (local supply/consumer flow management)
- Demand Side Management (DSM, consumer flow management)
- Fault elimination in substations (consumer flow management)

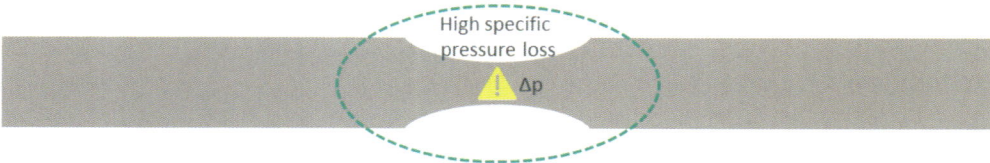

Figure 35. Seven measures for eliminating bottlenecks (Brange et al., 2017)

risks and opportunities affecting this viability should be accounted for in these calculations that are not necessarily present at the time of development (Brange et al., 2018). To include such subjects, (Brange et al., 2019) introduce a decision-making process providing a summary of the advantages and disadvantages of different factors and different bottleneck solutions, including a description of a real case demonstrating the application of this decision-making process.

4.4 Successful cases of temperature reduction in existing systems

According to district heating network analysis, it makes sense to identify supply areas where a reduction in supply temperature is possible based on the existing customer base and distribution structure. When analysing customers, it is necessary to determine the capacity and supply temperatures required on an individual basis based on specific requirements, such as space heating, hot water supply and process heat requirements. Available technical network data (e.g. heat losses, pipe dimensions, pressure ratios and flow parameters) and heat load conditions enable the determination of how high the supply temperature should be at the feeding point. This constitutes the minimum requirement for the network supply temperature. If necessary, it can make sense to further reduce the supply temperature and to decentralise supply to individual high-temperature customers using booster solutions, such as heat pumps (HPs) and electric boilers. Substations at the customer end would subsequently have to be converted to the lowest temperature for the area. This should be possible without major problems, given normal building conditions and heating surfaces. For domestic hot water preparation, substations could also be used with, if necessary, additional heating registers.

Viborg (DK, 210 GWh/a) is an exemplary model of this approach. By introducing a motivation tariff, smart meters, extensive communication, and optimisation measures, they reduced the annual supply temperature from 80 to 66 °C and the annual return temperature from 50 to 40 °C (both average annual values) between 2002 and 2018 (Diget, 2019), see Figure 36. This was achieved because the buildings in the system requiring the highest supply returns became more efficient; their consequent lower temperature demands allowed the whole district heating network to operate at lower temperatures. Further improvement potentials are predicted to lower the average supply temperature by another 4–5 °C over the coming years while still comfortably meeting the needs of all customers. However, return temperatures must also be reduced (Diget, 2019).

The lower temperature levels decreased heat loss, saving around 670 000 euro per year, producing an estimated cost reduction gradient (CRG) of 0.29 euro/(MWh·°C). By introducing a motivation tariff, Viborg has created a win-win situation for both customers and the operator. This motivation model costs the company around 270 000 euro per year, with the overall impact of the motivation tariff being savings of around 400 000 euro per year (Diget, 2019).

Another exemplary case study is **Borås** (SE; 611 GWh/a), where the annual average supply and return temperatures in the network were reduced from 96/68 °C (1985/86) to 79/47 °C (1996), see Figure 36. The ten-year temperature decrease was obtained through three main measures:

- Increasing the maximum pressure from 10 to 16 bars, enabling increased flows and lower supply temperatures;
- Introducing flow limiters in half of the substations (the largest substations); and
- Fulfilling a quality assurance program detecting malfunctions in substations.

The lower temperature level led to a more efficient energy system with higher electricity production within the combined heat and power (CHP) plant, lower costs due to less fuel usage and a higher coefficient of performance (COP) for the heat pump. The CRG for a lower network temperature level was estimated as 0.05 euro/(MWh·°C). This low value is due to the lower price and cost levels in 1995. Based on a temperature level reduction of 18 °C, the cumulative annual value was calculated to be about 500 000 euro. The total economic effort was about 2 million euro, giving a pay-back of about 4 years. This benefit was obtained without any major investments because only directed maintenance and lim-

> Network analysis allows identification of bottlenecks in a system and determination of the minimum supply temperature for the network. An easy approach is gradually reducing the supply temperature until the critical customer is adequately supplied.

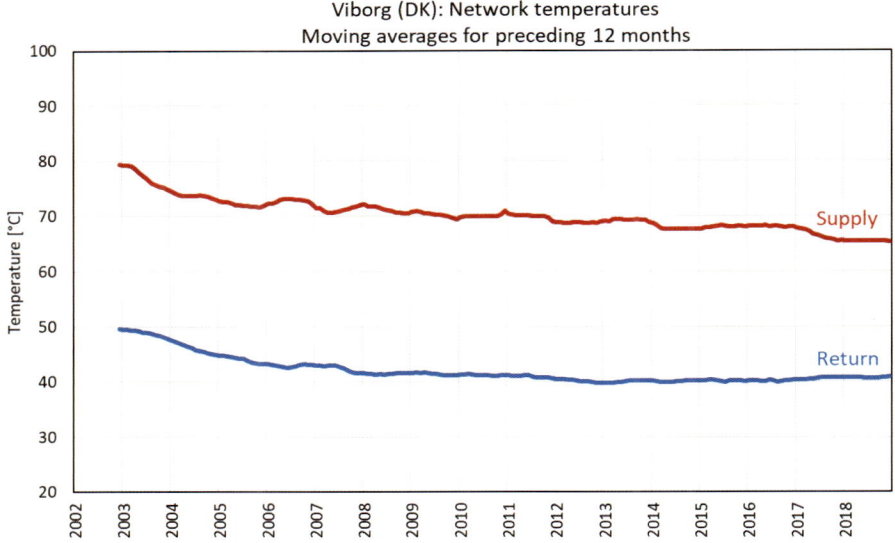

Figure 36. Supply and return temperature in the Viborg (DK) district heating network (Diget, 2019)

ited investments were used. So, the economic benefits of these temperature level reduction measures are obvious (Dahlberg & Werner, 1997).

Although both supply and return temperature reductions were achieved up to the end of the 1990s, these achievements were subsequently offset. Supply temperature increased from below 80 °C, returning to its 1989 level of 90 °C. Since 1994, the return temperature has been stable (just below 50 °C). The increased supply temperature can be explained by:

- Abandoned focus on temperature reduction;
- Installation of an absorption chiller requiring high supply temperatures during summers; and
- Connection of various single-family houses requiring greater circulation flow.

The Borås case demonstrates the importance of remaining committed to temperature reduction. Abandoning the priority can sacrifice the efforts made for previous improvements.

Meanwhile, the **Middelfart** (DK; 96 GWh/a) utility successfully lowered temperature levels from averages of 80.6/47.6 °C in 2009 to averages of 63.0/40.5 °C in 2019, see Figure 38. The district heating company took part in the development and testing of software tools that can help reduce the return temperature in district heating networks. Furthermore, this case study demonstrated a process that district heating companies can follow when working towards a low-temperature operation profile. During the process, network heat loss was reduced by 25%, with economic benefits estimated at approximately 560 000 euro per year (Sipilä & Rämä, 2016). Considering an average temperature reduction of 12 °C, the CRG is estimated as 0.49 euro/(MWh·°C). The Middelfart example demonstrates the possibility of including customer installations in the optimisation of a district heating system by monitoring the operating conditions of customer substations, providing service checks for customer installations and implementing a return temperature tariff that motivates customers to improve their own internal heat distribution systems to subsequently benefit from optimisation (Sipilä & Rämä, 2016).

The district heating network of **Gleisdorf** (AT; 6.5 GWh/a) was operated at temperature levels of about 90/60 °C from 2009 to 2016. Triggered by research projects, the service operator became aware of the importance of lowering system temperatures and the related technical and economic benefits. Consequently, starting in 2016, the district heating system reduced supply temperature stepwise to arrive at the current level of 83 °C. Meanwhile, return temperature was reduced to the current level of 49.5 °C (Figure 39) through comprehensive monitoring and various measures at the substation and customer levels (e.g. changing or cleaning heat exchangers, adjusting control setpoints and closing pri-

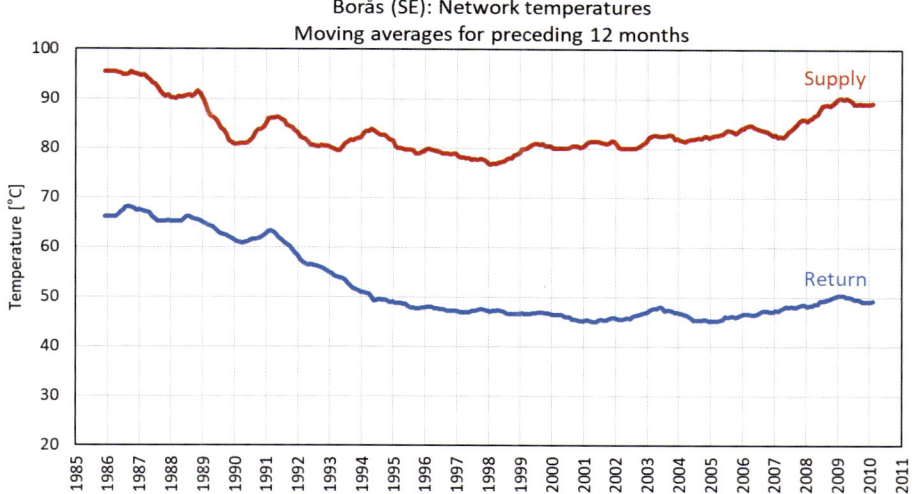

Figure 37. Supply and return temperature in the Borås (SE) district heating network. Source: (Dahlberg & Werner, 1997) complemented with later years.

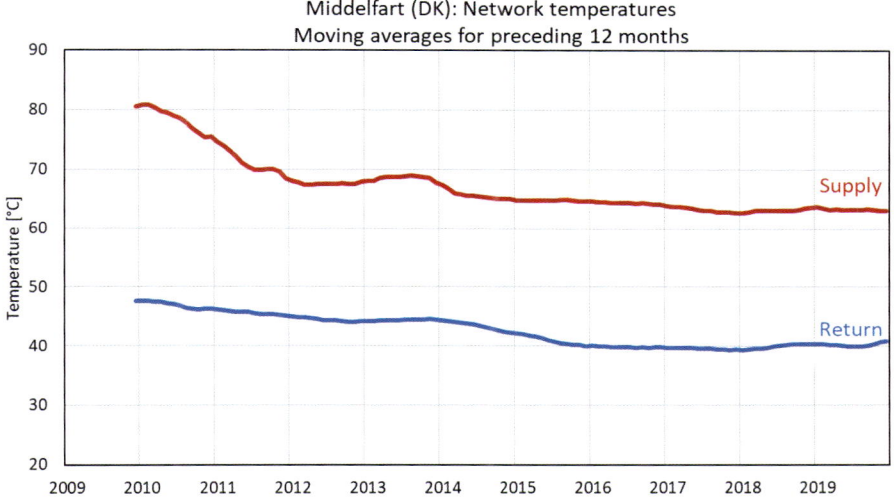

Figure 38. Supply and return temperatures for the Middelfart (DK) district heating network (Middelfart Fjernvarme, 2020).

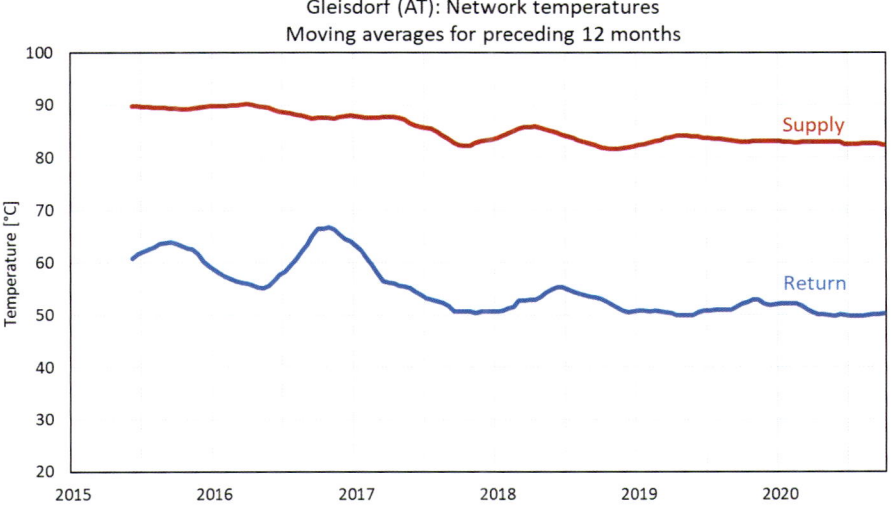

Figure 39. Supply and return temperctures for the Gleisdorf (AT) district heating network, according to AEE INTEC.

mary and secondary side bypasses). However, the operators remain convinced that further reductions of supply and the return temperatures are possible and are planning further actions to do so. Based on an average temperature reduction of 10 °C, a cost reduction of approximately 7 150 euro/a can be achieved through higher yields from solar plants and lower heat losses, producing a CRG of 0.11 euro/(MWh·°C). Although that does not sound like much, it is a relevant saving for a small system such as Gleisdorf's, and it is sufficient for financing and continuing the described optimisation measures. However, the most significant benefit is that lower temperature levels provided a foundation for integrating a new low-temperature heat source – an heat pump at a wastewater treatment plant – and a new network section with supply temperatures in the range of 70–75 °C. The heat pump's integration will lead to a significant increase of the CRG to 0.28 euro/(MWh·°C). Hence, economically and ecologically, the temperature reduction measures will substantially profit the Gleisdorf district heating system.

4.5 Subnetworks

To achieve an optimum at an overall level, distributed subnetworks can be established beyond the (existing) central structures to establish immediate (and local) actions to reduce system temperatures. Adjusting the temperatures through adapting heating systems, or adopting other technical measures, can be a challenge because most can only be implemented in the medium-to-long term, a function of the dominant building stock, the high investment costs and, sometimes, the difficult ownership structures. In such cases, subnetwork solutions can be applied, as highlighted in case studies 'Stadtwerk Lehen' (Salzburg, AT; see also Section 7.7) and 'Graz Reininghaus' (Graz, AT; see also Section 7.8).

Such (re)structuring may be conducted for parts of the existing network or for new supply areas, where lower temperature (and pressure) levels are integrated as so-called secondary networks or subnetworks. Such networks are hydraulically separated heat networks that are (partially) supplied by an upstream network. These networks are usually characterised by lower temperature levels compared to the upstream network, resulting in a transmission network with high temperatures securing the supply for all districts and a distribution network supplying connected areas with lower system temperatures. Through this division, heat can be fed into either the transmission or the distribution network, depending on the operating conditions (see Figure 40). The transmission network also serves as a backup in the event that the heat source is not delivered or the heat source's temperature is too low. The link to the transmission network reduces the risk of depending completely on a low-temperature heat source, which makes investment (at lower costs) and business decisions easier (Köfinger et al., 2016; Lygnerud et al., 2019).

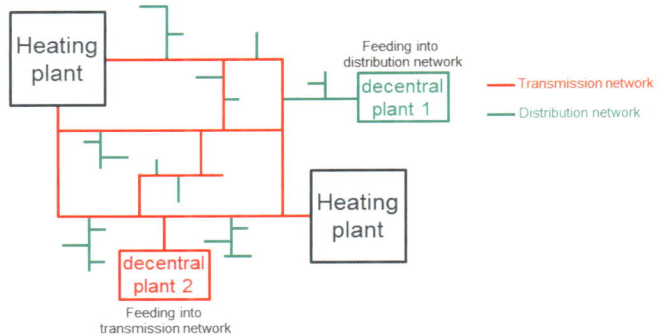

Figure 40. Feed-in points of distributed generators in a meshed heat network featuring transmission and distribution sections

Large urban district heating systems may be specially conditioned to favour higher temperatures in transmission networks, often a function of transmission capacities. However, it remains important to operate with lower temperatures in the distribution networks, with the Vienna district heating network providing a useful example of how this can work.

Example of district heating in Vienna

With supply temperatures between 80 and 160 °C (due to capacity) and return temperatures of approximately 60 °C, the Vienna district heating network is not a low-temperature network. A specific feature of the Vienna district heating network is its division into one major transmission network and several local distribution networks, which has the advantage of the networks operating according to different operating parameters (i.e. temperature and pressure), enabling connection of different district heating generations (e.g. 3GDH with

 Transmission systems supplying multiple distribution systems allow for the lowest temperature levels to be achieved locally. Such systems can be made on a large scale (e.g. Vienna) or a small scale (e.g. Stadtwerk Lehen and Graz Reininghaus).

4GDH). In some cases, subnetworks are already designed as low-temperature networks. For example, buildings are connected to heat networks in the area with the lowest network temperature (e.g. low energy houses) and, if necessary, to make the best possible use of local potentials, such as waste heat (Köfinger et al., 2016). Thus, Vienna serves as an example of how a large, high-temperature urban network can be successively transformed into a low-temperature networks (for more information on such transformations, see Chapter 8). Further information on the Vienna district heating network is provided in (Wien Energie, 2013) and (Wien Energie, 2020).

4.6 Cascading solutions

The basic principle of cascading is that a high return temperature customer supplies a low-temperature customer. For cascading solutions, existing technologies (i.e. 'classical' heat supply) can be used at the customer and system levels or even be combined with subnetworks (e.g. Vienna, as discussed in the previous section). Cascading connections of individual customers or customer clusters constitute a short term measure reducing return temperature. Here, the energetic use of the return pipe of high-temperature customers is the supply for low-temperature customers, which has a synergetic effect and can be either indirect (via the return pipe of the network) or direct. Options for the direct or indirect cascade connections of high- and low-temperature customers are presented in Figure 41, while Section 10.1 includes information about supply-to-supply and return-to-return connections for new customers or new distribution networks. Meanwhile, using buildings, building clusters or districts in a heat cascade requires analysis of the respective components according to their consumption and temperature profiles (Köfinger et al., 2016; Basciotti et al., 2016).

However, this solution has a built-in risk in the form of buildings that are energy renovated and have their temperature requirements reduced. On the one hand, this is beneficial to high-temperature supply buildings; on the other hand, if temperature decrease too much, low-temperature customers might not be supplied suffici-

 Cascading can be an intermediate solution, but it can also produce bottlenecks by lowering the subordinate system temperature. This can require new solutions, including reconnection.

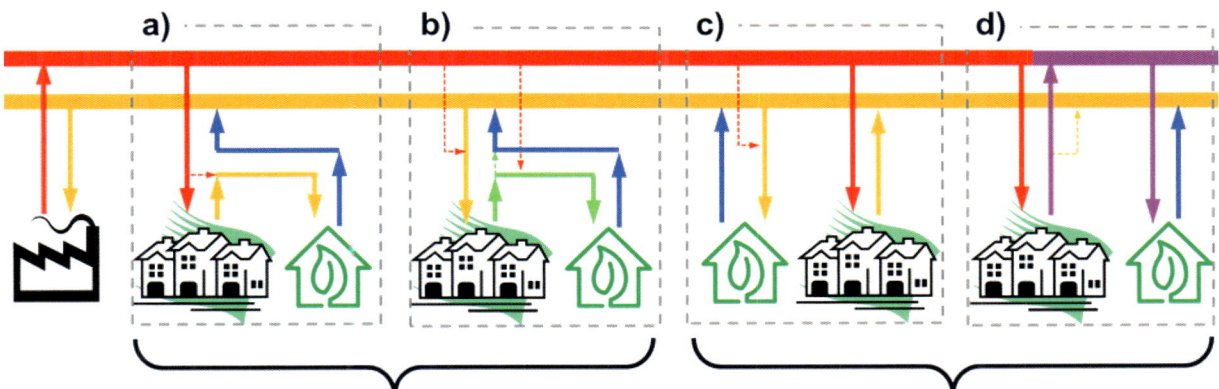

Figure 41. Options for direct or indirect cascading connection of high- and low-temperature customers: a) direct use of a high-temperature return; b) double cascade, the indirect use of a high-temperature customer and direct use with admixture; c) indirect use of the high-temperature return via the return pipe of the heating network; and d) indirect use via the supply pipe of the district heating network (Köfinger et al., 2016)

ently. If cascading stops functioning, solutions must be provided, which can require reconnection. Accordingly, cascading is a short term approach to reducing the return temperature.

The effects of cascading connections on return temperatures have been examined for different scenarios for two district heating networks: Vienna and Klagenfurt (AT). The modelled results can be summarised as follows (Köfinger et al., 2016; Basciotti et al., 2016):

- In Vienna, cascading could reduce the annual average temperature of the central heating system by between 0.7 and 1.5 °C. Depending on the scenario, a temperature reduction of up to 11.6 °C could be obtained for some network branches.
- In Klagenfurt, cascading could reduce the temperature of the central heating system in the considered pipe by an annual average of about 1.2 °C. The temperature reduction could be up to 9.6 °C for some network branches.

4.7 Installing heat pumps to address subsection demands

Numerous heat pump integration concepts have been described using external and internal heat sources (see Geyer et al., 2019). External sources are heat sources that bring outside energy into a system. These include, for example, ambient heat, industrial waste heat or heat from infrastructure (e.g. sewers and tunnels). Meanwhile, internal sources do not bring additional alternative energy directly into the system; instead, the district heating network is used as a heat source.

By integrating a heat pump into the return pipe, the temperature can be further reduced, as demonstrated by Figure 42. Increased cooling means other alternative energy sources can be either or both more easily and better integrated into the system, which can enable flue gas condensation if the return temperature in the network is too high. Furthermore, capacity is increased, and a subnetwork can be supplied. Different integration concepts are available on the IEA HPT Annex 47 project website (IEA HPT, 2020).

This concept has been realised by the 'District-Boost' project in Vienna, in which the return pipe of the primary district heating network is used as the heat source for the heat pump and has been connected via an additional hydraulic circuit. The heat sink inlet is connected to the return pipe of the secondary district heating network, and the heat sink outlet is connected to the supply pipe of the secondary district heating network. To guarantee that the heat sink's outlet temperature reaches the required supply temperature for the secondary district heating network, an admixing circuit is used. This circuit is hydraulically separated from the district heating network by thermal storage. The nominal heating capacity of the heat pump is about 255 kW, and the efficiency corresponds to heat source temperatures of 45/35 °C and heat sink temperatures of 63/75 °C, producing a COP of about 5.3 (Windholz et al., 2016; IEA HPT Annex 47, 2018).

In case study 'Stadtwerk Lehen' (Salzburg, AT; for more information, see Section 7.7), heat from roof-mounted

  Applications of heat pumps in district heating networks can:
- Increase transport capacity and supply of distribution or subnetworks
- Serve as a booster solution (temperature and capacity increase locally)
- Better discharge storage (increase storage capacity)

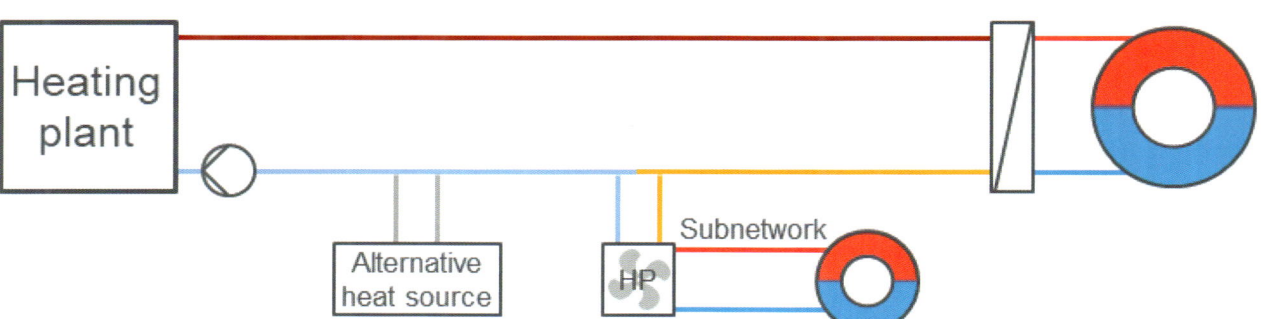

Figure 42. Integration of heat pumps with internal heat sources. On the one hand, this reduces the return flow; on the other hand, this increases capacity (modified from Geyer et al., 2019)

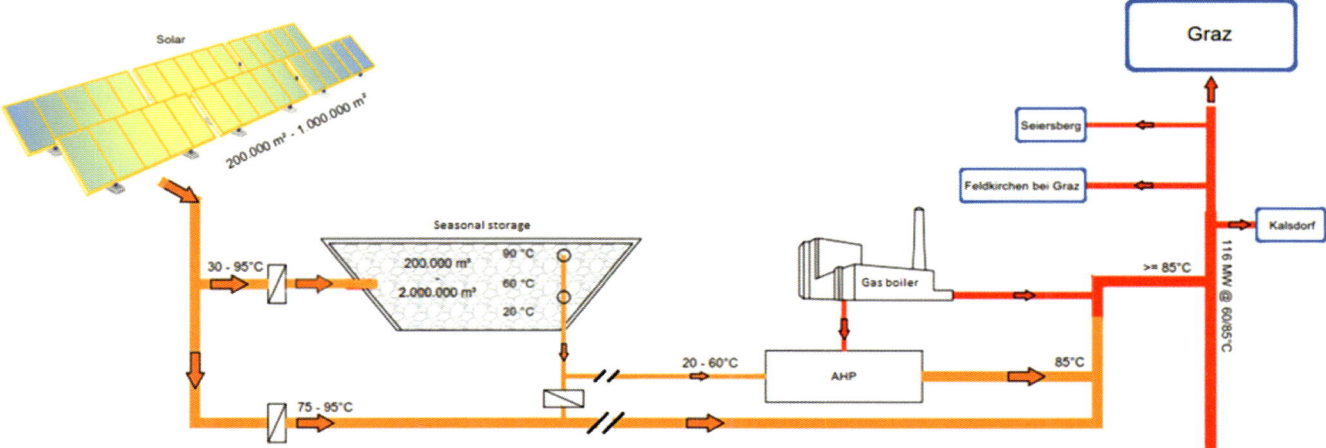

Figure 43. Scheme of the BIG Solar Graz concept (Reiter et al., 2016)

solar collectors (2,047 m²$_{gross}$) is stored in the central storage (200 m³), with an heat pump (160 kW) integrated to increase storage capacity. Additionally, the heat pump increases the yield of the solar system by up to 15%.

Elsewhere, the BIG Solar Graz concept is an innovative approach that bridges the gap between low- and high-temperature systems through heat pumps. Figure 43 indicates the low-temperature system in orange and the high-temperature system in red. Absorption heat pumps play a key role in this concept, improving both solar yields and storage capacity management. Gas boilers cover the residual load if more high temperatures are requested (Reiter et al., 2016).

4.8 Increased decentralised supply

Decentralised heat suppliers enable the integration of higher proportions of alternative heat sources. Given the close distance between supply and demand, heat loss can be relatively low. When a decentralised heat source belongs to a new stakeholder, close communication with the district heating operator is needed for each to understand the other's processes, including the volumes to be delivered, the temperatures needed and the value of the heat. Such discussions can constitute a barrier to investment in low-temperature heat sources, as recognised in both the high temperature industrial and the urban waste heat recovery context (Persson & Averfalk, 2018). Notably, such relationships require district heating providers to update their business logic to promote closer customer relationships in conjunction with a broader array of heat sources, especially where high-temperature sources correspond to an economy of scale and central production and lower temperature sources necessitate a decentralised and localised solution. Chapter 6 discusses this in greater details.

For the decentralised feed-in of heat into district heating systems, certain network-side requirements can ensure the safe and proper operation of the network. These requirements concern prevailing temperature and pressure levels, as well as the network control. Decentralised heat supply must consider two major output conditions: a stable required feed-in temperature and a feed-in heat rate equal to the heat output from the decentralised heat source. However, many installations cannot achieve the second condition due to variations in the feed-in heat rate. This problem can be resolved through different control concepts incorporating either temperature- or flow-control (Lennermo et al., 2019).

Additionally, network capacity and hydraulic limitations complicate the district heating network for decentralised suppliers. Figure 44 shows a main central feed, several customers and two decentralised feeders in one network segment. The red lines represent the supply pipes of a network, and the blue lines represent the return pipes. The distance between each supply and return pipe indicates the pressure level. The pressure level decreases as pipe length increases due to friction. A circulation pump is necessary to exploit a decentralised heat source and change the direction of

> The integration of decentralised suppliers enables a local capacity increase without increasing overall network temperatures.

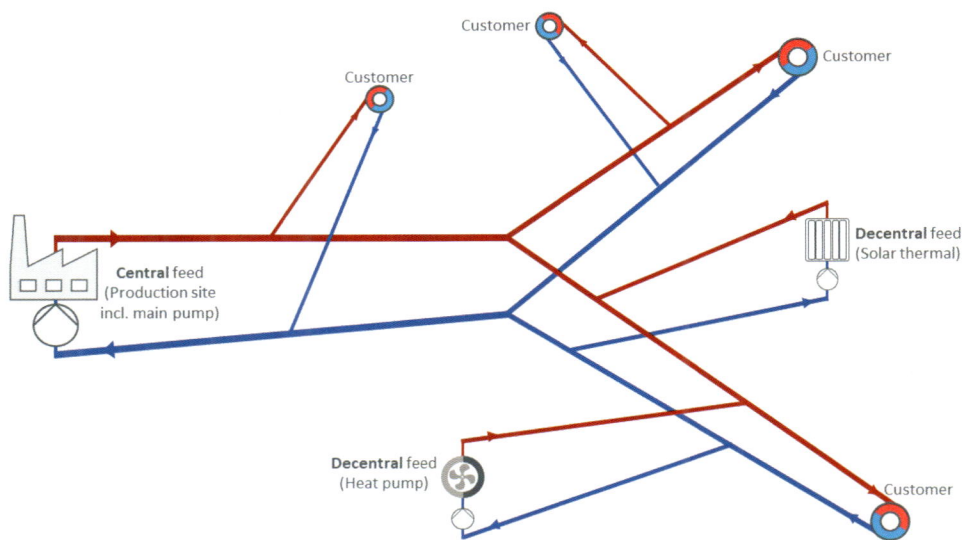

Figure 44. Illustrative pressure levels for a district heating network with centralised and decentralised heat sources.

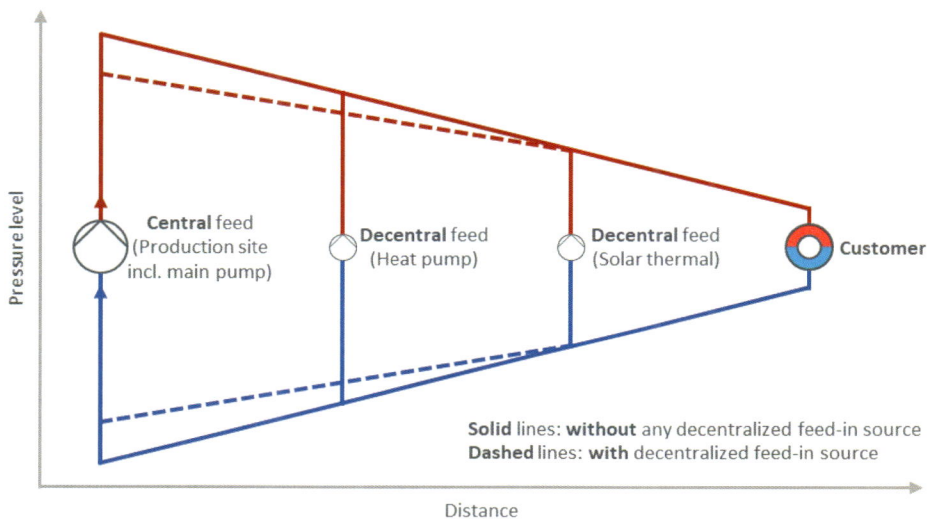

Figure 45. Visualisation of a district heating network showing centralised and decentralised heat sources and flow directions (modified from Nagler & Ponweiser, 2018)

the flow so that the heated water from a decentralised feed-in source can be pushed from the return pipe into the supply pipe. This is indicated by the increased distances between the supply and return pipes, where a pump is located and can increase the pressure level. Depending on the system control strategy, several different approaches are possible (further information is provided by Basciotti et al., 2017).

To demonstrate the challenges for pressure level control when decentralised suppliers are included in the district heating network, a simplified and generalised pressure level graph is provided in Figure 45. The solid red and blue lines represent the pressure level in the supply and return lines when there is no decentralised feed-in source. When a decentralised feed-in source is added in the return to supply configuration, with its own circulation pump, the pressure level distribution changes. Each distributed feed-in circulation pump needs to provide the pressure level sufficient for customers downstream in the district heating network. In the connection configuration presented in Figure 44, using distributed sources can allow the central circulation pump to promote a lower pressure rise because the feed-in circulation pumps provide the pressure increase necessary for downstream costumers. This lower increase and assumed pressure level distribution is represented by dashed lines in Figure 45. To ensure a

desirable pressure distribution and that the final customer receives sufficient pressure difference, the pressure of these distributed feed-in circulation pumps must be sufficiently increased.

4.9 Digitalisation opportunities

Digitalisation provides opportunities for sustainable district heating networks by increasing the share of both renewable and excess energies and enabling the reduction of temperature levels. According to the *Digital Roadmap for District Heating & Cooling* developed by Euroheat & Power (DHC+, 2019), digital technologies could make whole energy system smarter, more efficient and more reliable, as well as boosting the integration of renewables into the system. In the future, digital energy systems will enable district energy systems to fully optimise their plant and network operations while simultaneously empowering the end consumer. Digitalisation can optimise efficient use of connected infrastructures, generate scheduling according to forecast demand and enhance the integration of renewables. Although substantial digitalisation measures have already been implemented in some cases, this is not true for all or whole systems or applications. Regarding improvements, while considerable data is available for supply and distribution, data quality and quantity on the customer side could be deepened to allow optimisation measures to be employed at the system level.

Digitalisation abilities

The recently launched IEA DHC Annex TS4 'Digitalisation of District Heating and Cooling' (IEA DHC, 2020) aims to promote the opportunities for integrating digital processes into DHC schemes and clarify the role of digitalisation for different parts within the operation (and maintenance) of the DHC system. Furthermore, the programme will demonstrate the implementation of these technologies. However, new challenges need to be addressed, including data security and privacy concerns together with questions about data ow-

Table 3. Overview of digitalisation abilities in DHC systems divided by supply, distribution and demand (IEA DHC, 2020). The left side indicates where operational optimisation can be applied for what reasons. The right side lists methods that can be supported by analytics.

	Operational Optimisation	**Supported by analytics**
Supply	**Aim:** Dispatch optimisation/portfolio management of existing supply units for: • sector coupling/market participation of heat pump and CHP; • operation of small/large/seasonal storage; and • peak shaving.	**Aim:** Predictive maintenance using: • well-established tools for central supply units; and • a new approach using analyses of distributed prosumers. **Aim:** Evaluation of optimisation measures and operations (e.g. forecasts)
Distribution	**Aim:** Optimisation of network operational parameters such as • supply temperature, ΔT, flow rate, pressure distribution; • value settings for the opening of bypasses, redirecting flows, turn off network branches; and • handling distributed production/prosumers.	**Aim:** Analyses/predictive maintenance of: • historic segment load distribution (for assessing future extension options); • grid storage; • temperature and heat loss distribution; • leakages and bypasses; • material degradation/ageing; and • water quality;
Demand	**Aim:** Improve load management and supply and return temperature management by addressing • building management system; • direct control of substations; • balancing of secondary valves/hydraulics; and • storage in buildings (e.g. water, building mass).	**Aim:** Detection, diagnosis and correction of faults (e.g. high return temperature, peak loads, leakages and oscillations) for: • substations; • heat/utility meters; • building side controllers; and • valves, sensors and heat exchangers. **Aim:** Visualisation/analyses of consumption/ heat demand profile including • heat supply contract conformity; • user behaviour/efficiency; and • suggestions for improvements customer can make.

nership. Early results published in 2019 (IEA DHC, 2019) indicated that digitalisation could support operational optimisation and analytics of LTDH systems in the ways outlined in Table 3.

In the European Horizon 2020 project TEMPO (2021 [forthcoming]), the digitalisation of district heating networks plays a big role. Given there is as yet no univocal definition for 'digital heat networks', certain criteria or conditions need to be outlined. According to TEMPO, a digital heat network is a network featuring many sensors, as well as automated recording, transfer and storage of data, with data analysed automatically. Eventually, these data and analyses are used not only for billing but also for load forecasts based on consumption data, enabling optimisation of heating plants and storage scheduling within a short term prediction window. Analysis and visualisation of data is a critical feature of digital heat networks because raw data is of little interest. Instead, data must be transformed into comprehensible information that can be interpreted by humans. While engineers do this all the time manually (e.g. by translating numbers into graphs and tables), digital heat networks automate this step using machine learning or data mining algorithms. TEMPO (2021 [forthcoming]) exemplifies the development of different digital solutions for network optimisations, with solutions including:

- **A supervision ICT platform for detection and diagnosis of faults in district heating substations:** 75% of all district heating substations perform sub-optimally, producing an average 15–20 °C higher return temperature than necessary. The supervision platform automatically detects this suboptimal district heating substation behaviour.
- **Visualisation tools for expert and non-expert users:** These visualisation tools form the basis of a decision support system for expert users, which will become a powerful tool for detecting operational faults and deviations in the district heating network. The tools for non-expert users will provide knowledge on energy consumption and efficiency measures, empowering them as part of the energy chain.
- **A smart district heating network controller that can balance supply and demand and minimise return temperatures:** The controller balances heat demand and fluctuating renewable and residual heat sources, as well as further reducing return temperatures by influencing demand behaviour on the customer side and coordinating this at a network level.

Automatic fault detection using artificial intelligence and big data

Return temperatures in district heating systems have already been addressed in an IEA project (see Section 4.1), with the concepts of excess flow and target return temperatures being used to detect malfunctioning substations (Zinko et al., 2005). However, the diagnosis was limited to recognising the time of year malfunctions arise and deciding whether the issue concerned space heating or domestic hot water usage. While meter readings at district heating substations were previously performed manually for billing purposes, they are increasingly available at an hourly resolution, enabling inefficient behaviour to be detected proactively and cost-efficiently (Gadd, 2014). Meanwhile, (Gadd & Werner, 2015) used measurements of 135 Swedish district heating substations to identify three main types of faults, recognising minor and major faults in more than two-thirds of the substations. Although this analysis was completed manually, the researchers highlighted the possibility of automation. Based on that research, (Hamilton-Jones, 2020) developed an improved method using clustering algorithms for fast and automated detection of faults.

Automated fault detection and diagnosis for heating, ventilation and air-conditioning (HVAC) systems constitutes an active research area involving an array of methods (Kim & Katipamula, 2018). Despite the development costs for the model-based methods and the validation through measurements, the advantages are big as these models are/will be scalable and adaptable to any HVAC systems. To model the impact of faults producing higher district heating return temperatures, it is necessary to consider temperatures supplied by district heating networks, substations, customer heating systems and the buildings they serve. Existing simulation tools allow the execution of models accounting for the interplay between these systems and the provision of quantitative information regarding the resulting behaviour. By running sufficient similar simulations, training data can be produced for fault detection and diagnosis algorithms. If substantial measurements are directly available, methods such as big data and artificial intelligence can be used to detect errors and fault 'sig-

> Data-driven operational optimisation has enabled Assens Fjernvarme (DK) to lower its network temperature by 6–8 °C. Thus far, the utility has reduced its annual heat production by 2.5% and distribution loss by 12%. Additionally, it has been able to remove more than 100 bypasses around the network, with optimisation measures ultimately resulting in savings equivalent to about 30 euro per household (Kamstrup, 2020b).

natures' in real-time (for more insight, see Farouq et al., 2020). Applying these approaches could contribute to reducing system temperatures and optimising district heating networks, which could ultimately lead to transformations towards LTDH networks featuring higher proportions of renewable energies (Brès et al., 2019).

4.10 Design criteria for new systems

When planning new systems, it is necessary to consider not only locally available resources but also the needs of the customers. Possible cold supply and innovative supply concepts offer new solutions and services, with Section 10.1 providing an overview of possible network configurations.

Two key decisions for good planning

Heat loss increases as pipe diameter increases (due to increased surface area) and decreases with improved thermal insulation. Therefore, it is important to use the smallest possible diameter or, at most, one nominal size larger (Nussbaumer & Thalmann, 2016). Accordingly, during the design phase, key decisions must be taken regarding insulation performance (which impacts heat loss) and investment costs.

For ease of comparison, two networks have been used to demonstrate the differences. Network 1 features a 600 m pipe route length and pipe sizes up to DN65. Network 2 features a 1000 m pipe route length and pipe sizes up to DN100. For the larger network, an additional simulation swapped the larger pipes for double pipes sized up to DN65, demonstrating the significant decreases in heat loss. The comparisons presented in Table 4 and Table 5 provide an overview of the potential heat loss savings and related cost changes, which are crucial for individual project investment decisions.

The **first decision** concerns the use of **single or double** (supply and return within the same insulation) pipes. For flexible plastic solutions, double pipes are usually offered up to a size of DN65; for steel pipes, sizes can reach DN200. Table 4 summarises the effects of decisions regarding single or double pipe solutions, demonstrating heat loss savings up to almost 40%. Higher potential savings are especially possible for steel pipes.

Table 4. The first decision involves lowering heat losses through the use of double instead of single configurations of steel and flexible plastic pipes (Austroflex Rohr-Isoliersysteme GmbH, 2020)

Heat loss savings by…	Steel pipes	Flexible plastic pipes
using double instead of single pipes.	36 – 38%	22 – 30%
swapping only smaller pipes (up to DN65) for double pipes.	18 – 19%	10 – 15%

The **second decision** concerns **insulation class**. Insulation class describes the thickness of the thermal insulation around the carrier pipe. Flexible pipes with polyurethane hard foam (PUR) insulation demonstrated heat loss values similar to steel Series 2, reaching Series 3 levels with 'plus' insulation (see Figure 46). Based on catalogue values and project experience, Table 5 provides an overview of heat loss reduction compared to cost increases for Series 1, 2 and 3 steel pipes. Series 1 is the weakest insulation class, and Series 3 is the strongest, which corresponds to the least amount of heat loss. For flexible plastic pipes, there are no standardised classifications of insulation thickness. Thus, the flexible pipes used for this comparison used Basic to correspond to Series 1, Standard to correspond to Series 2 and Plus to correspond to Series 3. It is worth noting that other plastic solutions exist that have produced much greater heat losses, sometimes twice that of steel Series 2. Although heat loss savings are slightly less in comparison to steel, heat loss levels are slightly lower for plastic pipes with polyurethane insulation.

> The design phase is crucial for efficient and future-proof district heating networks. Here, two basic decisions regarding pipe configuration and insulation class must be made.
> Notably, lower system temperatures could decrease costs due to decreased insulation needs.

The incentive to emphasise insulation is lost in the context of reduced system temperatures, with insulation less relevant for cold district heating (CDH) networks, sometimes omitted because temperatures can be close to ground temperature, producing negligible heat loss. As such, high-temperature distribution systems need more insulation, and low-temperature systems need less insulation, meaning a lower insulation class can be chosen for lower excavation costs.

Table 5. The second decision involves comparing heat loss reduction to cost increases for steel and flexible plastic pipes (Austroflex Rohr-Isoliersysteme GmbH, 2020)

Comparison	Heat loss reduction	Network cost increase
Steel pipes		
Series 2 vs Series 1	16–18%	5%
Series 3 vs Series 1	26–28%	14%
Flexible plastic pipes		
Standard vs Basic	13–15%	6%
Plus vs Basic	21–25%	22%

Strategic focus on customer temperature requirements

Utilities services should also analyse and assess the technical situation on the customer side. Supplied heat should be used as efficiently as possible to achieve low system temperatures. Thus, from a strategic point of view, an operator should be careful to ensure that all consumers maintain the prescribed return temperatures, with new consumers only be connected if the connection does not require an increased supply temperature. Customer heat requirements can differ substantially (compare, for example, industry and household customers). Higher temperature requirements on the customer side can be offset by onsite booster solutions, with the district heating network providing the necessary baseload.

4.11 New innovative supply and distribution concepts

Given LTDH requires approaches overcoming the limitations of the business and supply logic of conventional district heating networks, in addition to the currently predominant two-pipe district heating systems with central pumping, various other concepts have been developed, built and proven. While heat loss is apparent in warm district heating (WDH) systems, it nearly vanishes in CDH systems. Thus, additional investments are required for additional heat supply to customer substations in CDH systems. However, such investments are not required in WDH systems.

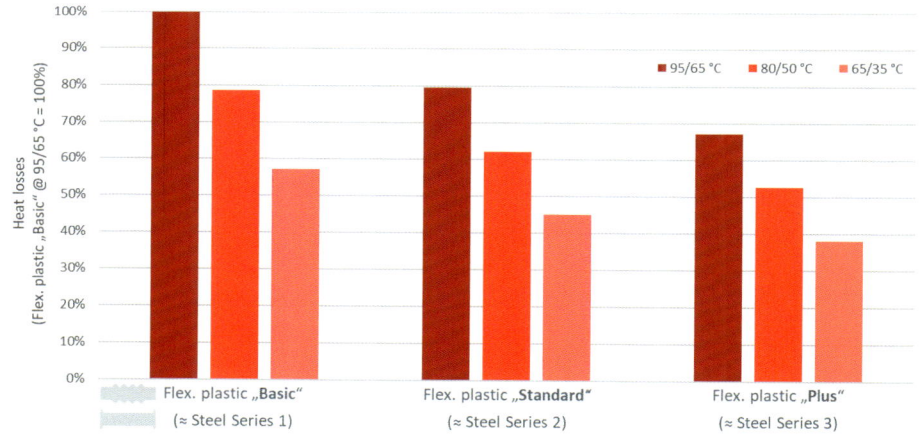

Figure 46. Comparison of heat loss values for flexible plastic pipes of different insulation classes at different temperature levels (Basic at 95/65 °C = 100%) (Austroflex Rohr-Isoliersysteme GmbH, 2020)

> When reducing heat loss from district heating pipes, there is a trade-off between the increasing pipe insulation costs and the associated heat distribution savings (Lund & Mohammadi, 2016).

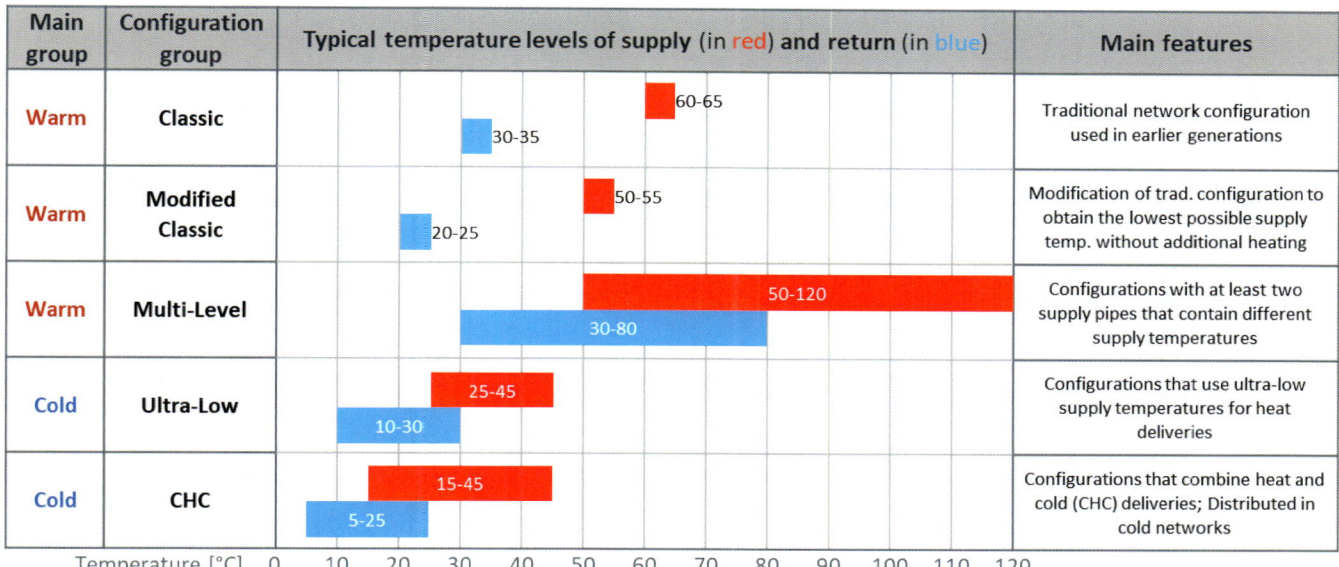

Figure 47. Overview of the network configurations and classifications this guidebook provides for low-temperature heat distribution, including typical temperature levels and principal features

 Warm district heating systems are preferred when excess heat is available at temperatures high enough to support the temperature demands of customers without additional heat supply. **Cold** district heating systems are applied when heat sources with temperatures below customer requirements are used as the main heat source.

An overview of the network configurations (classifications) used in this guidebook is provided in Figure 47. These configurations have been identified from projects performed by early adopters of low-temperature heat distribution. Notably, CDH and WDH networks are siblings in this configuration family. See Section 10.1 for detailed information on these configurations.

Heat sources with temperatures below customer requirements require an additional heat supply. An example is heat pumps, which can be applied in a centralised (in WDH systems) or decentralised (in CDH systems) manner. The choice depends on the heat density of the supply area and the heat loss associated with the cold or warm distribution.

Figure 48 compares the network heat loss for different supply concepts in a mixed-use neighbourhood in the city centre of Amstetten (AT) with about 590 inhabitants and 550 office workers. The neighbourhood has access to a conventional high-temperature district heating (95–55 °C; insulated steel pipes) system as well as both an LTDH (55–35 °C, insulated steel pipes) system and CDH (5–15 °C, plastic pipes without insulation) system for efficiency purposes. The comparison shows that the relative heat loss is approximately one-third lower for the LTDH system and two-thirds lower for the CDH system (Leppin, 2020).

Warm, or 'modified classic' district heating systems

This configuration group involves modifications to the traditional configuration enabling the lowest possible supply temperatures without any additional heating at the customer end. An example of a modified classic configuration is shown in the Section 10.1 as Figure 95; its corresponding characteristics are summarised in Table 25.

For heating and domestic hot water supply, three-pipe networks are a suitable option for lowering system temperatures. In the 4GDH projects TEMPO and TERMO, three-pipe networks have been designed. For TEMPO, this included one double pipe for supply and recirculation and one single pipe for common return. The third pipe fills a recirculation purpose and is usually two to three standard pipe sizes smaller than the supply pipe. The additional costs and savings resulting from the lower heat losses have been calculated and presented in Table 6. The additional pipe increases investment costs increase by approximately one third while reducing heat loss by 27%. Heat loss savings are calculated

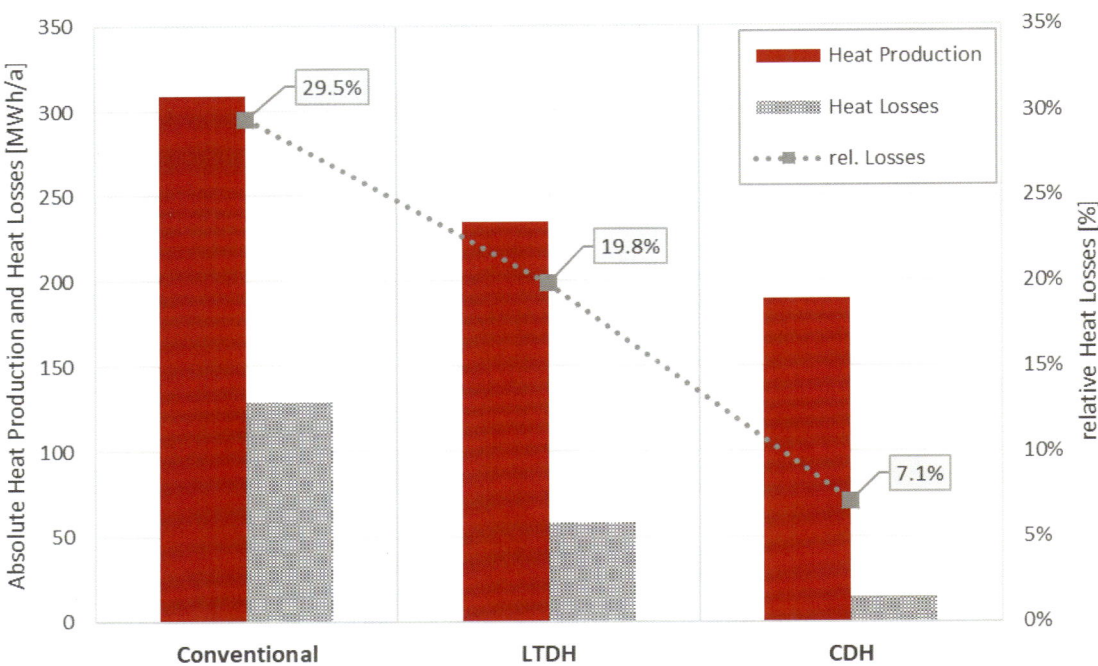

Figure 48. Comparison of different supply concepts and their network losses for a mixed-use neighbourhood in Amstetten (AT) with access to conventional high-temperature district heating (Leppin, 2020)

 Innovative **DHC network configurations** for heating and cooling supply are detailed in Section 10.1. Both conventional and new configurations are listed, which were collected within the scope of this project based on implemented examples. These should provide insights and suggestions on achieving the transformation to lower system temperatures. In each case, a graphical representation of hydraulic concepts is provided, as well as a description of the characteristics and typical temperature levels. While case studies are also mentioned, a more detailed description of the individual cases is found in Chapter 7.

according to the assumption that a two-pipe network operates at 70/40 °C because it cannot achieve the low return temperatures of a three-pipe network with 65 °C supply and 25 °C return (Engel, 2018).

Table 6. Comparison of a three-pipe and two-pipe network, including cost increases and heat loss savings (Engel, 2018)

Network comparison	Cost increase		Heat loss savings 65/25 °C vs 70/40 °C
	TEMPO	TERMO	
three-pipe vs two-pipe	34%	28%	27%

In other studies, the additional cost was lower, having foreseen one single pre-insulated pipe assembly with three pipes. This concept is only possible for small pipe sizes, such as those used for smaller house connections; the main pipes have to be double or single pipes (Engel, 2018).

To combine heating, cooling and domestic hot water, a four-pipe network is economically optimal.

Benefits of cold, or 'ultra-low', district heating systems

In contrast to WDH systems, which cover energy and temperature requirements directly, CDH systems represent a paradigm shift for grid-connected energy supply, using an additional decentralised heat supply to meet typical customer temperature demands. Importantly, labelling issues concerning CDH systems are addressed in Section 1.2.

Given these networks operate at very low temperatures – close to ambient temperature (5–25 °C) – plastic pipes without thermal insulation are used. Consequently, pipe investment costs are reduced. In contrast, given the normally low temperature spread of such systems, larger pipe diameters are required to limit flow and, thus, pumping costs. However, the space demanded by larger pipes is often not significantly higher because the pipes are not insulated, and the transport capacity is lower because of the additional decentralised heat supply units.

Notably, CDH systems allow the utilisation of a wide range of new and emission-free heat sources because temperature levels close to ambient temperature enable almost everything to become a heat source. However, such systems usually focus on low-temperature waste heat from industry and service sectors (e.g. all cooling processes), shallow geothermal energy, seas or lakes and wastewater. Depending on the seasonal availability of the heat source, some CDH systems use borehole thermal energy storage (BTES) to balance seasonal mismatches of supply and demand. From an economic perspective, however, the concept only works if inexpensive or free-of-charge heat sources are available, keeping the size of the BTES as small as possible. While conventional district heating systems provide a common temperature level to all heat customers, the decentralised heat pumps of CDH systems provide a tailormade temperature level for each heat customer. This enables high heat pump efficiency and lower heat loss.

By applying very low supply and return temperatures to a variant of the CDH system, the cold ultra-low district heating (ULTDH) system, the same temperatures used in district cooling systems can be obtained. That is, ULTDH systems provide district cooling using the same infrastructure and at the same time (e.g. one customer requires heat while others – for example, those with high internal heat loads – requires cooling). This provides an opportunity to supply both heating and cooling using the same distribution network. This synergy is known as combined heating and cooling (CHC), with the provision of cooling also a heat input in the ULTDH system context. However, to provide continuous cooling, system temperatures should be well balanced, not exceeding a certain temperature level.

The **successful implementation** of this concept was realised in **Zürich** in 2014 (see case study in Section 7.6). Here, cooling was provided for two data centres, with excess heat from that cooling process distributed within a CDH network combined with BTES at temperatures below 28 °C. Distributed heat pump stations used the CDH network as a source to provide heat to multiple secondary distribution networks operating at 40–68 °C. The mass flow in the system was generated by decentralised circulation pumps at each heat pump station. Although the CDH system operated with a temperature difference between the supply and return pipe of only 4 K, the power required for decentralised pumping was in the range of 1.5–3% (not including secondary distribution networks) of the heat delivered. Thus, it is in an acceptable range compared to conventional WDH and almost negligible compared to the power demand of the heat pumps.

At the Sydnes university campus in **Bergen** (Norway), a similar CDH network solution has been implemented. That system features one circulation pump that circulates seawater and several circulation pumps within the CDH network that circulates the water among the buildings. Some buildings use cold water directly for cooling, and some buildings have heat pumps for heating purposes. This CDH network has been extended since its establishment in 1994. Based on measurement data, the electricity use of the circulation pumps in the Bergen CDH network was 19% all energy supply to the systems, while the low-temperature heat from the sea was 45% and the electricity input to the heat pumps constituted 37%. Electricity is still relatively cheap, and district heating prices are regulated to be lower than the corresponding electricity cost. However, in recent years, electricity grid fees have increased significantly, which may motivate rethinking and optimising circulation pump operations.

Intermittent operation of heat distribution networks

Even if pipes are well-insulated, district heating networks with low heat densities struggle to be efficient. Especially during the low heat demand of the summer months, the substantial heat loss can promote low efficiency. A potential solution is the intermittent operation of (rather small) heat distribution networks, allowing the network to only operate for a few hours a day or week to charge storage(s) and enabling the district heating network to operate at a higher load for a short time before being switched off, thus reducing distribution losses. The storage facilities then supply customer needs, with the district heating network being restarted shortly before the storages empty (Clausen, 2009).

This approach was investigated by (Hammer et al., 2018) in the context of simulation studies of two district heating networks (3.5 and 5.8 GWh/a heat demand) in Austria, with the principle represented by Figure 49. By in-

The intermittent operation approach offers a possible solution for small district heating networks to reduce heat loss, especially during summer.
Intermittent operation means that district heating networks operate for only short periods during low load conditions (i.e. summer). During this time, storages are filled to meet customer demands. Heat loss can be reduced through these short operation periods.

stalling decentralised storages and shutting down the network during off-peak periods, distribution losses were reduced by up to 34%. Although increasing storage sizes seems advantageous at first glance, potentially reducing the required number of system initiation procedures, this would require longer charge times and greater charging capacities. Accordingly, important criteria include not only the downtime duration but also the network operation time. However, although heat losses can be reduced, sharp load changes during intermittent operational periods exert thermal stress on the respective installations and pipes. Thus, it should be noted that intermittent operations might impact the lifespan of the equipment, especially steel pipes, with plastic pipes better rated for increased on/off frequencies. However, to implement this operational strategy, specific situations must be considered in detail (Hammer et al., 2018).

In **Hjortshøj** (near Aarhus, DK), 20 low-energy houses were built between 2008 and 2014 and connected to an **intermittently operating district heating network**. Each house features a solar collector with a storage tank and is connected to the district heating network. The district heating plant provides hot water in pulses. During the heating season, the heat pulse heats the storage tanks in each individual house. During the summer, when there is no space heating demand, the district heating system is turned off for 3–4 months, and domestic hot water is delivered by the solar collectors, with electrical backup provided by the storage tank. For three heating seasons, measurements were taken at each house and at the supply networks, with the results demonstrating that heat loss can be reduced by up to 50% through intermittent operation (Olesen, 2015).

Meanwhile, the *TEMPO* project features a 'new rural district heating network in [the] Nürnberg region (Germany)' based on decentralised buffer storages and plastic pipes, which enable intermittent operation during summer. The district heating network is turned off when there is no heat demand (e.g. during sum-

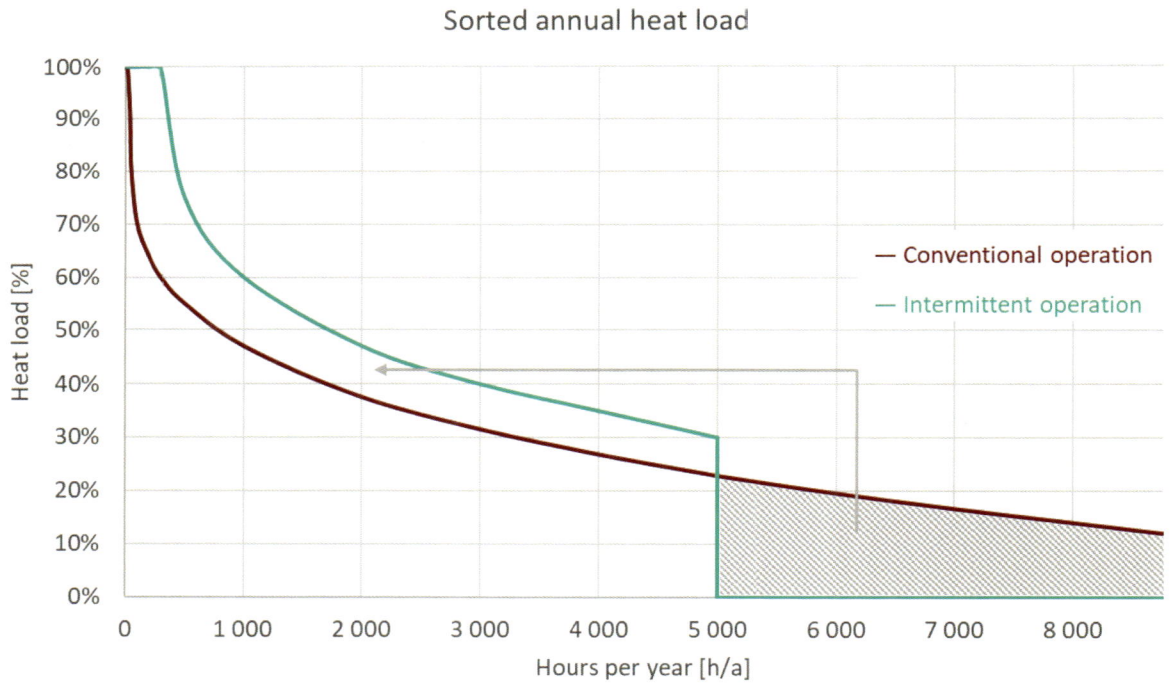

Figure 49. Qualitative visualisation of sorted annual heat load curves for conventional and intermittent operations (modified from Hammer et al., 2018)

mer nights), reducing both distribution heat loss and pumping energy. This concept works especially well for smaller and medium-sized networks of less than 100 connected consumers (TEMPO, 2021).

4.12 Lessons learned for obtaining lower system temperatures

The economic and ecological impacts of district heating networks can be significantly reduced through the broad implementation of low-temperature technologies. Excessive return temperatures and flows are typical in district heating systems, leading to system temperatures above design parameters. To achieve reduced system temperatures, it is necessary to take the multi-layered solution approaches that are available.

On the one hand, district heating operators must focus on problematic customer systems, those characterised by high heat demands, large mass flows and low cooling (identification methods are presented in Section 4.1; customer-side measures are presented in Chapter 3). On the other hand, measures must also be taken at the system level, especially in terms of generation and distribution. In the long term, efforts to achieve low system temperatures must not be limited to rare cases of network redesign but extended to existing heat distribution networks everywhere.

4.13 Major conclusions concerning low-temperature systems

The following conclusions constitute the most important take-aways from this chapter:

1. Before investing in any improvement measure, it is necessary to compare customer supply temperatures requirements with the primary supply side temperatures. In some cases, the critical supply temperatures needed by customers and those provided by the district heating network do not match. This happens because the district heating operator does not know the exact supply temperature required by customers and, thus, ensures that the temperature supplied is never below that needed to guarantee comfort.
2. When replacing existing substations or when designing new ones, heat exchangers with longer thermal lengths should be preferred as these will enable to obtain low supply and return temperatures
3. Many systems have already started their transformation to lower temperatures, demonstrating the wide range of possibilities and proven solutions that exist. However, it is important to maintain focus to avoid undermining improvement efforts.
4. All temperature reduction experiences should be utilised; this means taking advantage of the lessons learned by forerunners and the knowledge transfer that occurs within the district heating community.
5. Low-temperature systems contribute to the reduction of greenhouse gas emissions. Targeted policy

Important considerations when **beginning the low temperature transition** include:

- **Digitalisation** plays a decisive role in **identifying critical substations/customers** with **big improvement potential.** Continuous monitoring and analysis algorithms can help to keep system temperatures as low as possible. A helpful tool is the excess flow method; comparisons over different time horizons enables recognising what has changed and how. Using artificial intelligence and big data allows operators to complete analyses faster and automatically, processing vast amounts of data (more components/customers) in real-time.
- The utilisation of l**ocal heat sources** (see Sections 4.7 and 4.8) increases the security of supply and can create (local) added value. In combination with **subnetworks** (see Section 4.5), they are a suitable solution to local bottlenecks (see Section 4.3), keeping local supply temperatures at the lowest possible level. Furthermore, sub-sections in the district heating network can be successively transformed towards low-temperature networks.
- For **new systems**, the **planning phase** is essential for **efficient** and **future-proof networks**. When exploiting locally available alternative heat sources, innovative supply configurations (see Sections 4.11 and 10.1) enable the utilisation of the full potential of those sources.
- Finally, it is always helpful to learn from others. The **success stories** presented in Section 4.4 can be used to avoid initial mistakes, and lessons learned can be applied to convince decision-makers of the **obvious advantages** of **low-temperature** systems.

instruments and effective subsidies enable accelerated transformation. It is particularly important to raise awareness and sensitise political decision-makers to the necessity of low-temperature systems, especially given current energy policy frameworks hardly address their importance.

4.14 Literature references in Chapter 4

Austroflex Rohr-Isoliersysteme GmbH. (07. April 2020). Technisches Datenblatt (AustroPUR und AUSTROPEX). Retrieved from https://austroflex.com/download/

Averfalk, H., & Werner, S. (2018). Novel low temperature heat distribution technology. In Energy 145 (2018) (S. 526-539). ELSEVIER.

Basciotti, D., Köfinger, M., Marguerite, C., Terreros, O., Agugiaro, G., & Schmidt, R.-R. (2016). Methodology for the Assessment of Temperature Reduction Potentials in District Heating Networks by Demand Side Measures and Cascading Solutions. CLIMA 2016 - proceedings of the 12th REHVA World Congress. Aalborg: CLIMA 2016 - proceedings of the 12th REHVA World Congress.

Basciotti, D., Marguerite, C., Moser, M., Nagler, J., & Poier, H. (2017). heat_portfolio. Bericht zu hydraulischen Schaltungen und netzseitigen Regelstrategien zur Integration dezentraler Erzeuger in Wärmenetze in verallgemeinerungsfähiger Form. Wien, Graz: AIT Austrian Institute of Technology, TU Wien (IET), SOLID GmbH.

Bergsträßer, W., Hinz, A., Orozaliev, J., & Vajen, K. (6-7 October 2020). Lessons Learned from Excess Flow Analyses for Various District Heating Systems. 6th International Conference on Smart Energy Systems. Aalborg Denmark .

Brange, L., Englund, J., Sernhed, K., Thern, M., & Lauenburg, P. (2017). Bottlenecks in district heating systems and how to address them. The 15th International Symposium on District Heating and Cooling (S. 249-259). 10.1016/j.egypro.2017.05.072: Energy Procedia 116 (2017).

Brange, L., Sernhed, K., & Thern, M. (2019). Decision-making process for addressing bottleneck problems in district heating networks. International Journal of Sustainable Energy Planning and Management Vol. 20 2019 , (S. 37–50). http://dx.doi.org/10.5278/ijsepm.2019.20.4.

Brange, L., Thern, M., & Sernhed, K. (2018). Risks and opportunities for bottlenecks measures in Swedish district heating networks. 16th International Symposium on District Heating and Cooling (S. 380-389). 10.1016/j.egypro.2018.08.202: Energy Procedia 149 (2018).

Brès, A., Johansson, C., Geyer, R., Leoni, P., & Sjögren, J. (2019). Coupled Building and System Simulations for Detection and Diagnosis of High District Heating Return Temperatures. Proceedings of the 16th IBPSA Conference (S. 1874-1881). Rome: https://doi.org/10.26868/25222708.2019.210629.

Clausen, R. (2009). Undersøgelse af dynamiske forhold i et fjernvarmesystem mhp. at reducere varmetab (Masterthesis). Lyngby: Damnarks Tekninske Universitet (Institut for Mekanisk Teknologi).

Crane, M. (September 4-7, 2016). Individual apartment substation testing – development of a test and initial results. The 15th International Symposium on District Heating and Cooling. Seoul, Republic of Korea (South Korea).

Dahlberg, C., & Werner, S. (1997). Lower Temperature Level in The Borås District Heating System. Borås: EH&P-Unichal 1997.

DHC+. (2019). Digital Roadmap for District Heating & Cooling. Brussels: DHC+ Technology Platform c/o Euroheat & Power.

Diget, T. (2019). Motivation tariff - the key to a low temperature district heating network. In HOT|COOL, Journal No. 1/2019 (S. 19-22). Frederiksberg: DANISH BOARD OF DISTRICT HEATING.

Engel, C. (2018). 4th generation DH projects evaluated for H2020 projects. Waalwijk: Thermaflex International.

Farouq, S., Byttner, S., Bouguelia, M.-R., Nord, N., & Gadd, H. (2020). Large-scale monitoring of operationally diverse district heating substations: A reference-group based approach. Engineering Applications of Artificial Intelligence 90 (2020) 103492 (S. 16). https://doi.org/10.1016/j.engappai.2020.103492: ELSEVIER.

Gadd, H. (2014). To analyse measurements is to know! - Analysis of hourly meter readings in district heating systems (Ph. D. thesis). Lund: Lund University.

Gadd, H., & Werner, S. (2015). Fault detection in district heating substations. Applied Energy 157 (2015) (S. 51–59). http://dx.doi.org/10.1016/j.apenergy.2015.07.061: ELSEVIER.

Geyer, R., Hangartner, D., Lindahl, M., & Pedersen, S. (2019). Heat Pumps in District Heating and Cooling Systems. Vienna, Lucerne, Gothenburg, Aarhus: IEA Heat Pumping Technologies Annex 47.

Hamilton-Jones, M. (2020). Fault detection and optimization potential on the demand side of district heating systems. 6th International Conference on Smart Energy Systems. Aalborg Denmark.

Hammer, A., Sejkora, C., & Kienberger, T. (2018). Increasing district heating networks efficiency by means of temperature-flexible operation. Sustainable Energy, Grids and Networks 16 (2018) (S. 393–404). https://doi.org/10.1016/j.segan.2018.11.001: ELSEVIER.

IEA DHC. (24. August 2020). Annex TS4: Digitalisation of District Heating and Cooling. Retrieved 17 from https://www.iea-dhc.org/the-research/annexes/2018-2024-annex-ts4/

IEA DHC. (April 29th – 30th, 2019). Proceedings of the IEA DHC Annex Definition Workshop on: Digitalisation of District Heating and Cooling: Optimised Operation and Maintenance of District Heating and Cooling systems via Digital Process Management. Frankfurt am Main: IEA DHC|CHP.

IEA HPT. (08. April 2020). Annex 47: Heat Pumps in District Heating and Cooling systems. Retrieved from https://heatpumpingtechnologies.org/annex47/

IEA HPT Annex 47. (14. November 2018). District-Boost (Vienna). Retrieved from https://heatpumpingtechnologies.org/annex47/wp-content/uploads/sites/54/2018/12/annex-47-district-boost.pdf

Kamstrup. (24. August 2020a). Heat Intelligence - Intelligent analytics for the smart district heating system. Retrieved from https://www.kamstrup.com/en-en/heat-solutions/heat-analytics/heat-intelligence

Kamstrup. (29. October 2020b). Digitalisation delivers measurable results - Assens District Heating, Denmark. Retrieved from https://www.kamstrup.com/en-en/customer-references/heat/case-assens-district-heating

Kim, W., & Katipamula, S. (2018). A review of fault detection and diagnostics methods for building systems. Science and Technology for the Built Environment 24 (1), (S. 3-21). https://doi.org/10.1080/23744731.2017.1318008.

Köfinger, M., Basciotti, D., Schmidt, R.-R., Meißner, E., Doczekal, C., & Agugiaro, G. (2016). Low temperature district heating in Austria: Energetic, ecologic and economic comparison of four case studies. In Energy 110 (2016) (S. 95-104). ELSEVIER.

Köfinger, M., Schmidt, R.-R., Basciotti, D., Eder, K., Bogner, W., Koch, H., & Ondra, H. (2016). URBANcascade - Optimierung der Energie-Kaskaden in städtischen Energiesys-temen zur Maximierung der Gesamtsystemeffizienz und des Anteils erneuerbarer Energieträger und Abwärme. Wien: AIT Austrian Institute of Technology GmbH.

Kreisel, T. (04 November 2014). Fehleranalyse - Woraus resultieren hohe Rücklauftemperaturen? AGFW Seminar „Maßnahmen zur Erreichung niedriger Rücklauftemperaturen". Essen.

Lennermo, G., Lauenburg, P., & Werner, S. (2019). Control of decentralised solar district heating. Solar Energy 179 (2019) (S. 307–315). https://doi.org/10.1016/j.solener.2018.12.080: ELSEVIER.

Leppin, L. (2020). DeStoSimKaFe - Konzeptentwicklung & gekoppelte deterministisch/stochastische Simulation und Bewertung Kalter Fernwärme zur Wärme- & Kälteversorgung. e·nova - International Conference (S. 51-60). https://onedrive.live.com/?authkey=%21AJvEnIw6fU958MU&cid=B1F25D3854D3DDFA&id=B1F25D3854D3DDFA%214535&parId=B1F25D3854D3DDFA%214484&o=OneUp: Fachhochschule Burgenland GmbH.

Li, H., Dalla Rosa, A., & Svendsen, S. (September 5-7, 2010). Design of a low temperature district heating network with supply recirculation. 12th International Symposium on District Heating and Cooling, (S. pp. 73-80). Tallinn (Estonia).

Lund, R., & Mohammadi, S. (2016). Choice of insulation standard for pipe networks in 4th generation district heating systems. Applied Thermal Engineering 98 (2016) (S. 256–264). http://dx.doi.org/10.1016/j.applthermaleng.2015.12.015: ELSEVIER.

Lygnerud, K., Wheatcroft, E., & Wynn, H. (2019). Contracts, Business Models and Barriers to Investing in Low Temperature District Heating Projects. In Urban District Heating and Cooling Technologies (S. Appl. Sci. 2019, 9(15), 3142). doi:10.3390/app9153142: Applied Sciences.

Middelfart Fjernvarme. (2020). Operational measurment data of supply and return temperature. Middelfart: Middelfart Fjernvarme a.m.b.a.

Nagler, J., & Ponweiser, K. (28.02.2018). Hydraulische Schaltungen, netzseitige Regelungsstrategien zur Integration dezentraler Erzeuger in thermische Netze. Wien: Workshop: Nahwärmesysteme nachhaltig (um)gestalten @ AIT.

Nussbaumer, T., & Thalmann, S. (2016). Influence of system design on heat distribution costs in district. In Energy 101 (2016) (S. 496-505). ELSEVIER.

Olesen, G. B. (2015). Optimering og erfaringsopsamling fra pulsvarme til lavenergihuse ved andelssamfundet i Hjortshøj" ved Energiselskabet ved Andelssamfundet i Hjortshøj. Hjortshøj: Energiselskabet ved Andelssamfundet i Hjortshøj.

Oltmanns, J. J. (2020). Analysis and Improvement of an Existing University District Energy System (Dissertation). Darmstadt: Technische Universität Darmstadt.

Overhage, A. (04 November 2014). Einflussmöglichkeiten von Nutzern und Energieversorgungsunternehmen. AGFW Seminar „Maßnahmen zur Erreichung niedriger Rücklauftemperaturen". Essen.

Persson, U., & Averfalk, H. (2018). Accessible urban waste heat. Halmstad: ReUseHeat.

Reiter, P., Poier, H., & Holter, C. (June 2016). BIG Solar Graz: Solar District Heating in Graz – 500,000 m² for 20% Solar Fraction. In Energy Procedia (Volume 91) (S. 578-584). https://doi.org/10.1016/j.egypro.2016.06.204: ELSEVIER.

Sipilä, K., & Rämä, M. (2016). Low Temperature District Heating for Future Energy Systems. Subtask D: Case studies and demonstrations. Frankfurt am Main: IEA DHC|CHP Annex TS1.

TEMPO. (2021 [forthcoming]). TEMPO – Temperature Optimisation for Low Temperature District Heating across Europe. Mol: VITO | EnergyVille, NODA, AIT, Thermaflex, Solites, OCHSNER, Vattenfall, ENERPIPE, A2A, Halmstad University, Euroheat & Power | DHC+. Retrieved from https://www.tempo-dhc.eu/

TEMPO. (25. January 2021). Nurnberg Region: New Rural District Heating Network. Retrieved from https://www.tempo-dhc.eu/nurnberg/

Vojens Fjernvarme. (2020). Årsrapport for 2019. Vojens: Vojens Fjernvarme a.m.b.a.

Wien Energie. (August 2013). Technische Richtlinien - Technische Auslegungsbedingungen. Retrieved 17 January 2020 from http://www.wienenergie.at/media/files/2015/technische%20richtlinie%20tr-tab%20technische%20auslegungsbedingungen_140557.pdf

Wien Energie. (2020). Fernwärme. Retrieved 17 January 2020 from https://www.wienenergie.at/eportal3/ep/channelView.do/pageTypeId/67823/channelId/-47775

Windholz, B., Lauermann, M., Ondra, H., & Höller, M. (15. November 2016). District Boost: Einsatz von Wärmepumpen im Wiener Fernwärmenetz. 2. Praxis- und Wissensforum Fernwärme & Fernkälte. Wien.

Wirths, A. (23.-24. September 2008). Einfluss der Netzrücklauftemperatur auf die Effizienz von Fernwärmesystemen. 13. Dresdner Fernwärmekolloquium. Dresden.

Zinko, H., Lee, H., Kim, B.-K., Kim, Y.-H., Lindkvist, H., Loewen, A., . . . Wigbels, M. (2005). Improvement of Operational Temperature Differences in District Heating Systems. Frankfurt am Main: IEA DHC|CHP.

5 APPLIED STUDY: CAMPUS LICHTWIESE AT TU DARMSTADT

Authors: Johannes Oltmanns and Frank Dammel, TU Darmstadt

The Technical University of Darmstadt's Campus Lichtwiese is a perfect showcase for the systematic and consequent analysis and target-oriented optimisation of district heating system temperatures. The analysis presented in this chapter can be applied to any district heating system and the authors strongly recommend doing so.

5.1 Transferability of the applied study

The data necessary to calculate the metrics used in this chapter should be part of the standard measurement and monitoring system. If this is not the case, the respective monitoring system should be adjusted accordingly. Temperatures in district heating networks, on the supply as much as on the return side, depend heavily on a few critical buildings. The system supply temperatures may be high due to the requirement of one substation for a high temperature while all other substations might be able to ensure their users have comfortable temperatures at a lower level. On the return side, a substation with a high return temperature usually operates at a small temperature difference between supply and return and demands a high mass flow, resulting in an above-average influence on the system return temperature. On one hand, this means that a high return temperature in one substation can be decisive for the return temperature of the whole network. On the other hand, improvement measures taken in a few buildings with poor performance can have a great impact on the entire system. The metrics presented in the following sections serve to identify critical substations and determine which measures should be applied to achieve significant temperature reductions with minimal effort.

5.2 Background of the applied study

The Technical University of Darmstadt (TU Darmstadt) is one of the leading technical universities in Germany. In 2018, almost 26,000 students were studying in one of 113 study programs, with Computer Science, Mechanical Engineering, and Business Engineering being the most popular fields. The university is divided into four principal sites (Lichtwiese, Stadtmitte, Botanischer Garten, Hochschulstadion), with a total main usable area of 310 000 m², in 164 buildings. TU Darmstadt has pledged to fulfil the German national climate protection goals at the local level of its campus areas, namely a reduction of the area-specific CO_2 emissions by 80 %, compared to the 1990 level, by 2050 (Oltmanns, 2019). Energy systems are one of the core research areas at TU Darmstadt. Since 2016, an interdisciplinary team of researchers from Architecture, Electrical Engineering, and Mechanical Engineering has been working on the project "EnEff: Stadt Campus Lichtwiese", which is funded by the German Federal Ministry for Economic Affairs and Energy.

The project serves to support the Operations department at TU Darmstadt in its quest to achieve an energy transition at the local level at Campus Lichtwiese. So far, outputs of the project consist of an enhancement of the energy monitoring infrastructure, as well as the realisation of different implementation projects, such as data centre waste heat integration into the district heating network (Oltmanns et al., 2020) and a field study on temperature reduction and demand side management in the Architecture Institute building. The final deliverable of the project will be an energy concept for the future development of Campus Lichtwiese after 2030. This concept will set a path for the decarbonisation of the campus energy system, including generation, distribution and consumption of heat, cooling, and electric energy. In this process, a reduction of district heating network temperatures is crucial, as it represents a prerequisite for the integration of more renewable and waste heat, which is necessary to reach the university's own carbon dioxide emissions reduction goal.

Campus Lichtwiese is a typical university campus, built on the outskirts of Darmstadt from the late 1960s onwards once an expansion of the original campus in the city centre was not possible anymore. It unites different use cases such as lecture halls, laboratory buildings, office buildings, and the university dining hall in a self-contained area outside the city, representing a very good subject for an applied study demonstrating how to realise energy transition on a local level. The campus comes with several unique characteristics compared to other districts, namely that all buildings are non-residential, leading to low hot water demands and a high need for ventilation, especially in laboratory buildings.

All buildings, as well as the entire energy-relevant infrastructure, are owned by the university itself although generation facilities and distribution networks are operated in the context of a contracting agreement while the university's own Operations department manages the buildings. Heat and power for the university are currently supplied by combined heat and power (CHP) gas plants and backup heat-only gas boilers. A district cooling network supplies cooling energy and is connected to an absorption chiller and backup compression chillers. Table 7 gives an overview of the most important facts about the Lichtwiese district heating system.

Table 7. Parameters of the TU Darmstadt Campus Lichtwiese district heating system.

Number of buildings connected to district heating	32
Total heated floor area	150 000 m²
Annual heat supplied into district heating network (2018)	25.2 GWh/a
Annual heat demand buildings (2018)	22.6 GWh/a
Trench length in district heating network	4.2 km
Annual average district heating supply temperature (2017-2019)	88 °C
Annual average district heating return temperature (2017-2019)	58 °C
Annual average ambient temperature (2017-2019)	11 °C

In the following sections, the measures introduced in Chapters 3 and 4 are applied to Campus Lichtwiese. To reduce the temperatures in the network, first the return temperature reduction potential $\overline{\Delta T}_{R,j}$ (see Section 4.1) and the area-specific building heat demand (see Section 4.1) are calculated, to identify which buildings at Campus Lichtwiese have the biggest impact on the network temperatures. Subsequently, the underlying issues inside the buildings are identified for the three types of in-building heating circuits, space heating (Section 3.6), hot water preparation (Section 3.7), and ventilation (Section 3.8). For each type of heating circuit, different categories are defined based on the metrics presented in Section 3.5, and detailed results providing examples are presented in each category.

5.3 Monitoring data for identification of temperature reduction opportunities

Heat is supplied from the district heating network (primary side) to the in-building distribution network (secondary side) at a substation. Substations at Campus Lichtwiese are either designed with an indirect connection (heat transfer via heat exchanger, secondary side representing an independent hydraulic circuit) or a direct connection (no hydraulic separation between primary and secondary side), with or without an admixture of return flow on the secondary side to adapt the secondary supply temperature. Figure 50 illustrates the two general substation concepts with a direct (left) or indirect (right) connection to the district heating network.

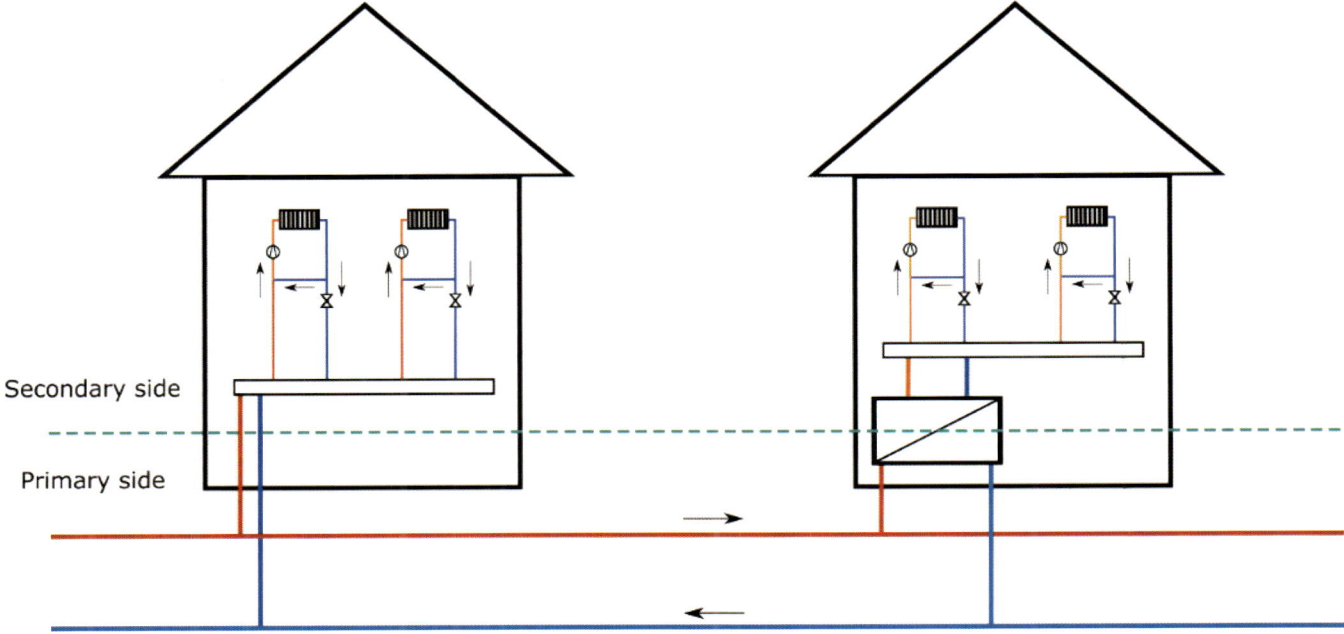

Figure 50. Directly and indirectly connected substations of the current system at TU Darmstadt Campus Lichtwiese.

Heat demand cannot be measured directly but is a function of the volume flow and the supply and return temperatures of a heating circuit.

The heat flow $\dot{Q}_{i,j}$ at time i to the customer from the heating circuit j is determined using a first law formulation for a stationary flow through the substation:

$$\dot{Q}_{i,j} = \dot{V}_{i,j} \cdot \rho_W \cdot c_W \cdot (T_{S,i,j} - T_{R,i,j}) \quad \text{Equation 16}$$

To calculate $\dot{Q}_{i,j}$, the volume flow $\dot{V}_{i,j}$ and the supply and return temperatures $T_{S,i,j}$ and $T_{R,i,j}$ have to be measured. The density ρ_W and the specific heat capacity c_W of water can be found in the literature (VDI-Gesellschaft Verfahrenstechnik und Chemieingenieurwesen, 2010). Figure 51 shows the two different heat measurement principles used at TU Darmstadt Campus Lichtwiese: in-line (left) and clamp-on (right) flow measurement, with temperature sensors being installed on the outer pipe wall or within the flow. Clamp-on heat meters are a common solution for retrofitting purposes because they do not require opening the pipe itself, thus facilitating the installation process in an existing heating circuit considerably.

Table 8 gives an overview of the uncertainty of the different values necessary for the heat flow calculation.

Table 8. Uncertainties in heat flow calculation.

Value	Uncertainty
Volume flow	between ±0.25% and ±1%
Temperature	Max. ±0,85 K (at 100 °C)
Specific heat capacity water	± 0.5%
Water density	± 2%

To identify where to apply which measure to decrease the network temperatures, both primary and secondary side heat monitoring data are necessary. Primary side data serve to determine which substations have the highest impact on the network return temperature and where the specific heat demand is highest. Secondary side monitoring makes it possible to go into more detail on the operation modes of different heating circuits inside buildings. Therefore, it helps to determine what kind of issue leads to high temperatures in a certain building and how complex it will be to solve the problem.

Primary side data are monitored by 29 heat meters for the 32 buildings connected to district heating (see Figure 52). Some buildings contain separately measured substations for space heating, ventilation, and hot water preparation, while in other cases, the heating circuits of several buildings are combined in one measurement point (see Table 9 for a list of the substations and the types of heating circuits represented therein). On the secondary side, 37 heating circuits in 13 different buildings are equipped with heat meters. It was not possible to install heat meters in every building and heating circuit; thus, buildings with high heat demand and primary side return temperatures were prioritised.

Additionally, some new buildings with low temperatures were equipped with heat meters, to serve as best practice examples. Figure 52 gives an overview of the distribution of secondary side heat meters at Campus Lichtwiese. The numbers indicate the names of the primary side heat meters for each building or group of buildings, and the colours show the number of heating circuits measured with secondary heat meters in each building. The Mechanical Engineering laboratory buildings, 3102 and 3106, are supplied from the same substation located in 3102, with heating circuits supplying both buildings at the same time. Accordingly, only one primary heat meter measures both buildings, and all

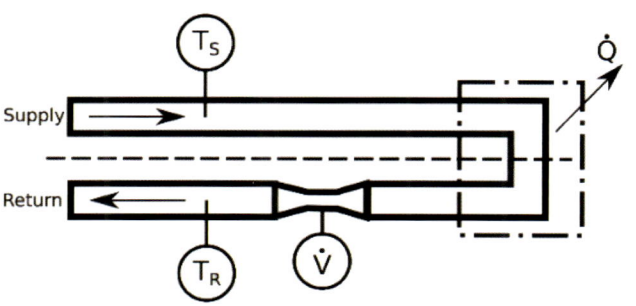 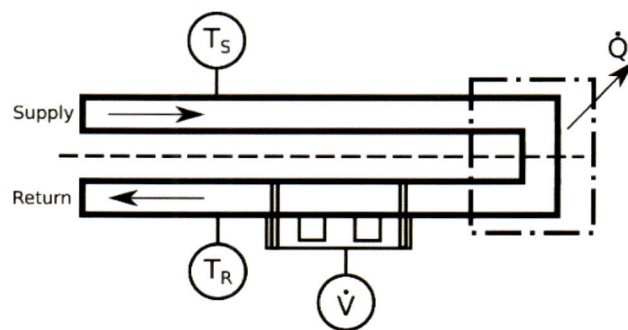

Figure 51. Different measurement principles used at TU Darmstadt Campus Lichtwiese.

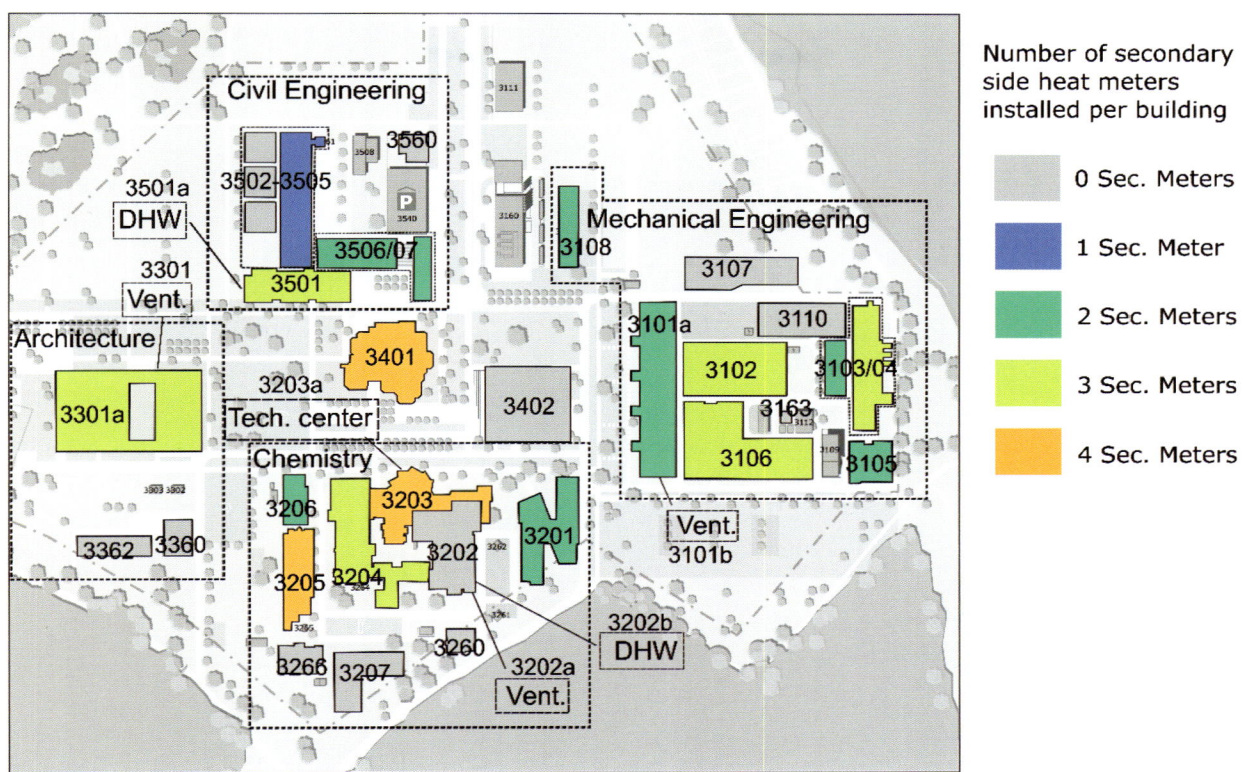

Figure 52. Primary and secondary side heat meters at Campus Lichtwiese.

Table 9. List of primary side substations at Campus Lichtwiese.

Substation name	Substation number	Space heating	Hot water preparation	Ventilation
Mechanical Engineering Institutes	3101a	x		
Mechanical Engineering Institutes	3101b			x
Mechanical Engineering Laboratories 1 & 2	3102/3106	x		x
Mechanical Engineering Laboratories 5 & 4	3103/04	x		x
Mechanical Engineering Laboratories 3	3105	x		x
Mechanical Engineering Laboratories 6	3107	x		x
Energy Centre	3108	x		x
Material Science Institutes	3201	x		x
Organic Chemistry Institutes & Laboratories	3202	x		
Organic Chemistry Institutes & Laboratories	3202a			x
Organic Chemistry Institutes & Laboratories	3202b		x	
Chemistry Lecture Halls & Library	3203	x		
Chemistry Lecture Halls & Library	3203a			x
Physical Chemistry	3204	x		x
Inorganic Chemistry	3205	x		x
Centre of Smart Interfaces	3206	x		x
M³ Laboratory Building	3207	x		x
Disposal Centre Chemistry	3266	x		
Architecture Institutes	3301			x
Architecture Institutes	3301a	x		
Day-care Centre	3360	x		
Kindergarten	3362	x		
University Dining Hall	3401	x	x	x
Lecture Hall & Media Centre	3402	x		x
Civil Engineering Institutes Old	3501	x		x
Civil Engineering Institutes Old	3501a		x	
Civil Engineering Laboratories 1-4	3502-3505	x		
Civil Engineering: Institutes New & Laboratories 5	3506/07	x	x	
Recycling Centre	3560	x		

the installed secondary side heat meters in 3102 cover both buildings at the same time.

Historic primary side monitoring data are available at TU Darmstadt going back to the end of 2016, which makes it possible to derive average values of the metrics presented below. The averaged metrics are based on time periods of complete years because supply and return temperatures depend on ambient conditions. Therefore, averaged metrics based on shorter time periods might be biased due to an overestimation of the impact of a specific season. The secondary side monitoring system was implemented only recently, and year-long time series are not yet available. Nevertheless, the characteristic operation modes of different substations can also be identified using the shorter time series available. The analysis of the performance of the secondary side heating circuits is based on week-long time series between March 13, 2020 and March 26, 2020. This time period represents the first week of the COVID-19 social distancing restrictions in Darmstadt, which led to exceptionally few students as well as faculty and staff members being present in the university's buildings. Therefore, hot water demand was significantly lower than during normal operations (see Section 5.6). The effect of the lockdown situation on space heating and ventilation heating is comparatively low because buildings need to remain heated and ventilated even if only very few people are present.

5.4 Impact of district improvement measures on the return temperature

To identify the most critical buildings in terms of return temperatures, first the return temperature reduction potential $\overline{\Delta T}_{R,j}$ for each substation j in the network is calculated (see Section 4.1):

$$\overline{\Delta T}_{R,j} = \frac{1}{n_{ts}} \sum_{i=1}^{n_{ts}} \frac{(T_{R,i,j} - T_{\text{target}}) \cdot \dot{M}_{i,j}}{\sum_{j=1}^{n_{sub}} M_{i,j}} \qquad \text{Equation 17}$$

Figure 53 shows a map of the buildings in the campus area. For each of the building substations, the return temperature reduction potential is calculated considering that temperatures at all substations are reduced simultaneously to an average target return temperature of T_{target} = 35 °C, which represents the desired future return temperature level for Campus Lichtwiese and a reduction of 23 K compared to the current temperature level of T_R = 58 °C. The colours in the figure indicate which buildings have a high return temperature reduction potential (red) and which have a low one (green).

Faultily controlled substations are highlighted, making it obvious that control errors, such as shortcut mass flows in ventilation heating circuits or hydraulically unbalanced space heating, are a major cause of high return temperatures in this district. Another cause is water tanks for hot water preparation. The specific sources of control errors at TU Darmstadt Campus Lichtwiese will be discussed in Section 5.6–5.8.

Figure 54 compares the return temperature reduction potential $\overline{\Delta T}_{R,j}$ to the mass-averaged return temperature of each substation $\overline{T}_{R,\dot{M},j}$:

$$\overline{T}_{R,\dot{M},j} = \frac{\sum_{i=1}^{n_{ts}}(\dot{M}_{i,j} \cdot T_{R,i,j})}{\sum_{i=1}^{n_{ts}}(\dot{M}_{i,j})} \qquad \text{Equation 18}$$

The size of the dots represents the average annual heat demand for each substation. The analysis shows that a high return temperature reduction potential is not only linked to substations with high absolute return temperatures but can also be a result of a very high mass flow at a moderate return temperature, due to high heat demand.

The five most critical substations are found in the Mechanical Engineering laboratories 3102/3106, the Organic as well as the Inorganic Chemistry buildings, 3202 and 3205, and the university dining hall, 3401. The two most critical substations have a return temperature reduction potential of more than 6 K, and the other three a reduction potential above 1.5 K while the reduction potential of all other substations is below 1 K. The five critical substations represent a return temperature reduction potential of almost 21 K. If only the return temperature from the most critical substations is lowered, the resulting district heating temperature reduction is somewhat smaller, due to the decrease in mass flow at the critical substations.

5.5 Comprehensive building renovation

High return temperatures are often a result of poor energetic performance of buildings; thus, in the second step, the heat demand Q_j of the substation j in relation to the total head demand Q_{DHN} is compared to the area-specific heat demand q_j (see Section 4.1).

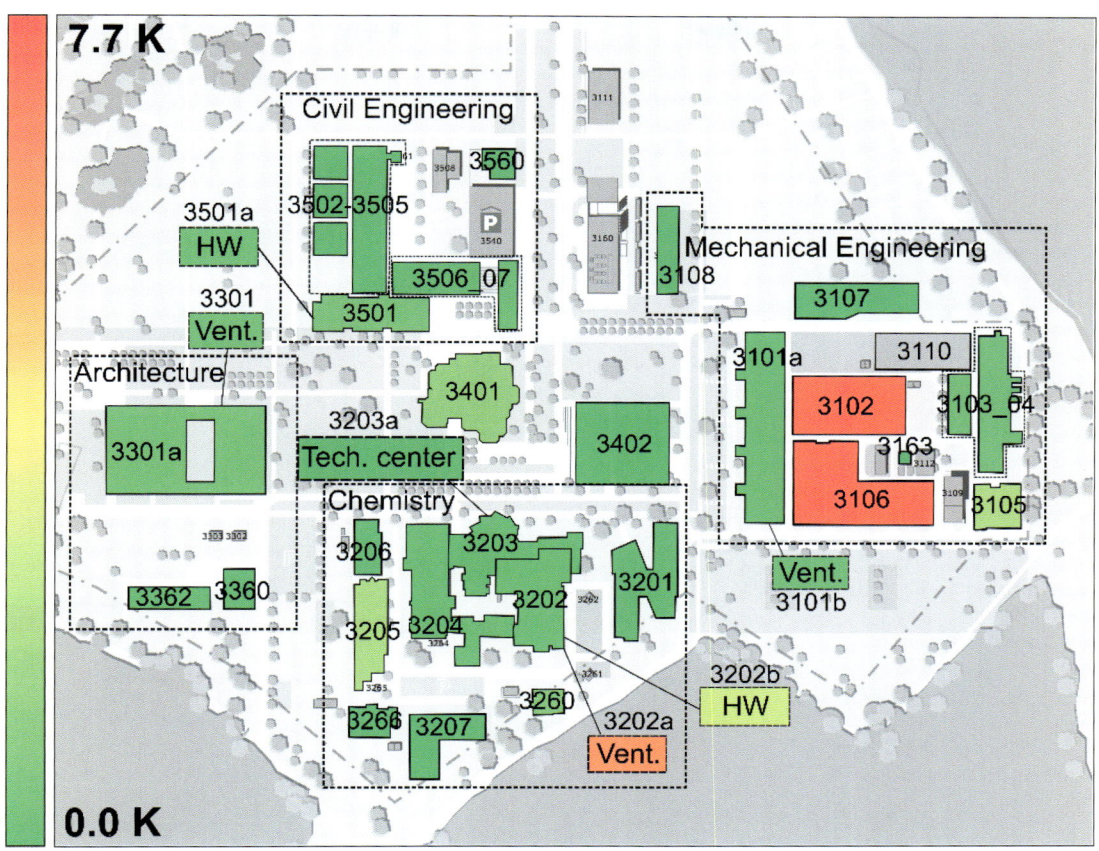

Figure 53. Return temperature reduction potential for each substation at Campus Lichtwiese.

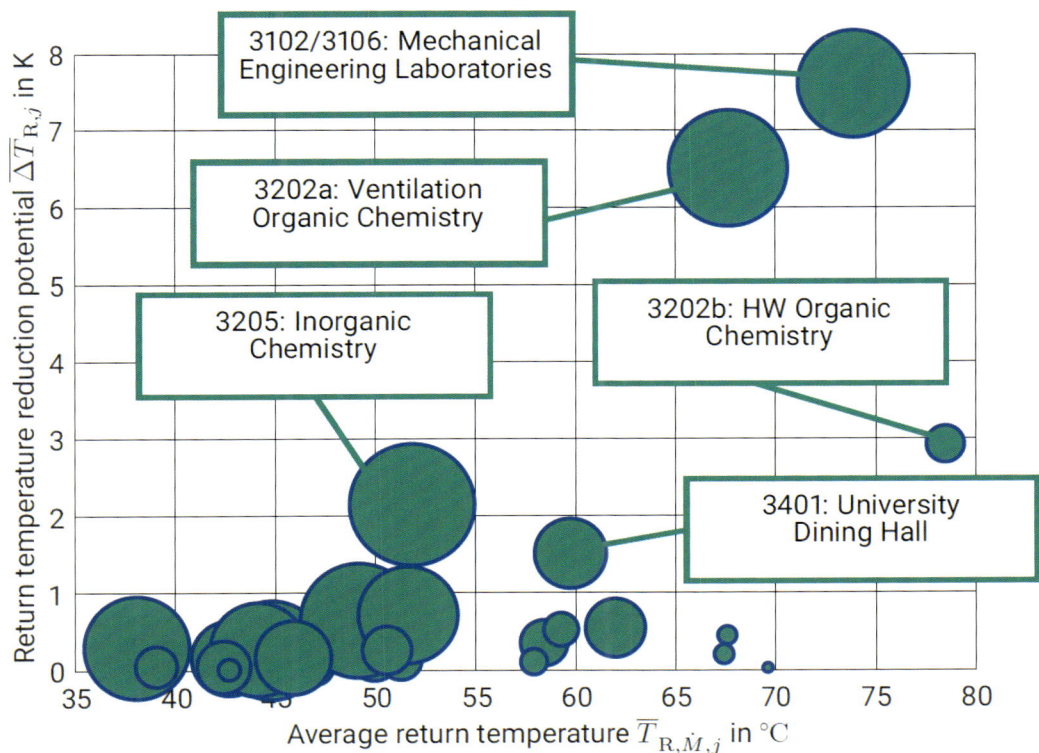

Figure 54. Return temperature reduction potential over average return temperature for each substation at Campus Lichtwiese.

APPLIED STUDY: CAMPUS LICHTWIESE AT TU DARMSTADT

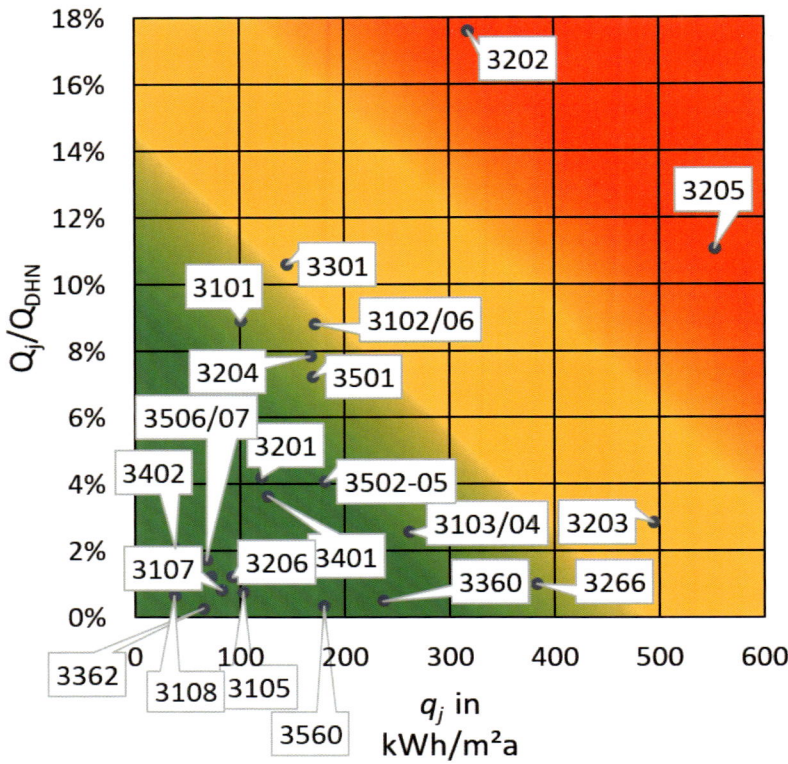

Figure 55. Heat demand Q_j of each substation j in relation to total heat demand Q_{DHN} vs. area-specific heat demand q_j

In Figure 55, the results of an analysis of the energetic performance of the buildings can be seen. For buildings containing several substations measured by primary side heat meters, the analysis was based on the aggregated heat demand of all substations inside the respective building. A few meters measure the combined heat demand of several buildings. For a list of the building names please consult Table 9.

Only a few buildings are within the range or below the reference values for specific heat demand in non-residential buildings in Germany (between 90 kWh/m²a and 140 kWh/m²a, depending on the building type) (Bundesministerium für Wirtschaft und Energie und Bundesministerium für Umwelt, Naturschutz, Bau und Reaktorsicherheit, 2015). While this demonstrates that most buildings should be renovated in the future, the first priority is the Organic Chemistry (3202) and Inorganic Chemistry (3205) buildings, where both specific heat demand q_j and share in total district heat demand $\frac{Q_j}{Q_{DHN}}$ are significantly higher than in all other buildings. In both cases, the ventilation system represents a major part of the heat demand. Another study carried out at TU Darmstadt concluded that the installation of a ventilation heat recovery system could save about 55% of the annual heat demand in the Organic Chemistry building 3202 and almost 70% in the Inorganic Chemistry building 3205. A few other buildings, such as the Chemistry lecture hall and library building (3203), as well as the waste disposal building (3266), also represent very high specific heat demands but a comparatively low share in the overall heat demand of the district. Thus, they are not the first priority for renovation measures.

5.6 Performance of hot water preparation

All buildings at Campus Lichtwiese are non-residential, which is why domestic hot water (DHW) preparation plays a minor role in the total heat demand. Nevertheless, some DHW heating circuits have a significant impact on the network return temperatures. DHW is needed mainly for coffee kitchens. In some cases, DHW is also used for laboratory purposes, and a few buildings are equipped with showers. Different technologies are used to supply DHW. In most buildings, hot water is not supplied centrally but by using in-line electric heaters. In a few cases, tanks including district heating heat exchangers inside the tank are used for domestic hot water preparation. Three domestic hot water preparation heating circuits on campus are equipped with meters, all three of which show different operation modes. These are presented in Figure 56.

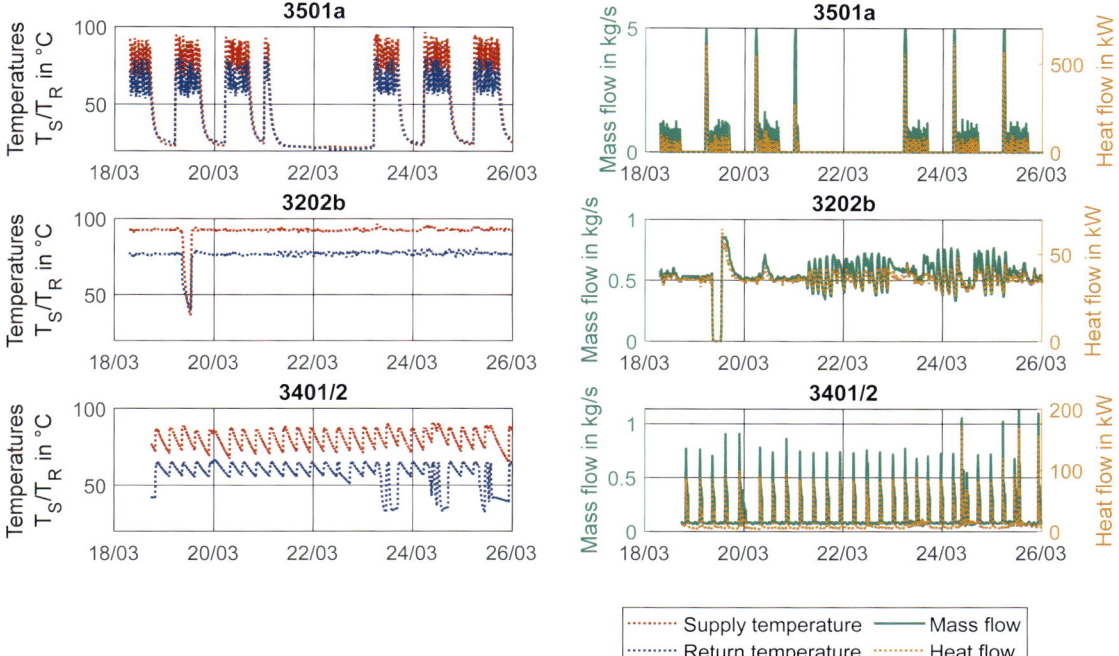

Figure 56. Temperatures, mass flow, and heat flow of the heat circuits supplying hot water.

The data used for this study were collected at the beginning of the first COVID-19 lockdown in Germany. This does not necessarily have an impact on space heating and ventilation heat demand but does impact the hot water heat demand significantly because the number of people present at the university was decreased sharply compared to normal operations. At the time of writing this guidebook, TU Darmstadt remains largely closed to students and faculty, with teaching mostly happening online, which is why it was not possible to analyse the performance of the DHW heating circuits under non-pandemic circumstances. The demand presented in Figure 56 is mainly the result of heat losses.

The first DHW heating circuit (3501a), located in the old Civil Engineering Institute building 3501 undergoes hourly heating cycles during workdays between 6:00 am and 6:00 pm and is turned off during nights and weekends. Mass flow and heat demand see high peaks in early mornings and modulate at a low level over the rest of the day. To avoid the early morning peaks in heat demand as well as high supply and return temperatures, the re-heating mass flow could be reduced, as explained in Section 3.7. The second heating circuit (3202b), supplying the Organic Chemistry building 3202, operates at roughly constant and very high supply and return temperatures. Heat demand and mass flow are low and see little fluctuation. The reason for the high temperatures in this example could be the fact that the hot water tank connected to this heating circuit supplies a very large building with long pipes that must be maintained at the required temperature. To guarantee the required hot water supply temperature at any point in the building, the temperature provided to the heat exchanger inside the hot water tank is significantly higher than necessary according to German regulations.

The third example, representing the DHW preparation in the university dining hall 3401, shows a saw-tooth profile for both supply and return temperatures, indicating regular heating cycles of the domestic hot water tank every 6 hours. While the tank is heating up, both its supply and return temperatures rise steeply. After a heating period of 1 hour, the temperature once again starts to decrease slowly. Due to a minimum mass flow of about $\dot{M}_{3401/2}$ = 0.1 kg/s in this heating circuit, the return temperature remains high even when no heat is used. At the end of the observed period, the return temperature in this heating circuit falls more steeply several times, suggesting that the characteristic saw-tooth profile is mainly due to heat losses during a period without demand for hot water in this heating circuit.

All three setups have a negative effect on the campus thermal energy system because they result in high return temperatures. While the first setup (3501a) has the lowest average impact on the return temperature, it comes with the disadvantage of high peaks in heat demand in early mornings. Only a minor part of the total campus heat demand is used for hot water preparation, thus switching to in-line electric heaters or heat pumps in buildings currently equipped with storage tanks would be the recommended solution.

5.7 Performance of space heating circuits

The operation of the space heating circuits at TU Darmstadt Campus Lichtwiese can be divided into four different categories, S1-S4. S1 to S3 result from well-controlled systems (see Figure 57), while category S4 pertains to faulty operations (see Figure 58).

S1. **No night setback and low temperatures:** The most typical operation mode for space heating is shown using the example of the heating circuit 3101/1. In this S1 category, supply and return temperatures depend on the ambient temperature. During night hours, heat demand and temperatures increase while they are lower during the day when the ambient temperature is higher. To reduce the heat flow supplied to the rooms when the ambient temperature increases, first the overall secondary side mass flow is reduced sharply, resulting in a decrease in both supply and return temperatures. Subsequently, the return flow admixture is increased, while the flow of primary supply water remains low, resulting in further decreasing secondary side temperatures. In the afternoon, when the ambient temperature starts to fall once again, the primary side mass flow is increased and reaches its maximum in the early mornings.

S2. **Night setback and low temperatures:** The second category, S2, pertains to the space heating circuits of the university dining hall 3401, as well as the Mechanical Engineering workshop building 3103. It has two different characteristic supply and return temperature levels, one during daytime and the other during night hours and on weekends, due to night setback. Nevertheless, this operation mode is considered as correctly controlled.

S3. **High supply temperature:** The third category, S3, which can be applied to the Inorganic Chemistry building 3205, is characterised by a constant supply temperature close to the primary supply temperature. The heat demand and the return temperature increase during cold night hours and go down during warm days. During a short period on March 19, 2020, the substation heat flow demand, as well as the mass flow in the heating circuit, dropped to zero, resulting in a sharp drop in primary and secondary supply temperature as well as secondary return temperature. This behaviour shows that there are no shortcut mass flows at low heat demand in this heating circuit and that it will thus not contribute to high network return temperatures during summer months.

Figure 58 displays category S4, characterised by fast fluctuations in temperatures during night-time. To gain a better understanding of these fluctuations, a close-up view is shown in the second row of the diagram.

S4. **Night-time fluctuations:** Category S4 is demonstrated using the example of the heating circuit 3102/6 in one of the old Mechanical Engineering laboratory buildings. During nights and early mornings, supply and return temperatures as well as mass flow oscillate significantly, leading to fluctuation of the adjusted heat flow between $\dot{Q}_{SH,3102/6}$ = 70 kW and $\dot{Q}_{SH,3102/6}$ = 200 kW. The second row of Figure 58 shows a close-up view of the night-time fluctuations on March 23, 2020. During this time, the temperatures showed a sinusoidal behaviour with a cycle duration of 1 hour and an amplitude of 10 K (return temperature) and 15 K (supply temperature). In this close-up view, the influence of system inertia can be seen, which is why adjusted temperatures and heat flow are calculated. This behaviour can be the result of a badly functioning return flow admixture thermostatic valve that is constantly opening and closing. The mass flow is high at all times, leading to a small temperature difference between the supply and the return line.

The highest $\overline{T}_{R,j}$ can be seen in the correctly controlled heating circuits of the Mechanical Engineering workshop building 3103, and the Inorganic Chemistry building 3205, possibly due to undersized radiators. The average secondary side supply temperatures differ significantly. In the Inorganic Chemistry building 3205, space heating is operated at almost primary side supply temperature while in the Mechanical Engineering laboratory building 3105, the average secondary side supply temperature is only $\overline{T}_{S,3105/2}$ = 37.8 °C. The combination of the average return temperatures and the investigation of operation modes gives an overview of which heating circuits should be addressed first to improve the functioning of the entire system.

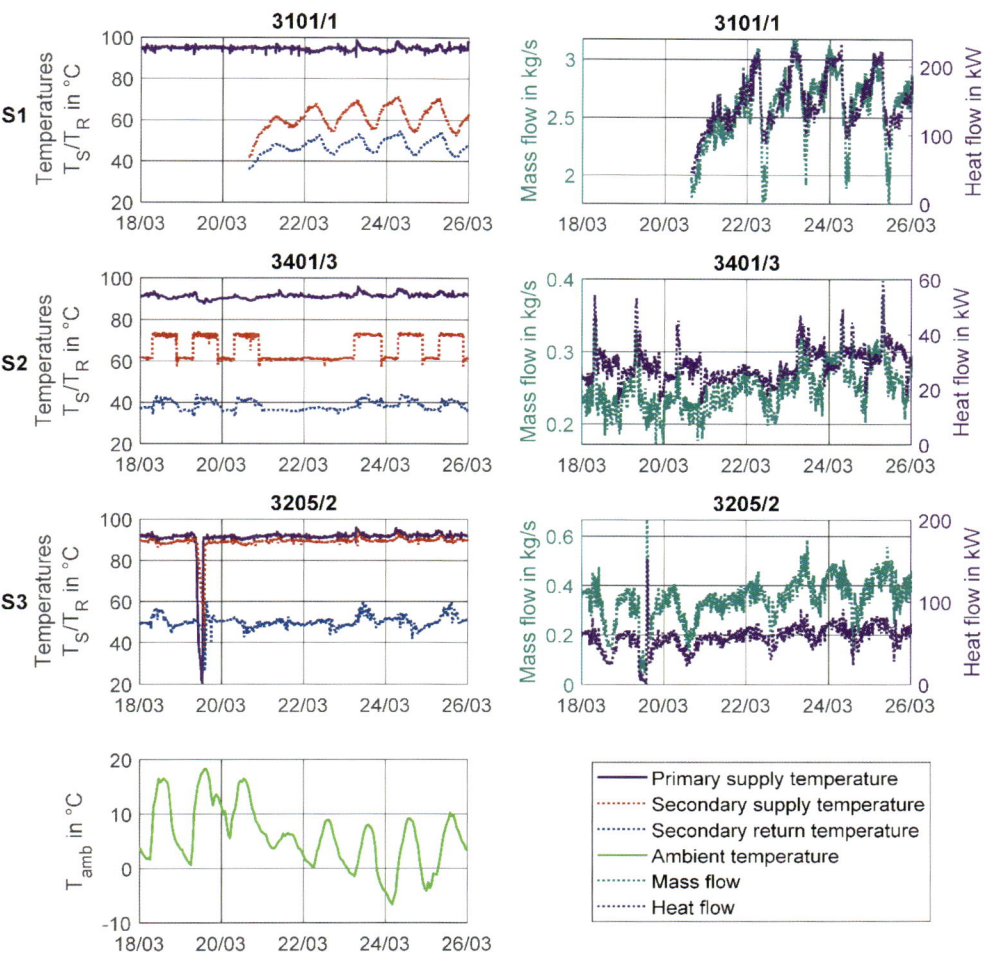

Figure 57. Temperatures, mass flow, and heat flow for the three categories S1-S3 of well-functioning space heating circuits

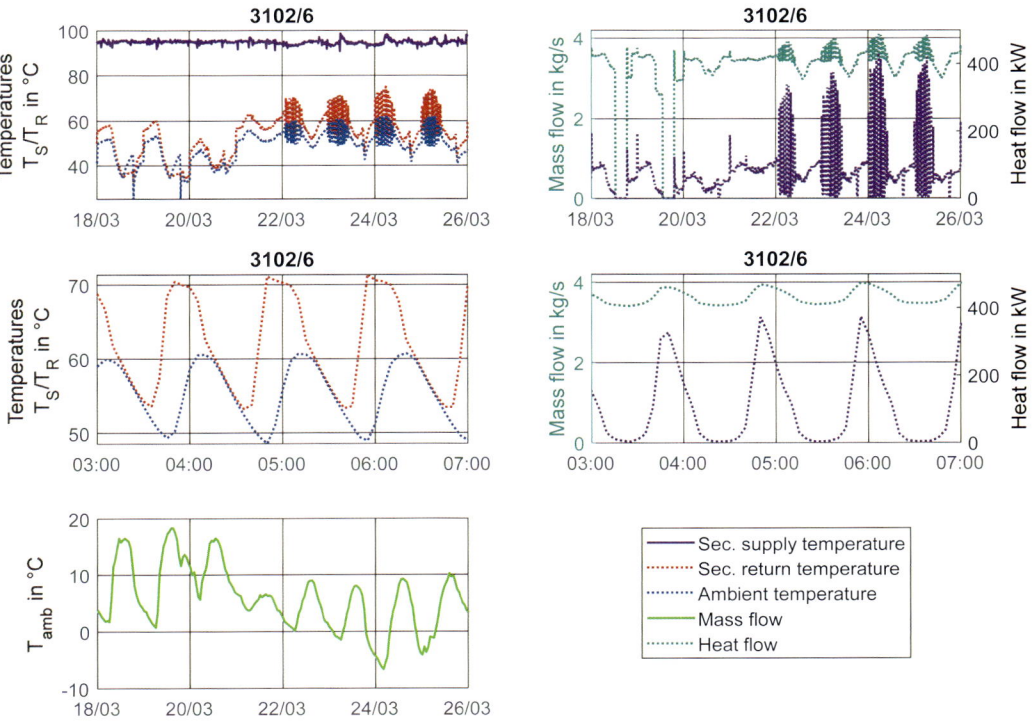

Figure 58. Temperatures, mass flow, and heat load of the faulty space heating circuit 3102/6.

Table 10 compares the performance of the different space heating circuits in terms of average supply and return temperatures (see Section 3.5):

$$\overline{T}_{S/R,j} = \frac{1}{n_{ts}} \sum_{i=1}^{n_{ts}} T_{S/R,i,j} \qquad \text{Equation 19}$$

Table 10. Comparison of the performance of the space heating circuits by time averages of supply and return temperatures (based on data collected between March 18 and March 26, 2020).

Heating Circuit	Operation Mode	$\overline{T}_{S,j}$ in °C	$\overline{T}_{R,j}$ in °C
3101/1	S1	61.1	47.5
3101/2	S1	55.4	45.7
3105/2	S1	37.8	32.6
3201/1	S1	54.1	47.1
3203/3	S1	49.0	40.0
3203/4	S1	48.8	41.3
3204/3	S1	45.7	36.4
3206/2	S1	42.2	33.6
3301/1	S1	55.4	46.4
3301/2	S1	53.3	42.4
3501/1	S1	50.6	41.5
3501/2	S1	63.4	51.4
3507/1	S1	38.4	29.7
3103/1	S2	72.4	54.9
3103/2	S2	72.0	58.1
3401/3	S2	66.3	38.5
3401/4	S2	66.4	44.0
3205/1	S3	90.0	55.6
3205/2	S3	89.0	50.2
3102/1	S4	53.8	48.8
3102/2	S4	55.1	48.3
3102/5	S4	53.9	48.0
3102/6	S4	53.1	48.3

5.8 Performance of ventilation heating circuits

The performance of a ventilation control system can be determined by comparing the heat demand $\dot{Q}_{vent,j}$ to the return temperature $T_{R,vent,j}$ of the ventilation heating circuit using a simple linear regression function (see Section 3.5):

$$\dot{Q}_{vent,j} = \beta_{0,j} + \beta_{1,j} \cdot T_{R,vent,j} + \epsilon_j \qquad \text{Equation 20}$$

The ventilation heating circuits can be divided into five categories V1-V5. The first three categories V1-V3 represent correctly controlled heating circuits while V4 and V5 are a result of errors in the ventilation control system. Figure 59 shows the time series of the primary and secondary side supply temperature, as well as the secondary side return temperature (left), the time series of the heat flow and the mass flow (centre), and the regression of the heat flow over the return temperature (right) of the three categories of well-functioning ventilation heating circuits.

V1. **High supply temperature and fluctuating mass flow:** The first category V1 can be shown using the example of the ventilation heating circuit of the Material Science building 3201/2. The supply temperature in this heating circuit is close to the primary supply temperature while the mass flow, as well as the return temperature, fluctuate with the heat demand. The regression of the heat demand over the return temperature yields a high positive gradient of $\beta_{1,3201/2} = 9.5$ kW/K (coefficient of determination $R^2_{3102/2} = 0.79$). A reduction of the primary supply temperature would lead to an increase in the return temperature from this type of heating circuit. To decrease the return temperature, increasing the size of the ventilation heat exchanger would be necessary.

V2. **Low supply temperature and fluctuating mass flow:** The ventilation heating circuit 3206/1 of the modern „Center of Smart Interfaces" (CSI) constructed in 2011 serves as an example for the second category V2. In this category, the secondary side is hydraulically disconnected from the primary side via a substation heat exchanger, and the ventilation heating supply temperature is significantly lower than the primary side supply temperature. As for the Material Science building ventilation system, the mass flow is closely related to the heat demand. Also in this category, the regression of the heat demand over the return temperature yields a positive gradient of $\beta_{1,3201/2} = 7.0$ kW/K ($R^2_{3102/2} = 0.75$). Both the first and the second categories are examples of a control system defined by a constant secondary side supply temperature (see Section 3.2). While the Material Science building 3201 is directly connected to the district heating system, the CSI building 3206 includes a substation heat exchanger, thus reducing the supply temperature before it enters the ventilation heating circuit.

V3. **Low supply temperature and constant mass flow:** The third category, V3, is an example of a control system defined by a constant mass flow and varying supply temperature depending on the heat demand (see Section 3.2). It can be found in the ventilation heating circuit of the new Civil Engineering laboratory building 3507/2. Both supply and return temperatures are adapted to the heat demand while the mass flow is constant, except when the heat demand falls to $\dot{Q}_{\text{vent }j}$ = 0 kW and the circulation pump is turned off. The regression of the heat demand over the return temperature results in a positive gradient of $\beta_{1,3201/2}$= 1.5 kW/K ($R^2_{3102/2}$= 0.67). This operation mode yields both low supply and return temperatures and is a good option for LTDH applications. When no heat demand, and consequently no mass flow exists, the supply and return temperatures are determined by the ambient temperature inside the substation and are almost equal, except for measurement uncertainties.

Figure 60 shows the results for the two categories V4 and V5 of faulty ventilation control systems occurring at TU Darmstadt Lichtwiese.

V4. High supply and high return temperature:
Several ventilation heating circuits show a behaviour comparable to the one in the old Mechanical Engineering laboratories buildings 3102/3106 (displayed here using the example of ventilation heating circuit 3102/4). In this example of category V4, the secondary side supply temperature is close to the primary side supply temperature, as in the ventilation heating circuit of the Material Science building 3201/2, but the mass flow is very high, resulting in a return temperature only about 10 K lower, and fluctuating between 80 °C and 85 °C. While the heat demand in the well-functioning ventilation heating circuit 3201/2 and the faulty circuit 3102/4 are in the same range, the mass flow is about five times higher in the latter. The gradient of the regression of the heat flow over the return temperature is negative ($\beta_{1,3201/2}$= -3.8 kW/K), but the correlation between the two variables is low ($R^2_{3102/2}$= 0.13). The most probable reason for the excessive return temperature from this heating circuit is an undesired bypass mass flow, but it could also be a result of an undersized ventilation heat exchanger.

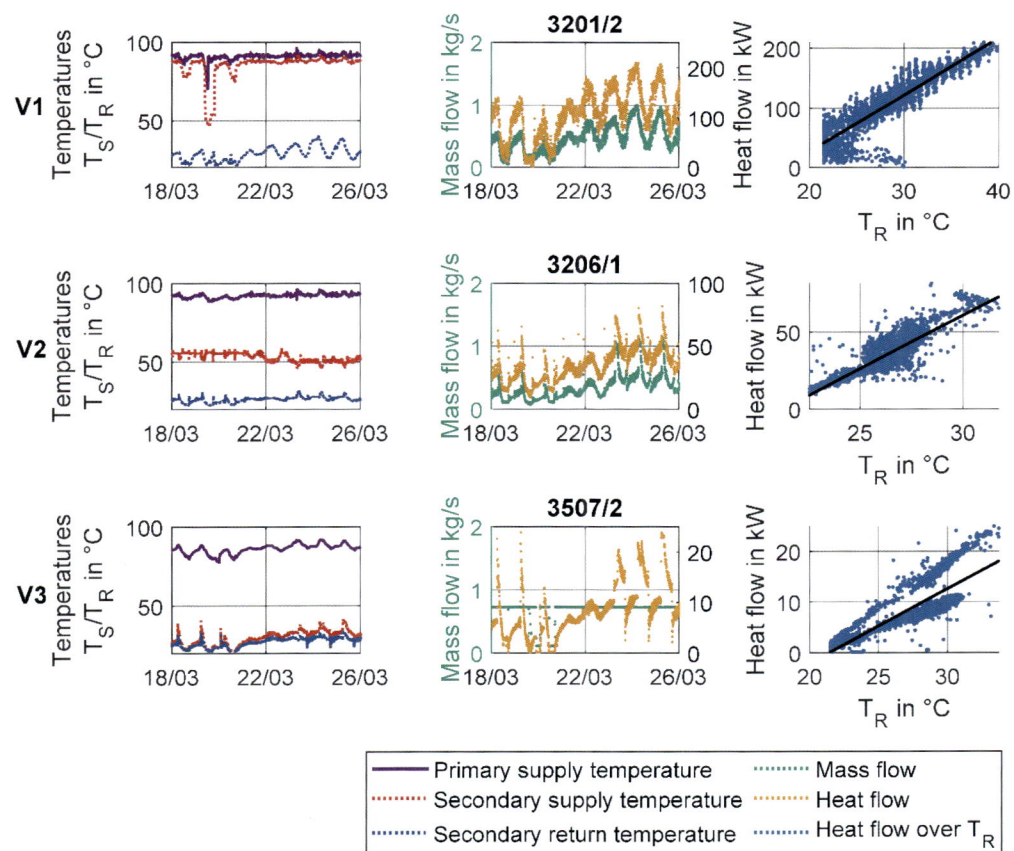

Figure 59. Temperatures, mass flow, heat flow, and heat flow over return temperature for the categories V1-V3 of well-functioning ventilation heating circuits.

V5a. High return temperature at low heat demand:
A typical error in ventilation heating circuits at TU Darmstadt Lichtwiese is high return temperatures due to recirculation mass flows, here shown using the example of the ventilation heating circuit in the university dining hall 3401/1. In this example of category V5, an excessive recirculation mass flow at low heat demand leads to a negative gradient of the regression of the heat demand over the return temperature $\beta_{1,3401/1}$ = -2.2 kW/K ($R^2_{3102/2}$ = 0.89). The factor between the maximum and minimum heat flow $f_{\dot{Q}_{vent}}$ = 19 is almost five times higher than the factor between the maximum and minimum mass flow $f_{\dot{M}_{vent}}$ = 4.

V5b. High return temperature at low heat demand:
Another example representing category V5 is the ventilation heating circuit 3205/4 in the Inorganic Chemistry building. At first glance, it seems comparable to the one of heating circuit 3201/2 ($\beta_{1,3205/4}$ = 4.7 kW/K compared to $\beta_{1,3201/2}$ = 9.5 kW/K), but the coefficient of determination is very low ($R^2_{3205/4}$ = 0.08). Figure 60 shows that the behaviour of this heating circuit is essentially a mix of categories V1 and V5a. As long as the heat demand is sufficiently high, the heating circuit behaves like a V1 heating circuit, but during nights with low heat demand, the mass flow is not reduced sufficiently, leading to high return temperatures as in the case of 3401/1. Since the heat demand in this building is generally high, critical low heat demands are seldom reached, leading to an overall positive regression gradient $\beta_{1,3205/4}$. The regression makes this aspect visible, showing two different characteristic areas, one with a high positive gradient as in category V1, and one with a negative gradient typical for category V5. Even though this heating circuit yields comparatively low return temperatures during the time period considered here, it will contribute to high return temperatures during the summer season.

Table 11 compares the performance of the different ventilation heating circuits in terms of the secondary side supply temperature $\bar{T}_{S,j}$, the secondary side return temperature $\bar{T}_{R,j}$, and the gradient β_1 of the regression of the heat demand $\dot{Q}_{vent\,j}$ over the secondary side re-

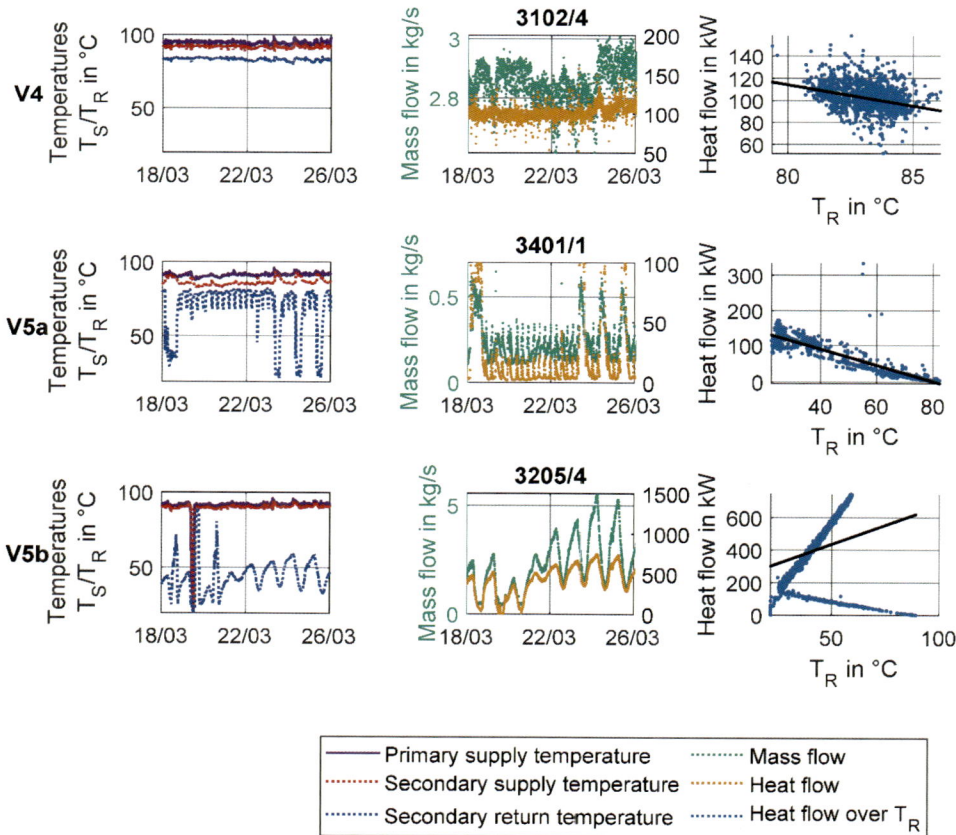

Figure 60. Temperatures, mass flow, heat flow, and heat flow over TR for the two categories V4 & V5 of faulty ventilation heating circuits, including a distinction between categories V5a and V5b.

turn temperature $T_{R,vent,j}$ < 30 °C. Well-functioning ventilation heating circuits in categories V1-V3 reach low average secondary side return temperatures, except for the heating circuit 3501/3 $\bar{T}_{R,3501/3}$ = 38.5 °C, even though it is also considered to be operating correctly. Ventilation heating circuits with control errors show characteristic high return temperatures ($\bar{T}_{R,j}$ > 54 °C, except for the two heating circuits 3205/3 and 3205/4 in category V5b). The average secondary supply temperature $\bar{T}_{S,j}$ is in the range of the primary supply temperature, except for the heating circuits in categories V2 and V3.

Table 11. Comparison of the performance of the ventilation heating circuits (based on data collected between March 18 and March 26, 2020).

Heating Circuit	Operation Mode	$\bar{T}_{S,j}$ in °C	$\bar{T}_{R,j}$ in °C	β_1 in kW/K	R^2
3201/2	V1	85.2	28.7	9.5	0.79
3501/3	V1	90.0	38.5	2.7	0.43
3206/1	V2	53.1	26.2	7.0	0.75
3203/2	V3	31.7	24.8	0.1	0.03
3507/2	V3	29.8	27.0	1.5	0.67
3102/3	V4	91.9	74.5	-4.4	0.58
3102/4	V4	91.9	83.4	-3.8	0.13
3105/1	V4	89.8	81.6	-3.9	0.43
3203/1	V5a	85.7	56.0	-0.6	0.17
3301/3	V5a	86.8	54.5	-2.7	0.75
3401/1	V5a	86.4	69.4	-2.2	0.89
3205/3	V5b	88.7	27.9	0.5	0.08
3205/4	V5b	89.3	44.3	4.7	0.08

5.9 Reduction of the district heating supply temperature

As can be seen in the examples shown above, many heating circuits, especially for space heating, do not make use of a high primary side supply temperature. For ventilation, supplying the heat exchanger at primary supply temperature is a lot more common, but a well-functioning ventilation heating circuit operates at very high temperature differences and would still be able to reach reasonable return temperatures if the supply temperature were slightly lower. Figure 61 shows the impact of a reduction of the supply temperature on the two well-functioning ventilation heating circuits operating approximately at primary supply temperatures (in the Material Science Building 3201 and the Old Civil Engineering Institute Building 3501). A reduction of the primary supply temperature by ΔT_S = 8 K would lead to an increase in the return temperature by about 5 K on average in these heating circuits.

In well-functioning heating circuits with high temperatures, a local increase in the supply temperature via a heat pump instead of retrofitting radiators or ventilation heat exchangers might be a solution. If there exists an error in the control of a heating circuit, this error should be corrected first, rather than investing in an additional heat pump. A booster heat pump only makes sense for space heating in the Inorganic Chemistry building 3205 (see category S3). The heating circuits in this building come with constantly high supply and return temperatures, but their control is functioning correctly.

5.10 Recommendations for actions

Based on the analysis described in this chapter, the following measures were identified as the most pressing issues to lower district heating temperatures and heat demand at TU Darmstadt Campus Lichtwiese:

1. Correction of ventilation control errors: The first priority is to reduce recirculation mass flows in the ventilation heating circuits 3102/3, 3102/4 and 3105/1 of the Mechanical Engineering laboratory buildings as well as in the Organic Chemistry ventilation system 3202a. Additionally, it makes sense to attend to the high return temperatures at low heat demand in the ventilation heating circuits 3203/1, 3205/3, 3205/4, 3301/3 and 3401/1.

2. Hot water preparation: The hot water preparation tank in the Organic Chemistry building (heating circuit 3202b) should be replaced with an in-line electric heater, and the operation mode of the hot water tank in the university dining hall 3401 should be adapted as follows. If the rules for water hygiene allow it, the tank could be heated only during operating hours of the dining hall, as was implemented in the case of the hot water tank in the Civil Engineering institute building 3501 (heating circuit 3501a). This is not an ideal solution, but it is an easy way to reduce the average return temperature from this heating circuit with no additional costs.

3. Building renovation: Comprehensive building renovation should be carried out in the Inorganic Chemistry building 3205. In this case, the specific heat demand and the temperatures are high even though the control of the space heating circuits is functioning correctly, suggesting that the heat demand increased after the initial installation of the heating system. To lower the high specific heat demand as well as the temperatures, the

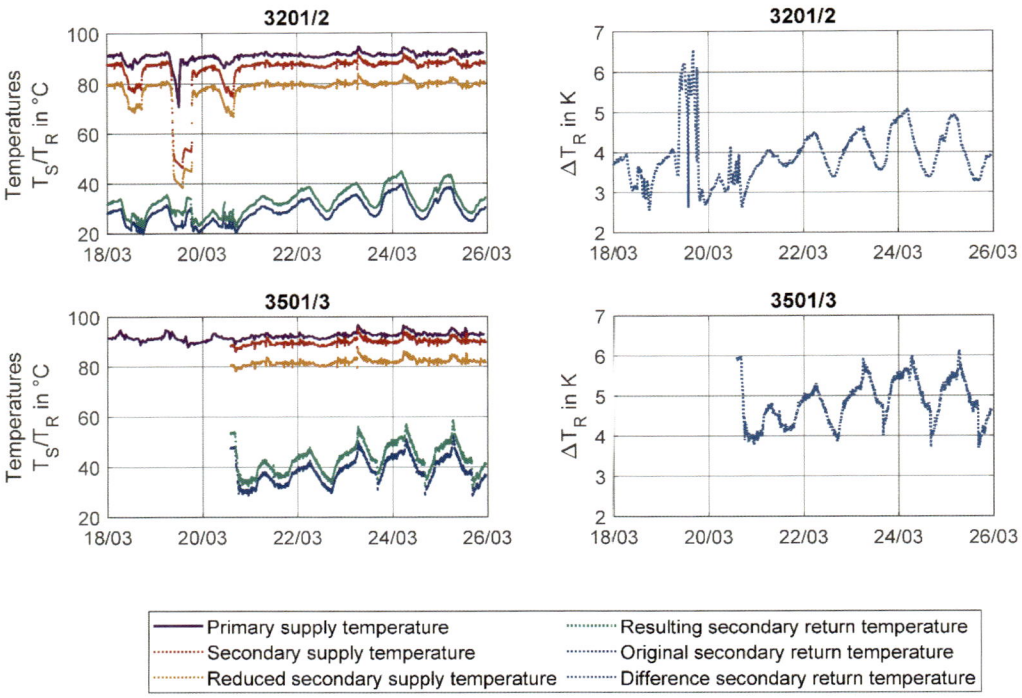

Figure 61. Temperatures before and after supply temperature reduction for the heating circuits 3201/2 and 3501/3

installation of a heat recovery system for the ventilation is a high priority. Additionally, surface heating would be an option to reduce space heating temperature demands. Even though this building is one of the more recently erected buildings on campus (erected in 1995), it shows construction issues, such as a leaking roof. Therefore, user comfort and conservation of the building structure are further arguments for a comprehensive renovation of this building. The Organic Chemistry building 3202, which also shows a very high specific heat demand, is already being renovated. Preparing the building for low-temperature heating is a crucial aspect of this process.

4. District heating supply temperature reduction: Many heating circuits do not require the current high district heating supply temperature level but operate at significantly lower secondary supply temperatures. Those which do make use of the primary supply temperature often show control errors. After resolving control errors in the buildings 3102/3106, 3105, 3202, 3203, 3205, 3301 and 3401, the supply temperature can be reduced step by step. In the ventilation heating circuits 3201/2 and 3501/3, such a reduction will lead to a slight increase in return temperatures. Instead of installing surface heating in the Inorganic Chemistry building, increasing the supply temperature locally using a booster heat pump would also be an option to avoid exceptionally high return temperatures from those substations.

For TU Darmstadt Campus Lichtwiese, significant construction activities are projected for the upcoming decades; thus, the low temperature heat supply should also be considered when new buildings are being planned. This includes high standards for the thermal envelope of the buildings, surface heating with differential pressure control, appropriate sizes for ventilation heat exchangers and decentralised hot water preparation. In new buildings, thermal energy monitoring on the primary and secondary side of the substation should become standard equipment.

5.11 Energetic, ecologic and economic comparison of the proposed actions

In the following section, the energetic, ecologic and economic impacts of an implementation of the measures proposed above will be presented. Scenario (heat recovery) considers a reduction of the return temperature in the five most critical substations (3102/3106, 3202a, 3202b, 3205 and 3401) to an average return temperature of 35 °C and an installation of ventilation heat recovery in the Organic and Inorganic Chemistry buildings (3202 and 3205). This is compared to the reference scenario, representing the current (2018) setup of the

TU Darmstadt energy system. First, the annual time series of the resulting return temperatures, network heat losses and changes in total heat supply are presented. Subsequently, results regarding final energy supply, CO_2 emissions, and costs are discussed.

In Figure 62, the impact of different measures on the campus return temperature, network heat losses, and the change in total heat supply is shown. The median return temperature reduction in scenario is . The temperature reduction is slightly higher in summer months than during wintertime, because return temperature errors, especially in ventilation heating circuits, are more frequent in the summer than they are during the winter season (see category V5 in Section 5.7). The second graph shows the total heat loss from the district heating pipes, which can be reduced by 5 % on average in scenario . The third graph makes it possible to understand the influence of the proposed measures on the total heat supply of the campus. In scenario , the yearly heat supply to Campus Lichtwiese is reduced by 15.5 %, mainly due to savings created by the installation of ventilation heat recovery in the critical buildings.

In Figure 63, the yearly final energy supply to Campus Lichtwiese in scenario S_{HR} is compared to S_{ref}. In this diagram, results for the different types of energy relevant to the TU Darmstadt energy system are presented: heat from heat-only boilers, heat and electric energy from combined heat and power plants (CHP), cooling energy from absorption chillers and compression chillers, electric energy supplied from the grid, and electric energy for generation purposes, such as circulation pumps (electric energy contractor).

The yearly final energy supply to Campus Lichtwiese is reduced by about 6% based on the measures presented above. This is mainly due to a reduction of the heat supply from heat-only boilers by 22% (3000 MWh/a). CHP heat supply is also reduced by 9-10%, leading to an increase in the power supply from the public grid by 9%.

Carbon dioxide emissions are calculated based on a emission factor of $f_{CO2,el,2018}$ = 0.474 t_{co2}/MWh for external electric energy, and $f_{CO2,gas}$ = 0.202 t_{co2}/MWh for natural gas used as input to CHP plants and heat-only boilers. Figure 64 shows that the total carbon dioxide emissions can be reduced by 4.5% while the heat supply related emissions (Heat-only boiler-HOB and Heat CHP) see a reduction of 19%.

Figure 65 illustrates a comparison of the capacity-related, demand-related, and CO_2 emissions costs in the different scenarios. While demand-related and CO_2

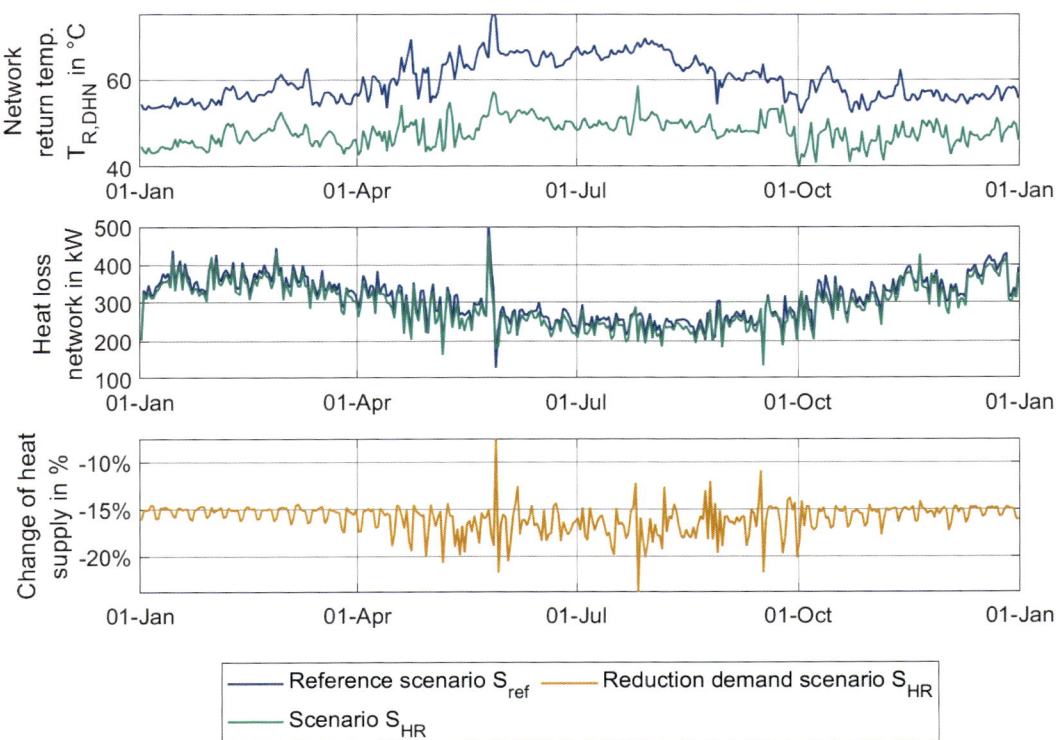

Figure 62. Impact of the proposed temperature reduction measures on network return temperatures, heat losses, and heat supply.

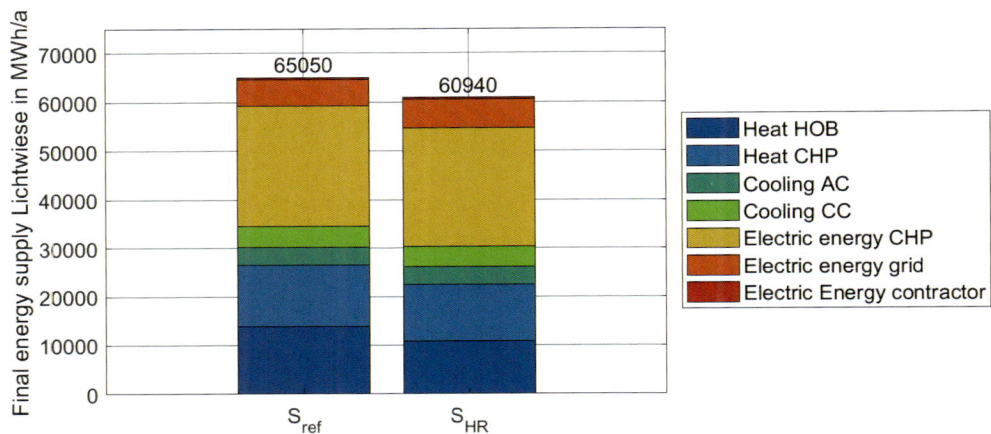

Figure 63. Impact of the proposed temperature reduction measures on final energy supply.

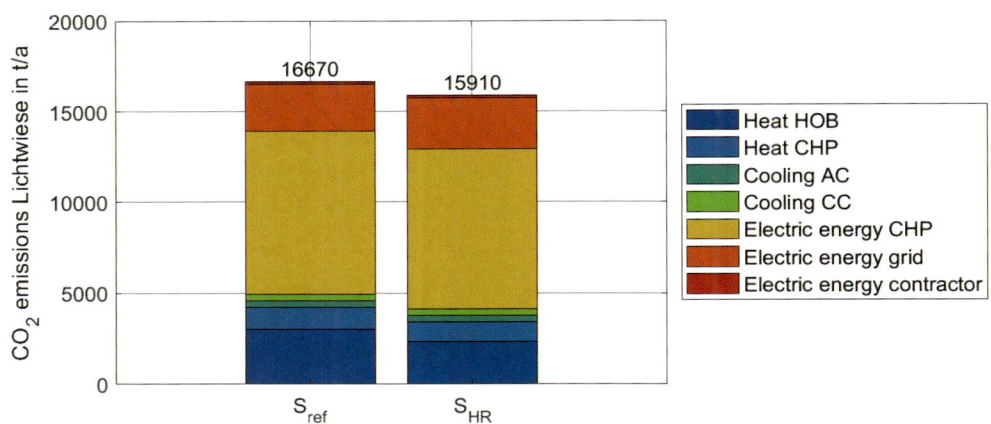

Figure 64. Impact of the proposed temperature reduction measures on CO_2 emissions.

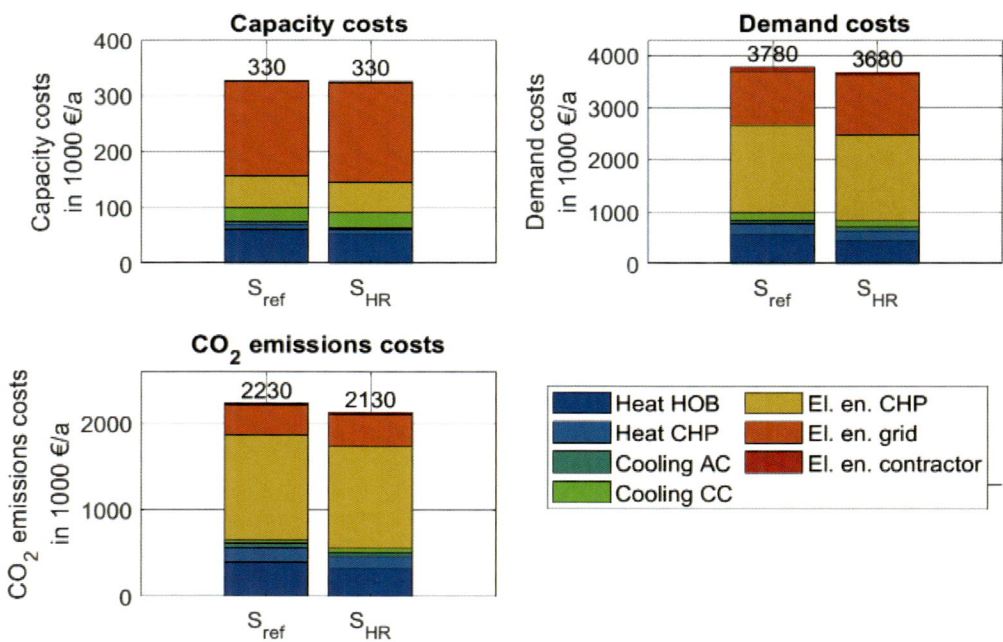

Figure 65. Impact of the proposed temperature reduction measures on the capacity-related, demand-related, and carbon dioxide emission costs.

emissions costs can be decreased in S_{HR} as compared to S_{ref}, capacity-related costs remain constant. Since a reduction in heat demand leads to a reduction in CHP electric energy generation, grid electric energy demand increases, leading to higher fractions of costs for electric energy supply.

Figure 66 compares the total annuities, which can be lowered by 200 000 euro/a in S_{HR} compared to S_{ref}. Investment and operation costs of the measures proposed were not included in this comparison. Instead, the maximum initial investment C_0 can be calculated, taking into account an annuity of the investment A_{inv} and operation-related costs A_{op} equal to the decrease in capacity-related, demand-related and CO_2 emissions costs $A_{inv} + A_{op}$ = 200 000 euro/a:

$$C_0 = \frac{A_{inv} + A_{op}}{f_a \cdot (1 + f_{op} \cdot f_b)} = 1.2 \text{ Meuro} \qquad \text{Equation 21}$$

In this equation, $f_a = \frac{q-1}{1-q^{-t}}$ represents the annuity factor, including a nominal interest rate of q = 1.05 and an observation period of t = 10 a. The observation period is short compared to the average lifetime of the facilities in the district energy system, but decision-makers are often reluctant to base their investment decisions on very long observation periods; hence, a shorter observation period is reasonable in this context. The factor f_{op} combines the cost factors for maintenance, servicing and inspection. In *Fundamentals and economic calculation VDI 2067* (Verein Deutscher Ingenieure, 2012), detailed factors for each facility are listed. In the context of this guidebook, a simplified approach is used and an average factor for operation-related costs f_{op} = 1.03 is considered for all facilities. The price-dynamic cash value factor is represented by $f_b = \frac{1-(\frac{r}{q})^t}{q-r} = 8.39$, including a price change factor r = 1.2 Meuro.

Based on the results obtained in this section, the proposed measures are economically feasible, as long as the total investment does not exceed 1.2 Meuro.

5.12 Major conclusions from the applied study

The applied study of temperature reduction in the district heating network at TU Darmstadt's Campus Lichtwiese serves as a showcase of how to reduce temperatures in an existing district heating system and is equally applicable to many other district heating systems. It makes evident that operational errors within the building heating infrastructure lead to considerable increases in the network temperatures and shows that individual problems in a few buildings can have a significant impact on the entire network. At the same time, fixing the most critical issues helps to reduce network temperatures considerably, especially on the return side.

The study also reveals a major barrier for reducing network temperatures: as long as the generation in the district heating system is realised via CHP plants and boilers, the immediate benefit of reducing network temperatures is low because many of the economic benefits presented in Chapter 2 do not apply to a fossil-based energy system. At the same time, renewable heat sources, such as geothermal, solar thermal, or local waste heat are low-temperature heat sources, which are neither economically nor energetically feasible in high-temperature district heating systems. In a high-temperature district heating network, low-temperature renewable heat can only be integrated using a heat pump

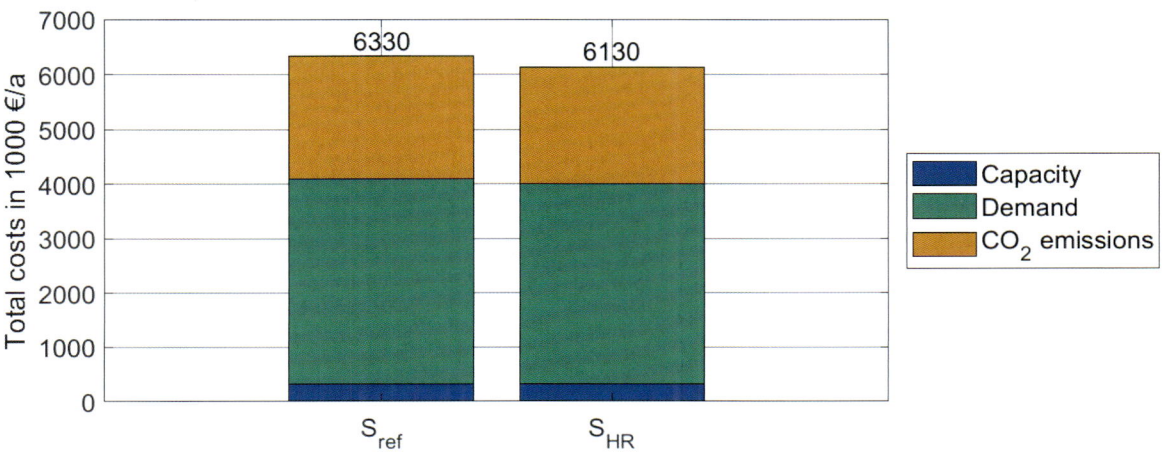

Figure 66. Impact of the proposed temperature reduction measures on the total annual energy-related costs.

at low efficiencies, resulting in high electric energy demands for the heat pump. To be able to realise a transition from fossil-based to renewable district heating, it is thus necessary to first implement a transition from 2GDH or 3GDH to LTDH. Nevertheless, a reorganisation of business models will be necessary in order to make LTDH economically feasible on a larger scale. This aspect will be discussed in more detail in the following chapter.

5.13 Literature references in Chapter 5

Bundesministerium für Wirtschaft und Energie, Bundesministerium für Umwelt, Naturschutz, Bau und Reaktorsicherheit (2015). Bekanntmachung der Regeln für Energieverbrauchswerte und der Vergleichswerte im Nichtwohngebäudebestand. URL: http://www.coaching-kommunaler-klimaschutz.de/fileadmin/inhalte/Dokumente/StarterSet/BMVBS_Energieverbrauchskennwerte_und_der_Vergleichswerte_im_Nichtwohngeb%C3%A4udebestand.pdf (visited on January 26, 2021).

Oltmanns, J. (2019). Auf dem Weg zu einem energieeffizienten Campus: Projekt EnEff:Stadt Campus Lichtwiese - Weiterentwicklung eines Energiesystems auf Quartiersebene. Hoch3 (3), 4–6. URL: https://www.tu-darmstadt.de/universitaet/aktuelles_meldungen/publikationen/publikationen_archiv/einzelansicht_10624.de.jsp (visited on January 26, 2021).

Oltmanns, J., Sauerwein, D., Dammel, F., Stephan, P., Kuhn, C. (2020). Potential for waste heat utilization of hot-water-cooled data centers: A case study. Energy Science & Engineering, 8(5), 1793–1810. DOI: https://doi.org/10.1002/ese3.633.

Verein Deutscher Ingenieure (2012). Economic efficiency of building installations (Fundamentals and economic calculation, VDI 2067). Berlin: Beuth Verlag.

6 COMPETITIVENESS OF LOW-TEMPERATURE DISTRICT HEATING

Authors: Kristina Lygnerud and Helge Averfalk, Halmstad University

Increased operational efficiency from lower system temperatures, optimised technical configurations for low-temperature heat distribution and the resulting economic benefits (see Chapter 2) are frequently discussed. Less discussed are additional competitive advantages of low-temperature solutions. Therefore, in this chapter, how a stand-alone LTDH solution, or a combination of conventional district heating and low-temperature solutions can increase the overall competitiveness of the district heating business case are addressed. The discussion is first addressed from an overall business model perspective and a national viewpoint. The discussion is concluded with a more detailed analysis of the heat distribution cost in the LTDH context.

The shift from high-temperature heat sources and a centralised heat supply to low-temperature heat sources and a decentralised heat supply requires different business logic than that in the conventional district heating context. Hence, the differences are reviewed for the conventional versus low-temperature business model to understand how the LTDH business model combined with the conventional model can increase the overall district heating competitiveness (Sections 6.1–6.4). Turning to the national level, the national setting affects district heating and the potential of competitive LTDH solutions is reviewed (Section 6.5). Finally, the parameters of the heat distribution cost are addressed, which is vital to district heating companies interested in LTDH investments because it is an essential indicator of the competitiveness of any district heating system (Sections 6.6–6.9).

6.1 Traits of business models in early LTDH installations

According to economic theory, investment decisions are made to maximise the long term market value of a company (Myers, 1984), which is achieved by optimising the discounted net present value of future cash flows (Cyert & March, 1963), maximising the utility of the capital invested (Edwards, 1954). New ways of conducting business are either associated with uncertainty or risk. In the latter case, the probability of events can be estimated (Knight, 1921).

Investments in LTDH are different from investments in 3GDH solutions. One main difference is that the heat supply shifts from sources of high temperature (allowing a district heating operator to acquire and use it where it is most efficient) to a combination of several low-temperature heat sources that are limited in size and tied to a specific location (waste, geothermal, waste heat, and ambient heat). The LTDH context necessitates an increased focus on distribution and storage of heat and using it when it is most cost-efficient, which means that the usual method of conducting district heating business (business logic) must be upgraded.

However, how the business model must be adjusted to account for LTDH and the resulting business logic shift is not often discussed. This was further evident in the search for demonstrators to showcase in Chapter 7. For none of these, a description of business model was provided. Instead, focus was on technology. If a decision-maker makes the investment decision based on technical information only, a risk exists that the investment

> **Business model - what is it?**
>
> Business models describe how different pieces of a business fit together but do not account for choices made to meet competition and are not to be confused with the business strategy. Business models are also different from products, companies, industries, networks, technology, internal organisations, and value chains. Common features of business models are that they provide insight on customers (value, relationships, segments, and channels), resources (activities, resources, and partners) and the cost/income structure resulting from customer and resource choices, which are generic elements of business models. The business model generates foreseen cash flows on which investment decisions are made (such as NPV analysis). The business model is built along the lines of a certain logic, reflecting a core value (such as carbon neutrality) or a core condition (such as large volumes of high-temperature heat).

decision may not maximise the market value or the utility of the invested capital.

To understand the differences between high-temperature and low-temperature business models, a paper co-financed by the TS2 project was written based on the analysis of six LTDH cases (Lygnerud, 2019). This paper addresses the following research question: *Do district heating companies that implement low-temperature solutions develop their business models at the same time as they make the shift in technology?* The answer to this question is no. The main conclusion in the paper is that none of the six studied cases upgraded the business model or logic. Instead, the high-temperature context was applied to the low-temperature solution, leading to the loss of the potential value created in the low-temperature context.

In this book, the goal is to support LTDH implementations. Therefore, findings in the paper mentioned above point out how low-temperature investments can increase the competitiveness of the district heating customer offer. First, the readers are provided with an understanding of the characteristics of the cases analysed in the paper. Then, important findings from reviewing the cases are summarised, and the questions to ask to facilitate the inclusion of LTDH investments in the district heating portfolio are listed.

6.2 The cases

The cases were chosen to understand the business model shift between low and high temperature. The cases are interpreted as forerunners from which the district heating industry can learn how to maximise the effect of the technology shift from the third generation to LTDH. The cases were selected on four criteria.

The first criterion was project initiation before the end of 2019. The second criterion was that the ownership of the district heating company undertaking the case should be a mixture of private and public companies (three companies are privately owned, and three are publicly owned). The third criterion was that the system heat sources must have a lower temperature than the sources in conventional third-generation systems. The fourth criterion was to feature cases from both the frontrunner countries in district heating technology where a tradition exists of large district heating systems and countries with a tradition of local district heating systems or less mature district heating markets. Interviews were conducted with the project managers for the six low-temperature implementation projects.

Case 1: A city in Germany

In Germany, the site owner does not want the location disclosed to avoid interfering with public procurement processes. It is a developing area with good possibilities to challenge traditional thinking and form an area that addresses changing living behaviour, construction, interaction, and energy system innovation. The project aims to focus on energy system innovation and development from an environmental and primary energy perspective, taking advantage of the existing district heating network (using return heat) and locally available heat from the sewage system. The case provides heating and cooling for an area equal to the size of around 200 000 m^2. The project is a collaboration between the site developer and the local privately owned energy company. Electricity produced onsite (from renewable biomethane), excess heat from the sewage system and heat from the low-temperature return heat in the district heating network will be combined. With an innovative, fourth-generation district heating piping system, the project aims for a low primary energy demand using low system temperatures that decrease energy loss. The heat-source temperatures range from 15 °C (sewage water heat in wintertime) to 40 °C (return temperature of the district heating system).

Case 2: Darmstadt, Germany (also featured in Chapter 5)

The case is located on the Lichtwiese campus of Technical University in Darmstadt. The university aims to contribute to the German energy transition using waste heat from a high-performance computing data centre located on campus. The project is a collaboration between a research team consisting of architects, electrical engineers, mechanical engineers, and administration from the Institute of Technical Thermodynamics and the local district heating supplier (a private company owned by the city of Darmstadt). The campus is a typical university campus erected in the 1960s which later expanded on several occasions. The university currently contains heat from three gas engines in combined heat and power plants (total of 7 MW$_{th}$) and six gas boilers (55.8 MW$_{th}$). Since 2017, there has been a district cooling network onsite supplied by absorption cooling (1 W$_{th}$). In addition to the heat and power station, a data centre serves as a low-temperature heat source (supplying 360 kW$_{th}$, a small fraction of what is needed in total). The heat-source temperature ranges from 40 °C to 45 °C coming from high-performance computer servers. Because this temperature is too low to be used directly for heating purposes, the temperature of the data centre waste heat is upgraded via a heat pump from 60 °C to 70 °C and integrated into the return line of the district heating system.

Case 3: Albertslund, Denmark

Albertslund is a town west of Copenhagen in Denmark that invested in district heating in 1964. Denmark is a mature district heating market, and district heating covers 97% of the heat demand in Albertslund. Within the next 10 years, depending on the willingness and speed among building owners to refurbish, approximately 5000 dwellings are expected to be renovated in a city of 28 000 inhabitants. The heat is purchased from a district heating transmission company and then distributed through the locally owned district heating distribution network (owned and operated by the municipality). The purchased heat is primarily generated by CHP from biomass and waste incineration, and waste heat through a heat pump from a computer centre is included in the distribution system.

The city has ambitious environmental goals and annually presents its green accounts (where the targets of CO_2 neutral heat and electricity by 2025 and 100% reduction of CO_2 emissions by 2050 are important). One measure to reach the goals is to lower temperatures in the district heating distribution network. The target was launched in 2016, and by January 2026, the supply temperature level in the system should be 60 °C. The target will be met by energy savings in buildings and building adjustments for LTDH, and the distribution system will be updated to allow for lower temperatures. Work to partition the distribution system into low-temperature systems has been initiated.

Today, five local islands have installed a shunt to mix hot and cold water for the area. The buildings in these areas are new constructions or are social housing with substantial refurbishment efforts. Currently, more than 1000 households are supplied with LTDH. The project is a collaboration between the municipal district heating network operator and the building owners in the city. There is a dialogue forum called the 'User Council' with representatives from 50 different housing areas. In this council, all aspects related to the local supply of heat, water, sewage, outdoor lighting, garbage, and re-use are discussed, and local politicians address the council's suggestions. In the User Council, low temperature in district heating has been agreed upon, and a low-temperature action plan is updated annually. The interest in preparatory work for low-temperature installations has increased over time from the house owner side. The Albertslund district heating supply offers the end users the option to rent a new heating unit with service. All 7500 heat meters have been changed to smart meters. For the coming years, the focus will be on the big data analysis of meter and weather data to provide the end users with a more exact consultation to prepare homeowners for lower temperatures.

Case 4: Vallda Heberg, Sweden

This case is in Vallda Heberg, a residential area in the city of Kungsbacka south of Göteborg in Sweden. The area is newly built and was completed in 2013. The houses are passive energy houses and encompass a retirement home, service sector (gym and offices), single-family houses, and apartment buildings. The installation was made by the local housing company EKSTA Bostads AB, owned by the city. This company is the local real estate company but also provides heat to its own buildings. The company is a forerunner in Sweden, in that they use solar in district heating whenever possible. Because the company owns the buildings and provides them with heat, they understood that lower temperatures in the network would make solar energy more efficient. They decided to conduct a test in the Vallda Heberg area (the heat source is solar with a temperature range of 70 °C to 80 °C). The company has considered the whole area rather than the building level when setting the system boundaries of the implementation. One way to secure a good summer load is to connect the washing machines and dishwashers to the building heating system. The heat is delivered to the building heat exchanger (e.g., no apartment substations were installed). The project results from the company being able to affect both the energy system and building construction in the area. All customers in the area chose to connect to the district heating network. The installation is a small network that will not be connected to the large-scale district heating network in the locality.

Case 5: Nottingham, United Kingdom

The fifth case is in the city of Nottingham in the UK. It is the first LTDH network installed in the UK. There is an existing, second-generation district heating network in the city (planned in the 1960s and in operation 1973) where the heat source is waste incineration combined with gas boilers. The district heating market in the UK is fragmented but entering a growth phase. The heat source feeding the LTDH island is the return heat from the district heating system (at 60 °C to 65 °C).

The local district heating company is municipally owned, which is also the case for the Nottingham City Homes. The implemented scheme is that Nottingham City Homes has its own energy service company (commonly reffered to as a ESCO) for the low-temperature area. Moreover, Nottingham City Homes is a separate company that owns 30 000 homes in Nottingham and is owned by the local council. An area of 94 flats belonging to Nottingham City Homes was substantially refurbished and attached to a newly built low-temperature grid (also owned by the housing company). The work was done within the realms of the Renovation Model for URBAN regeneration (REMOURBAN) project (H2020 financing) as part of the EU Smart Cities and Communities grant, and the installation was completed in March 2019.

The local university has been actively engaged in the energy installation, providing knowledge and installation advice and monitoring the installations. The consumption is measured per apartment, and the gas boilers previously used in the apartments were substituted with an apartment substation. The small volumes of water standing left in the heat exchangers do not allow *Legionella* bacteria growth and are not a problem.

Case 6: Wörgl, Austria

In Austria, the city of Wörgl has a local district heating company, which is new to the district heating market (it was installed in 2013). The heat is generated by biomass and industrial waste and residual heat from a dairy factory (with waste and residual heat flows at a temperature of 45 °C to 80 °C). The district heating system is in the growth phase, and one strategy for growth is to build small heat islands connected to a large network. The city has been building an area of social housing targeting young families. When building the houses, the city decided to go for LTDH, which is well suited for floor heating, primarily using waste heat from the dairy factory. The city engaged a team that included the city planner, developer, architect, and pipe manufacturer at an early stage to identify an efficient solution for the area. As a result, a prefabricated pipe solution was designed to save on piping costs and space used in the area. The *Legionella* aspect is treated locally with booster heat pumps for heating the water.

In Table 12, a summary of the companies in the case study is provided. The respondents in the cases are the project managers of the low-temperature implementations. This target group was deemed most apt to provide an understanding of the implementations ranging from the implementation purpose to the technical constraints. The applied analysis model is the 'business model canvas' (please view the full paper for further information). The interviews were semi-structured, and the questions were elaborated based on the business model canvas components. The relevance of the questions was tested with practitioners from three district heating companies in three countries (Italy, France, and Germany), and their input was incorporated into the interview guide. The interviews lasted between 45 minutes and 1 hour. During the interviews, notes were taken, and the interviews were documented in conjunction with each interview occasion. The analysis identified whether or where development in the business models could be detected in the cases where a technology shift occurred, compared with the conventional district heating business models.

6.3 Results from case reviews

Business models describe how different pieces of a business fit together but do not account for choices made to meet competition and are not to be confused with the business strategy. Business models are also different from products, companies, industries, networks, technology, internal organisation, and value chains. Common features of business models provide insight into customers (value, relationships, segments, and channels), resources (activities, resources, and partners) and the cost/income structure resulting from customer and resource choices, which are generic elements of business models.

The analysis model used in the study is the 'Business Model Canvas' (Osterwalder & Pigneur, 2010). The model addresses customer-oriented activities (the customer segment, the value perceived by the customer, the relationship with the customer, and the communication channel used to communicate with the customer) and resource-oriented activities (key activities, resources, and partnerships). How the value is delivered to the customer leads to both costs and income, two additional parts of the canvas. The main results are pre-

Table 12. Characteristics of heat supply options for district heating.

Case	Implementation status	Ownership structure of district heating company	Heat sources and their temperature	Mature heat market (Y/N)
1	Feasibility study completed	Private	Sewage water (15 °C in winter) and return of district heating (40 °C)	Y
2	In progress	Private	Waste heat from high-performance computers (40 °C to 45 °C)	Y
3	In progress	Public	District heating network (60 °C)	Y
4	Completed	Private	Solar (70 °C to 80 °C)	Y
5	Completed	Public	District heating network return (60 °C)	N
6	Completed	Public	District heating network using different heat sources (45 °C to 80 °C)	N*

*The Austrian district heating market is mature, but the Tyrol region where the demonstration site is located is characterised by local and small bio-fuelled heating centres.

sented for the customer side, resource-oriented side, and cost/income aspect.

(1) Customer orientation (customer value, segment, relationship, and channel)

Customer value reflects the value that the customer perceives from using a product or consuming a service compared to other alternatives. The conventional customer value of district heating is heat and hot water delivered. In the six cases, the possibility of engaging in activities that lower the environmental impact combined with saving energy was important to the partners. For example, using waste heat and renewable energy resources or merely better using the heat in the grid (return heat or sectioning the system) was a key driver, generating a selling point strengthening the customer value. For example, a green heating alternative can be offered. In greenfield investments where new buildings are being erected, the green aspect of LTDH is an important selling point. However, in new areas being built in cities with a district heating system, the green aspect of the LTDH island is a competitive advantage.

In different contexts, high-temperature district heating has played an essential role for reduced carbon dioxide emissions (often discussed in Sweden where a shift from fossil fuels to waste and forestry residues has been occurring since the 1970s). However, the element of green has traditionally not been put forward as a key selling point. Considering the goals for 2050 and that LTDH can contribute significantly to reaching them, it appears that it is time to reinforce the green value and use it as a selling point.

The *customer segment* reflects the target group of the offer. The conventional segment for district heating is large, professional customers. From the studied cases, half are a mixture of customer segments, including private homes. In addition, in half of the cases, a strong prosumer presence is actively engaged to facilitate heat recovery. Using a local heat source increases the possibility to engage in local activity, supporting local trade and industry.

Regarding *customer relationships*, the features of local engagement and, for prosumers, capitalising on an otherwise lost resource leads to the potential to establish loyal and long term relationships. In the days of increasing transparency and the possibility of comparing alternatives for the most cost-efficient solution, the link to the local community and the possibility of engaging in a long term and stable relationship can be critical factors for competing in the heat market.

Customer channel reflects the interaction with customers. It will not significantly change for any customer other than the prosumers for whom the LTDH investment must be tailored.

(2) Resource orientation (activities, resources, and partnerships)

The *resources* needed to undertake LTDH are similar in the six cases. The district heating system permits low supply temperatures, heat pumps (in four out of six cases), a local heat source, and staff with the capabilities to engage in the necessary dialogue and contract writing with prosumers.

Activities of importance are establishing efficient systems for heat recovery (either establishing a system including waste heat and a heat pump) or improving the current district heating system. Apart from the technical activity, communication and win-win creation with prosumers are imperative.

Partnerships reflect close collaboration with partners that facilitate the business to be undertaken. In the fossil fuel context, the providers of fuels were important partners. In the LTDH context, the prosumer and heat-source owner become the most critical partner.

(3) Cost/income aspect

In the six cases addressed above, investments are needed in new resources: primarily in heat pumps, prosumer relationship building, and the network. For further input on cost estimates for LTDH investments, more information is included in Sections 6.6 to 6.9, where the heat distribution cost is addressed. Moreover, LTDH can significantly lower the costs compared to the average capital cost of European district heating (based on an assessment of 37 early LTDH implementations).

Turning to income, in the six LTDH installations, the heating price charged to customers was kept identical to the conventional district heating price. Hence, no change was made to capitalise on the green value added to the district heating offer in the LTDH area. It is understandably difficult to increase the district heating pricing for existing district heating customers. However, it should be possible for a district heating provider to differentiate the price charged to customers in different contexts. In a newly built area, where no district heating tradition exists, it should be possible to come in with a premium price tag for green energy. Admittedly, this can be further complicated by the existing price regulations where differentiation is only possible for professional customers.

Motivations for LTDH

The trend in Europe is to upgrade price models by including different motivations for lowering the return temperatures. The driver in this is the increased efficiency and increased margins in the district heating system by lowering the temperature. In an LTDH context, the trend is relevant to address. In a study of 200 district heating companies in Sweden, it was identified that approximately 55% of them have a flow demand component linked to motivation. However, the effect of such motivation appears to be minor, considering the conventional high-temperature context (Petersson & Dahlberg Larsson, 2013). One main challenge when designing a motivational tariff is to find an appropriate balance between bonus and cost. As the buildings become increasingly efficient, a bonus system must be phased out to ensure that the bonuses paid to customers do not erode the operational gains from lower return temperatures. Within the TS2 team, this aspect was addressed several times, and it was identified that the bonus system must be upgraded to reflect the investments made. Over time, as the system efficiency is upgraded, the motivational levels must be adjusted accordingly.

According to the review of known cases of price model upgrades across Europe (Leoni, Geyer, & Schmidt, 2020), important aspects to consider are the conflict of actors (should the building owner or the district heating provider undertake the upgrade?) and the financing of the necessary activity. One solution is that the district heating provider finances the investment and the district heating customers lock-in to a determined district heating price for a given period (10 years or more). No established consensus exists on the most effective form or level of motivational tariff to lower temperatures in existing district heating systems. Moreover, limited information exists on how to value prosumer relationships bringing low-temperature heat sources into the heat portfolio, which is most likely a consequence of the limited prosumer engagement in district energy to date.

6.4 Questions to address when expanding to LTDH

In places where there are existing district heating systems, it is possible to include LTDH as a complement. The high-temperature grid can be kept as a backbone for heat distribution into which heat is supplied from

---

The price model - what is it?

A price model is designed to generate income and reflects how the component income structure in the larger business model context is generated. There is often confusion between the business model and its price component. The price model is not identical to the business model; it is only one part of it.

several decentralised heat sources. With such an approach, it is possible to compete with other heat alternatives based on unique selling points of LTDH (the value of green, long term customer relationships and tailored relationships with prosumers) allowing for further densification of the district heating system in the city. In markets with low district heating maturity, LTDH is competitive because it can fit the planned district and use available heat sources in a cost-efficient and green way. In a greenfield context, it is easier to build the LTDH business case around its actual conditions. In an existing district heating system context, the business case is more difficult because the logic of the existing business case is not efficiently transferable to the LTDH business case. Below, five example questions are listed to address when expanding the district heating portfolio to encompass LTDH.

6.5 National context and potential for LTDH to increase district heating competitiveness

Companies 'are born and learn to compete' in the context of the nations where they operate (Porter, 1990). In the diamond theory of national advantage, Porter addressed four cornerstones that jointly make up the 'diamond': (1) firm strategy, structure, and rivalry; (2) demand conditions; (3) related and supporting industries; and (4) factor conditions. For decision-makers on LTDH investments, it is relevant to understand how the cornerstones of nations, on an overall level, affect LTDH and how it can improve the competitiveness of district heating (addressed in Sections 1 to 4). The section concludes with examples of the LTDH context in five countries (two consolidating district heating markets: Sweden and Finland, two expanding markets: Norway and Canada, and one new market: Belgium).

(1) Firm strategy, structure, and rivalry

The strategies and structures applied by firms are a result of legislation, incentives (political), and competition. The European district energy landscape is heterogeneous, and the European district energy landscape can be split into four categories: consolidation, refurbishment, expansion, and developing countries (Aronsson & Hellmer, 2011). The strategy in consolidation countries is to densify the existing distribution network and fend the competition off to defend the current market share. In refurbishment countries, the strategy is to increase the efficiency of the current infrastructure and improve the security of the supply, ensuring that customers remain with the district energy solution. In expansion countries, the strategy is, of course, expansion, and in the countries developing their district heating market segment, the strategy is to learn about district heating and identify which solutions to implement and how to implement them.

In terms of structure, mature district heating markets and consolidation markets are characterised by 3GDH

Questions to address before the expansion of the district heating portfolio to encompass LTDH

1. **What customer value is in demand?**
   Is there any possibility to capitalise on the green value inherent in using low-temperature heat sources? What is the competition; is it a greener solution?
2. **Is it possible to engage the customer long term?**
   The LTDH solution can lead to a local engagement among building owners, motivating them to enter long term agreements. What is the situation in the district where LTDH is being considered?
3. **What prosumers are there in the foreseen district? How is a win-win created with them? How many of them can be secured? Can a long term relationship be established?**
   Are new resources or activities needed?
4. **What investments need to be made? Are heat pumps needed? Is there a capability in the current staff to combine low-temperature waste heat sources and heat pumps into efficient heat recovery systems?**
   The LTDH necessitates a customer dialogue with prosumers. Do the employees have the skillset to identify and secure low-temperature heat sources?
5. **Should the price component of the business model be upgraded?**
   Motivational tariffs are discussed in the district heating sector. With LTDH, low system temperatures are imperative. Depending on the desired improvements in the district heating system, the level of an efficient motivational tariff will vary.

networks whereas 2GDH and 3GDH networks are found in refurbishment countries. The structure in expanding markets and new markets is not necessarily linked to a high-temperature system. Solutions that are apt for the areas being exploited are being implemented (demand drives expansion), which means that high-temperature or low-temperature solutions can be equally feasible. Some countries have successfully chosen to invest in ultra-low temperatures since the early 1990s. For example, these include Switzerland and Italy, and the ultra-low-temperature system in Bergen, Norway, dates back to this period. Later in Canada, the Vancouver area experienced an increase in district heating systems since 2008 when municipalities in British Columbia required energy planning. The first installations were high temperature, but there has been increasing interest in ambient heat recovery from sewage water and boreholes that can supply heating or cooling depending on the demand.

The structure of an industry is closely linked to political incentives. In a study identifying the existence of explicit LTDH investments in European countries Persson, Averfalk, Nielsen, and Moreno (2020), only two countries made a dedicated low-temperature effort. In Denmark, a 4DH research centre was active between 2012 and 2018 with contributions from several Danish and international universities with many Danish district heating companies. In Germany, Wärmenetzsysteme 4.0 was launched in 2017 and provided 100 million euro for funding feasibility studies and pilot projects related to LTDH.

Turning to rivalry, (Porter, 1990) assumed that domestic rivalry triggers innovation and efficiency. In Europe, the main competition for district heating came from gas in the past and will come from heat pumps in the future. As a result of the competition for heat customers, district heating companies in mature and consolidating countries are actively investing in making the existing structures as efficient as possible. A critical aspect for doing so is to lower system temperatures, creating an interest in these district heating markets to understand LTDH investments. In addition, district heating systems in refurbishment countries in the Eastern parts of Europe necessitate efficiency improvements to make LTDH relevant. In expansive markets and new district heating markets, demand shapes the investments. Depending on the will of the investor to invest in a fossil-free solution (meeting the 2050 targets) or in a conventional solution (locking in the current practice of high-temperature district heating) will guide the investment decisions made in these countries.

(2) Demand conditions

The district heating providers provide heat and hot water for their consumers, meeting a basic human need. As such, customers tend to focus limited attention on their heat and hot water supply until it does not work efficiently.

However, the increasing demand for renewable energy from the user and regulators (at the local, regional, national, and EU levels) creates monetary incentives for renewable energy sources. Moreover, the district heating industry was built on the business idea of using heat that is otherwise wasted (be it from the production of electricity in a CHP or from an industrial process or urban infrastructure). A lack of incentives for waste heat recovery was found in a study surveying LTDH stakeholders across eight European countries (Leonte, 2019), coupled with incentives for other competing technologies. This lack is a problem because alternative technology incentives exist, such as those for high-efficiency CHPs, which suggests that, until incentives are offered on low-temperature heat recovery, it will be challenging for this new technology to compete with other existing technologies in terms of attracting investment. One explanation for this absence of incentives is that future heat solutions are being evaluated against the current norm (high-temperature solutions). Such assessments are oblivious to the fact that, in the future, the high-temperature possibility will be limited (no fossil fuels, no biomass, and no waste to incinerate). A narrow-mindedness exists regarding what constitutes a sustainable energy solution because incentives exist for renewable energy sources but not for other alternatives (e.g., LTDH).

Another driver of demand is awareness; today, limited knowledge about LTDH installations exist across Europe (Wheatcroft, Wynn, Lygnerud, Bonvicini, and Leonte, 2020). Until the knowledge that LTDH investments are feasible and that heat and hot water can be supplied by tapping the local sewage water system, service sector buildings, metro systems, or datacentres, these installations will be in low demand.

(3) Related and supporting industries

The related and supporting industries refer to clusters of industries that need and support each other. Further, district heating companies necessitate heat sources and, depending on the ownership configuration, operators of the central heat plant and distribution system. In essence, district heating companies need providers of pumps, heat switches, pipes, maintenance and control systems, and entrepreneurs for different forms of groundwork. In the context of LTDH installations, heat pump providers and fitters of the foreseen solution are needed. However, a limited number of fitters and installers can take the challenge to implement an unconventional heat recovery, creating a bottleneck. As no standardised solutions exist, there are also no standardised permits for undertaking low-temperature heat recovery. As a consequence, there are no standardised contracts with limited technical standards that can be applied to identify the conditions and value of the heat between the district heating company and the prosumer (Lygnerud, Wheatcroft, & Wynn, 2019).

(4) Factor conditions

Factor conditions refer to such aspects as the national infrastructure, labour force, land, and natural resources. Turning to the context of district heating, the most visible factor is the presence of different natural resources. For example, in Sweden and Finland, biomass is abundant and dominates the fuel mix of district heating companies. In central and southern Europe, the same is applicable for gas. A recent study has shown that a potential exists for urban waste heat (e.g., low-temperature waste heat) from sewage, datacentres, service sector buildings, and metro systems, which corresponds to 10% of Europe's current total heat demand (Persson et al., 2020). Turning to 2050 and the forecasted heat demand in EU28, according to Figure 1, these volumes represent one-quarter of the heat demand.

(5) The context of LTDH in five countries

To conclude this section, the situation of district heating and the low-temperature context in five countries have been reviewed within the TS2 project. Interviews with representatives from the International Energy Agency's District Heating and Cooling Executive Committee were conducted to understand the context of LTDH in five countries. The countries reflect consolidating district heating markets (Sweden and Finland), markets where the district energy technology is expanding but market share is still at a limited implementation level (Norway and Canada), and a new district heating market (Belgium). Semi-structured interviews were conducted to address 13 questions on two major topics: the current context of LTDH and incentives for LTDH investments. The definition of low for LTDH for the interviews was 50 °C supply and 20 °C return, which are technically feasible to achieve as a yearly average when each substation obtains a supply temperature of at least 50 °C (Averfalk & Werner, 2018).

Sweden and Finland

In the market context in the two consolidating heat markets, district heating is not expanding, apart from densifying activity in areas where a district heating system exists. As a result of the status quo, there is low market maturity for LTDH in Sweden, whereas it is not even a market in Finland. In both countries, however, a long tradition exists of recovering heat using large heat pumps. The primary heat source is sewage water, but ambient heat from the sea or industries has also been recovered. In the 1980s, electricity was very cheap in Sweden, and installations corresponding to 1500 MW were built (80% are still in operation). This heat is upgraded and used in high-temperature district heating networks, for example, with temperatures of around 85 °C or above (Averfalk, Ingvarsson, Persson, Gong, & Werner, 2017). A recent trend is to recover smaller low-temperature heat sources than those previously recovered, and heat from datacentres and grocery stores is being discussed.

In Sweden, there has also been activity to lower system temperatures in existing district heating networks. One early example is that of Mälarenergi in Västerås. Examples also exist of pure LTDH systems based on low-temperature sources (e.g., the system in Brunnshög using excess heat from two nearby research institutes). The interest in low-temperature heat sources is currently relatively limited in both Finland and Sweden due to building standards and the existing heating infrastructure.

In addition, LTDH is not foreseen to play a leading role in making the national district heating fuel mix fossil-free or increasing the energy efficiency in either country. In Sweden, LTDH is expected to grow and become one of several parts of the future Swedish district heating fuel mix. The low relevance of low temperature in these countries is reflected in a low knowledge level of LTDH in both countries.

In Sweden, there is a nascent discussion about future access to fuels, and it has been acknowledged that, to become fossil-free, several sectors will have an increased demand for biomass, which will increase the competition for biomass in the upcoming decades. Today, the part going to energy conversion is mainly waste wood from the forest industry or demolition wood. In addition, the challenge issued by the city of Helsinki in 2020 reflects an objective of phasing out fossil fuels from the district heating network and replacing them with options other than biomass. Hence, the current understanding that LTDH is of low relevance can change in the future when biomass has become a resource in high demand.

Finally, combined heat and power production is needed in Sweden to balance the power peaks in the electrical grid. A benefit of LTDH is enabling low-cost energy for the periods where CHPs are not operating. However, the benefits are often not harvested because the tradition of replacing the current sites with common high-temperature sites is pronounced.

Regarding incentives for district heating overall and LTDH in particular, support schemes for CHP were implemented during the 1990s in Sweden, but in the last years, the Energy Agency has started to focus on the overall energy system, where district energy is one of many components. A research programme is dedicated to district energy solutions but takes the energy system approach where district energy is one part of a smart city. Moreover, district heating is considered a mature technology with limited need for investment support, leading to excluding low-temperature solutions from targeted subsidies. Thus, they compete with incentivised renewable energy sources solutions. In terms of business model development for LTDH, there are pilot sites in Sweden (one is much discussed in Brunnshög, Lund), where the LTDH solutions can be profitable. No consensus yet exists on the LTDH business model. In terms of the value of green energy from harvesting low-temperature heat sources, the value is acknowledged but not yet capitalised in business models.

The incentives for RES are limited in Finland, but subsidies exist for pilot projects (where LTDH installations are found), and the decarbonisation of existing CHPs can be subsidised. The value of green heat sources is generally not capitalised on, but some customers, like those interested in certifications, such as BREEAM, can be willing to pay for a green and locally harvested low-temperature heat source.

Norway and Canada

In the market context, in Norway and Canada, district heating is expanding but is still at limited implementation. In Norway, hydropower electricity dominates, and both the building code and energy declaration code are disadvantageous for district heating compared with other heating alternatives. However, this might change with the ban to use fuel oil for heating (some exceptions remain) as of January 2020. In Canada, low-carbon electricity and gas also dominate. Many of the existing systems are small, and two-thirds of the systems are operated on university campuses or military and other government premises. Most larger and older systems are operated with steam or high-temperature water.

The potential for low-temperature solutions exists by resorting to electricity-driven heat pumps due to Norway's low electricity price. The low-temperature solutions are apt for new passive house buildings primarily through floor heating in bathroom areas, a comfort detail in high demand in the country. In Canada, British Columbia has experienced an increase in district heating systems since 2008 following legislation requiring municipalities to undertake an inventory of greenhouse gas emissions, create a plan to reduce their levels, and implement the plan. This legislation led to a significant increase in interest in renewable energy use and district heating. The first installations were high temperature but increasing interest has been expressed in ambient heat recovery from sewage water and boreholes that can supply either heating or cooling depending on the demand. For both countries, the LTDH market maturity and knowledge level on the topic are very low.

Both countries aim for electrification, but the available electricity sources are insufficient, making district energy an attractive flexibility provider in energy systems. In Canada, the distances between waste heat sources, such as the Alberta Industrial Heartland area near Edmonton, and existing district heating systems have tended to be too long for cost-efficient solutions. The lack of a general district heating infrastructure in Canada limits the opportunity for the significant use of industrial waste heat.

Regarding incentives for district heating overall and LTDH, infrastructure funds were available for district heating network development around the millennium in Norway. This kind of support has decreased over time, but interest has increased in waste heat usage because it plays a vital role in a circular economy. Customers have increasingly demanded green energy sources. Business model development for LTDH (according to the possibilities in Section 6.1) is not detectable.

Canada supports clean energy projects limiting greenhouse gas emissions through local regulations, such as the one mentioned for British Columbia (above), rather than national support schemes. How to use artificial intelligence through machine learning and how it can be used in an overall greening of business models has been discussed. However, this is a slow process. There has been no development of LTDH business models (according to the possibilities in Section 6.1).

Belgium
In the market context, Belgium is a market where gas is used to meet the heat demand with a few exceptions of district heating networks dating back to the 1970s. The gas has traditionally been imported from the Netherlands, but for a decade, the need to limit the gas export from the Netherlands to mitigate earthquakes from gas extraction has been discussed and recently implemented, cutting natural gas volumes in half. Therefore, a window of opportunity exists for district energy solutions to replace the existing gas networks. The new installations have been high temperature, but recent interest has been expressed in geothermal (shallow and deep) and waste heat recovery. The use of waste heat is limited but is being explored by the city of Antwerp, for example, where a petrochemical cluster is generating waste heat now recovered in a newly built district heating network as part of the city development projects. An LTDH market does not yet exist, and the knowledge level on LTDH is very low.

Regarding incentives for district heating overall and LTDH, regional support exists for district energy in Belgium, but the installations are high temperature, apart from (very few) pilot installations with low temperatures.

> In summary, a LTDH investment can bring additional value for the customer and can lead to a long-term loyal customer relationship, which are aspects of great importance for the overall future competitiveness of district heating. The inclusion of LTDH in the district heating mix can increase the competitiveness of district heating in any country.
> The extent of the increase in competitiveness depends on the context (e.g., the four parts of the diamond outlined above) of implementation. Discussed in Sections 6.1 to 6.4, the business model of LTDH creates a possibility to engage in a closer and long-term relationship while generating added value (e.g., green energy or the possibility to become a prosumer). These are important aspects for long-term district heating competitiveness overall, such as when fully exposed to the competition (e.g., in consolidation countries), when needing to retain customers (e.g., such as in refurbishment countries), in demand driven expansion, and in starting up a future energy system.
> Although LTDH should be relevant to any district heating country, the current maturity of LTDH markets is low. Several initiatives have been conducted worldwide (see more in the chapter on demonstrators), but no consensus exists on standards, permits, and contracts, which is a hurdle for LTDH implementation. Furthermore, in countries with a long district heating tradition, a lock-in effect exists in the high-temperature business logic, which is applied to the LTDH investment without modification (eroding the competitive advantage that LTDH investments can generate). Due to the low maturity in the LTDH market segment, little focus on business model development has occurred; instead all attention has been focussed on the technical feasibility, which is less efficient than developing both sides simultaneously.
> In countries with a more recent district heating tradition, there appears to be a possibility for district heating overall, including LTDH, to be a complementary solution to the main systems (e.g., electricity in Norway and Canada). In Norway, the ban on fuel oil use for heating and the desire for a circular economy can support LTDH expansion. In Canada, regional policies (such as the one in British Columbia) and long physical distances can be beneficial to local LTDH designs. In Belgium, the need to transition from gas to other energy alternatives is making district heating solutions of all sizes and temperature levels appealing.

6.6 Heat distribution characteristics in early implementations

This second part of this competitiveness chapter contains an analysis of characteristic parameters for heat distribution to explore the competitiveness for some early implementations of LTDH. This analysis is required because the features of these new low-temperature systems must be related to the density or sparsity of these heat distribution areas. The purpose of this analysis is to separate important low-temperature features from features belonging to the concentration of heat demands in heat distribution areas.

Hence, a proper assessment method is required. The ambition is to benchmark new heat distribution areas concerning their estimated heat distribution costs. The assessment method should preferably consist of parameter data that can be collected by the area developer or owner with relative ease. This second part details a basic method for assessing new heat distribution schemes and relates these to specific heat demands, concentrations of heat demand (Section 6.7), and temperature levels (Section 6.8). The characteristics in these sections are used to estimate heat distribution costs (Section 6.9).

The assessment is conducted using data from 37 networks where initiatives for lower heat distribution temperatures have been identified. This group of networks consists of implemented, planned, or simulated cases. The temperature levels in these networks vary from high to low levels. This assessment method was initially applied by Dalla Rosa et al. (2014). Input and output parameters are presented in Table 13, and the relationships between these parameters are presented in Table 14. The systems included in the analysis are presented in Table 15.

The specific heat and cold demand for the 37 cases are presented in Figure 67, where the cases are sorted according to country and location, as presented in Table 15. These specific demand cases consist of the annual space heating and cooling demand, domestic hot water demand, and heat loss from domestic hot water circulation over the total building floor area. In the case of cold demand, only five systems are included. They re-

Table 13. Nomenclature for the characteristic parameters of heat distribution.

Symbol	Definition	Units	Parameter	Type
A_L	Total plot area (for connected buildings)	m²	1	Input
A_S	Total building space area	m²	2	Input
L	Total trench length	m	3	Input
Q	Annual heat and cold supplied (input at heat plants)	MWh	4	Input
Q_S	Annual heat and cold delivered (output at substations)	MWh	5	Input
t_s	Annual average supply temperature	°C	6	Input
t_r	Annual average return temperature	°C	7	Input
t_a	Annual average ambient outdoor temperature	°C	8	Input
d_a	Average outer media pipe diameter	m	9	Input
a	Annuity factor	-	10	Input
C_1	Average construction cost constant	euro/m	11	Input
C_2	Average construction cost coefficient	euro/m²	12	Input
Symbol	**Definition**	**Units**	**Parameter**	**Type**
q	Specific heat and cold demands	kWh/m²	1	Output
e	Plot ratio	-	2	Output
q_L	Heat and cold densities	kWh/m²	3	Output
Q_S/L	Linear heat density	MWh/m	4	Output
w	Effective width	M	5	Output
G	Degree time number	°Ch	6	Output
Q_{hl}	Annual distribution heat loss	MWh	7	Output
q_{hl}	Annual average relative distribution heat loss	-	8	Output
K	Heat transfer coefficient	W/(m² °C)	9	Output
A_p	Outer media pipe surface area	m²	10	Output
C_d	Average distribution capital cost	euro/MWh	11	Output

Table 14. Relationships between the characteristic parameters.

$$\text{Specific heat demand: } q = \frac{Q_s}{A_s}$$

$$\text{Plot ratio: } e = \frac{A_s}{A_L}$$

$$\text{Heat density: } q_L = \frac{Q_s}{A_L}$$

$$\text{Linear heat density} = \frac{Q_s}{L}$$

$$\text{Effective width: } w = \frac{A_L}{L}$$

$$\text{Degree time number: } G = \int_0^{1\ year} \left(\frac{t_s + t_r}{2} - t_a\right) d\tau$$

$$\text{Annual distribution heat loss: } Q_{hl} = Q - Q_s$$

Annual average relative distribution heat loss:

$$q_{hl} = \frac{Q_{hl}}{Q}$$

$$\text{Heat transfer coefficient: } K = \frac{Q - Q_s}{2\pi d_o L \times G}$$

$$\text{Pipe surface area: } A_p = 2\pi d_a L$$

$$\left[\text{Average pipe diameter: } d_a = \frac{A_p}{\pi L}\right]$$

Average distribution capital cost:

$$C_d = \frac{a \times I}{Q_s} = \frac{a\left(\frac{I}{L}\right)}{\left(\frac{Q_s}{L}\right)} = \frac{a(C_1 + C_2 \times d_a)}{\left(\frac{Q_s}{L}\right)}$$

Table 15. The 37 district heating systems in the analysis as examples of early implementations of lower heat distribution temperatures together with their chapter and section appearances in this guidebook.

Location, system identifier	Country	Appearance
Gleisdorf, system	AT	7.3, 10.2, 10.3
Salzburg, Lehen	AT	7.7, 10.2, 10.3
Villach, Landskron	AT	10.2, 10.3
Vienna, Krieau-Viertel 2 plus	AT	10.2, 10.3
Okotoks AB, Drake Landing	CA	10.3
Geneva, Lac Nation	CH	10.3
Rotkreutz (Zug), Suurstoffi	CH	10.3
Visp, Visp-West	CH	10.3
Zürich, ETH Hönggerberg	CH	10.3
Zürich, FGZ Friesenberg	CH	7.6, 10.2, 10.3
Aachen, Karl-Kuck-Strasse	DE	10.2, 10.3
Bamberg, Lagarde campus	DE	7.12, 10.2, 10.3
Braunschweig, Rautheim	DE	7.9, 10.2, 10.3
Darmstadt, Campus Lichtwiese	DE	5, 10.3
Freiburg, Gutleutmatten	DE	10.3
Kassel, Zum Feldlager	DE	7.11, 10.2, 10.3
Mannheim, Benjamin-Franklin	DE	10.2, 10.3
Munich, Ackermannbogen	DE	10.3
Neuburg, LTDH	DE	10.3
Oldenburg, ENaQ	DE	10.2, 10.3
Recklinghausen, LTDH	DE	10.2, 10.3
Aarhus, Lystrup	DK	10.3
Höje Tåstrup, Sönderby	DK	10.3
Dublin, Tallaght	IE	10.3
Trondheim, NTNU Campus	NO	10.3
Halmstad, Ranagård	SE	10.3
Linköping, Ullstämma	SE	10.3
Linköping, Contemporary	SE	10.3
Linköping, Future	SE	10.3
Lund, Brunnshög	SE	7.10, 10.2, 10.3
Stockholm, Hjorthagen	SE	10.2, 10.3
Västerås, Bergsgrottan	SE	10.3
Västerås, Bjärby	SE	10.3
Västerås, Kaptenen	SE	10.3
Västerås, Kassel	SE	10.3
Västerås, Spinnakern	SE	10.3
Slough, Greenwatt Way	UK	10.3

present systems that apply combined heating and cooling. When the cooling demand is located outside heated buildings, the cooling demand is still related to the building space for heating to illustrate the proportion between the heating and cooling demand. The main argument for this inclusion of the external cooling demand is that the heat and cold demand should share capital distribution costs. The annual specific heat demand is, on average, 70 ± 37 kWh/m^2, including one standard deviation. The cases could be considered to exhibit good thermal performance because most cases have a specific heat demand below 100 kWh/m^2. The minimum and maximum values were 6 kWh/m^2 and 189 kWh/m^2, respectively. The lowest demand appears in the Geneva case, where the heat demand is only related to the heat subtracted from the return pipe in the district cooling system based on lake water but refers to the total building space. The highest heat demand appears in the Zürich FGZ case, giving a strong indication that the thermal performance for this case can be improved considerably in the future. However, the secondary heat distribution losses are included in this heat demand.

The five annual specific cold demand values vary between a minimum value of 21 kWh/m^2 and a maximum value of 432 kWh/m^2. The two lowest values appear in the Vienna and Visp cases. The highest value appears in the Zürich FGZ case, where the cooling demand appears in two data centres that provide heat recovery to be used by heated buildings.

6.7 Concentration of heat demand

This section concerns the variables considering the concentration of heat demand. These variables include the plot ratio, heat density, linear heat density, and effective width. The plot ratio is an independent variable in this context, whereas heat density, linear heat density, and effective width are the dependent variables. These are presented graphically for the studied group of cases in Figure 68, Figure 69, and Figure 70. In the heat distribution context of DHC, higher concentration yields higher cost-effectiveness in heat distribution as reported by (Persson & Werner, 2011) for some major European cities and by (Persson, Wiechers, Möller, & Werner, 2019) for the entire EU land area.

The plot ratio assesses the concentration of buildings on a given land area and determines the building density. Table 14 lists this dimensionless output parameter, which is determined by the total building space area over the total plot area. Hence, a value above 1 indicates a quite densely built environment. For the 37 analysed objects, the average plot ratio is 0.62 ± 0.60, including one standard deviation, whereas the difference between the minimum value of 0.1 for Visp and the maximum value of 3.27 for Vienna is relatively large. Hence, the densest area has a plot ratio that is 33 times higher than the sparsest area.

Heat and cold densities assess the concentration of heat and cold deliveries (output at substations) on a given land area. This output parameter assesses the to-

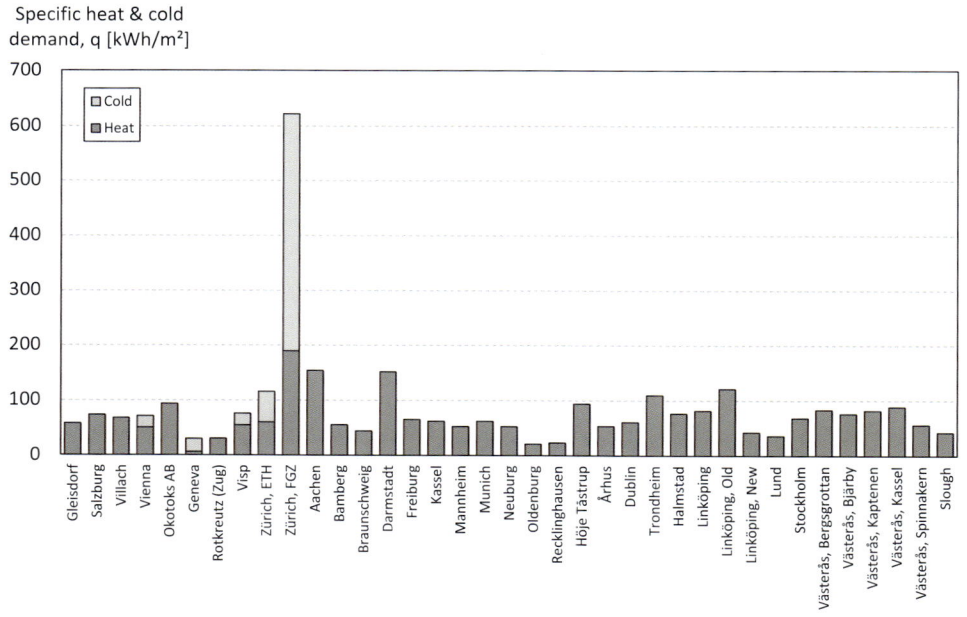

Figure 67. Specific heat and cold demand values for the 37 cases in the study.

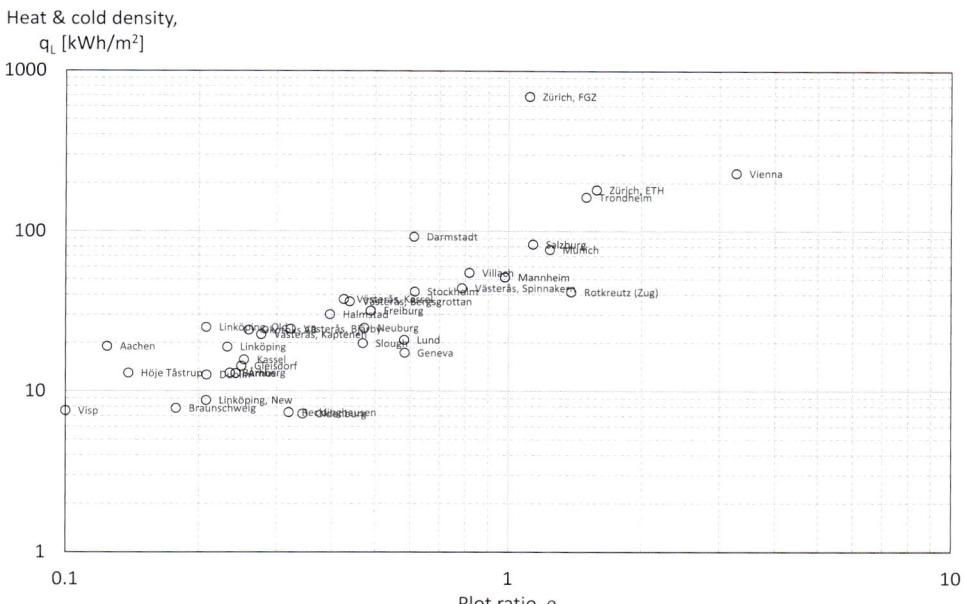

Figure 68. Combined heat and cold density as a function of the plot ratio.

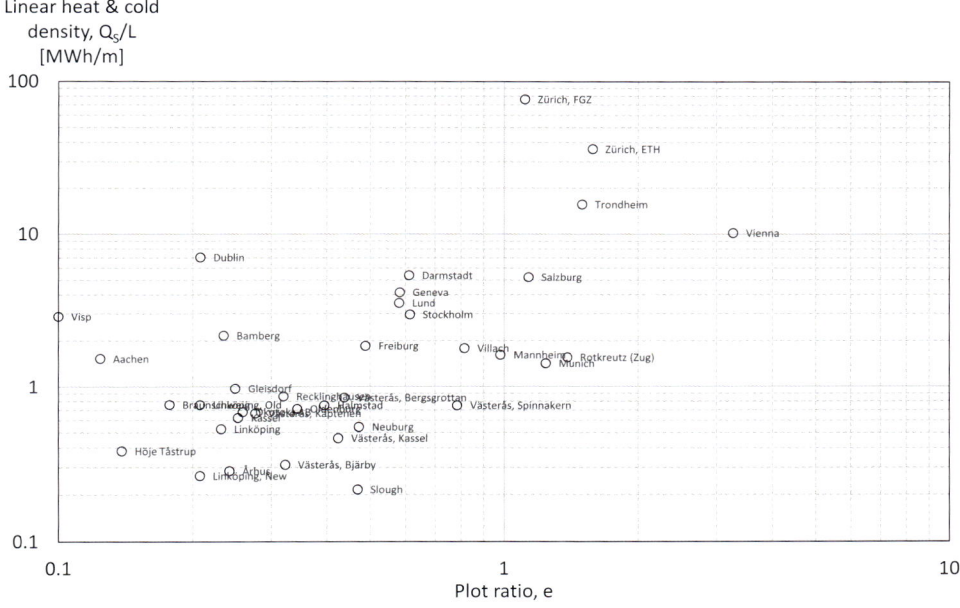

Figure 69. Combined linear heat and cold density as a function of the plot ratio.

tal annual heat and cold delivered in the total plot area. A clear positive linear relationship between the plot ratio and heat and cold density can be observed in Figure 68. One of the studied objects, Zürich, stands out with a very high cooling demand because its cold deliveries are associated with two major external data centres. Five heat and cold densities are lower than 10 kWh/m^2, which is a threshold typically considered too low for DHC areas.

The combined linear heat and cold density assesses the heat and cold delivery concentration (measured as input to the substations) for a given trench length. This is the most influential parameter for assessing the feasibility of the expansion of DHC networks (Frederiksen & Werner, 2013).

The average linear heat and cold density for the study group is 5.2 ± 13.5 MWh/m, including one standard deviation. For the combined linear heat and cold density, a large variation exists between the minimum value of 0.2 MWh/m for Slough and the maximum value of 77 MWh/m for Zürich FGZ. However, this high linear heat and cold density refers only to the cold network and do not include the secondary heat distribution network, according to Section 7.6. Hence, the concentration assessed by the total linear density is 350 times higher in the Zürich case than the Slough case.

The effective width is a measure assessing what trench pipe length is required for a specific land area. This output parameter denotes the relationship between the total plot area for connected buildings over the total trench length of the distribution network, according to Persson and Werner (2010). The DHC distribution is effective when the effective width is long. Short effective widths normally appear in sparse areas with detached single-family buildings.

For the studied group, the average effective width is 85 ± 109 m, including one standard deviation. The effective width also appears to be associated with a considerable variation between the minimum value of 11 m for Slough and the maximum value of 560 m for Dublin.

However, real effective widths beyond 100 m are rare in ordinary DHC systems. High width values are often associated with input errors from either wider plot areas or missing trench lengths. A typical error is that areas without any buildings, such as park areas, are included in the plot area inputs. Concerning trench lengths, a typical error is that only distribution pipes are included, omitting all service pipes.

6.8 Temperature levels

In addition to parameters related to concentration, parameters related to the applied temperature levels are also of interest. Two main parameters to be assessed are the annual degree time number and the average heat transfer coefficient for the heat distribution loss.

The degree time number for heat distribution assesses the temperature component in the DHC distribution. This output parameter denotes the time integral for the temperature difference between the annual average distribution temperature and the annual average ambient outdoor temperature. In this context, the related time is the annual number of hours; hence, the parameter unit is annotated °Ch (degree-hours).

For the studied case group, the average degree time number is 286 000 ± 160 000 °Ch, including one standard deviation. This low average reveals that many low-temperature networks appear in the analysed group. The minimum value of -90 000 °Ch appears in the Bamberg case, whereas the maximum value is 537 000 °Ch in the Darmstadt case. A negative degree time number is obtained when heat distribution temperatures are lower than the ambient temperature. Hence, the heat distribution loss becomes a heat gain from the ambient ground. According to the key findings in Chapter 5, it would be possible with low effort to lower the temperature level in the Darmstadt case to 360 000 °Ch. Current 3GDH systems have temperature levels between 450 000 and 600 000 °Ch, whereas the tempera-

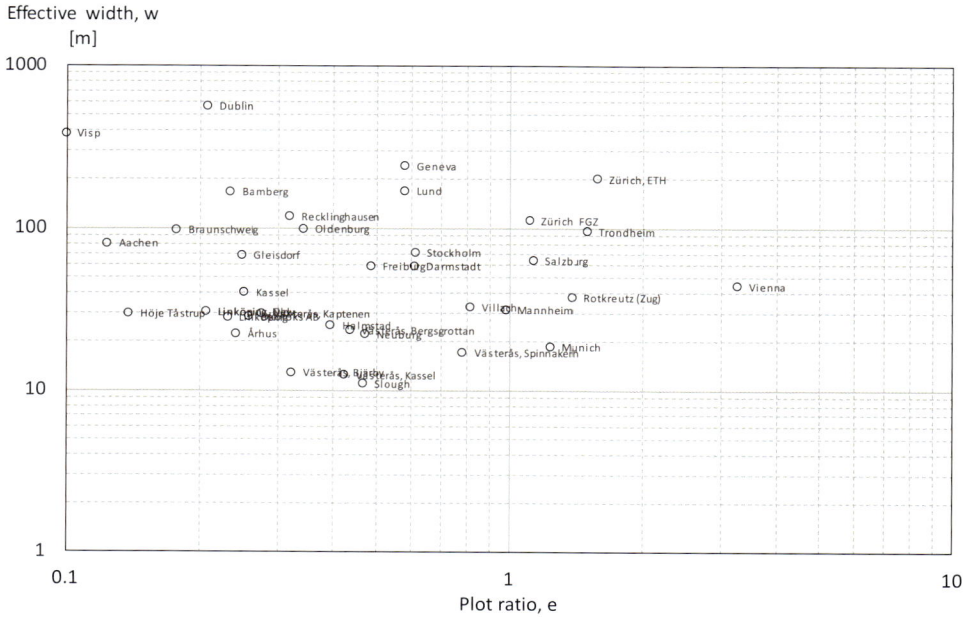

Figure 70. Effective width as a function of the plot ratio.

ture level interval for 2GDH networks can be estimated to 600 000–800 000 °Ch.

The annual average relative distribution heat loss assesses the magnitude of the heat distribution loss. This output parameter denotes the difference between the annual heat and cold supplied (input at heat and cold plants) and annual heat and cold delivered (output at substations) divided by the annual heat and cold supplied (input at heat and cold plants). The average value was 9% ± 8%, including one standard deviation, with the minimum value of -1% in the Bamberg case and the maximum value of 32% in the Västerås Kassel case. However, heat losses below 1% are not reproduced in Figure 71 and Figure 72 because a logarithmic scale is used for the relative heat distribution loss.

The variation of the relative heat distribution loss is presented in Figure 71 with the variation in the applied degree time number. Surprisingly, the correlation between these two parameters is quite low. The explanation is that a parameter other than the temperature level has a higher influence on the relative heat distribution loss. This fact is presented in Figure 72, where the variation of the relative heat distribution loss is presented with the variation in the combined linear heat and cold density. It is evident from this figure that low linear densities create higher relative heat distribution losses. Hence, it is more important to have high linear densities than low heat distribution temperatures to obtain low relative heat distribution losses. Thus, the linear density is the most decisive parameter for low distribution losses.

The heat transfer coefficient assesses the effectiveness of the pipe insulation. This output parameter denotes the heat transfer from the pipes with warm water through the pipe walls, insulation, and ground to the ambient temperature. The heat transfer coefficient is related to the outer pipe surface without insulation and the average temperature difference between the distribution pipes and the ambient temperature. The total pipe surface must be estimated, which can be achieved by collecting the trench lengths for each standard pipe dimension (described by DN, 'Diametre Nominal' numbers) in the studied distribution area to estimate the heat transfer coefficient.

Average heat transfer coefficients can then be estimated for each case by dividing the annual heat distribution loss by the total pipe surface area and annual degree time number. Corresponding heat transfer coefficients can also be estimated for commercial prefabricated district heating pipes from information about the manufacturer's estimated heat loss. Estimated average heat transfer coefficients for the studied cases and estimations for commercial district heating pipes from one manufacturer are presented in Figure 73.

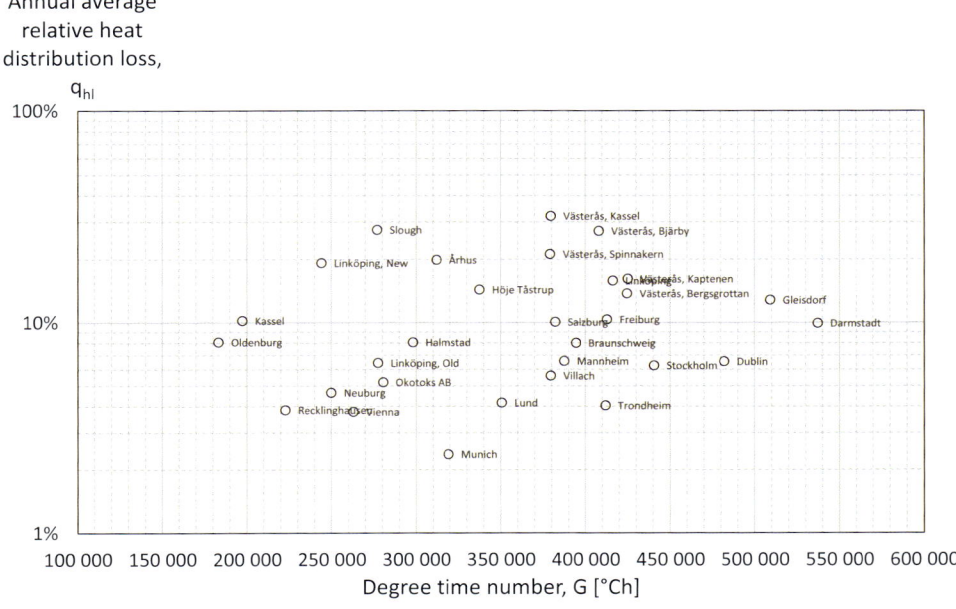

Figure 71. Relative annual heat distribution loss as a function of the temperature level expressed with the annual degree time number.

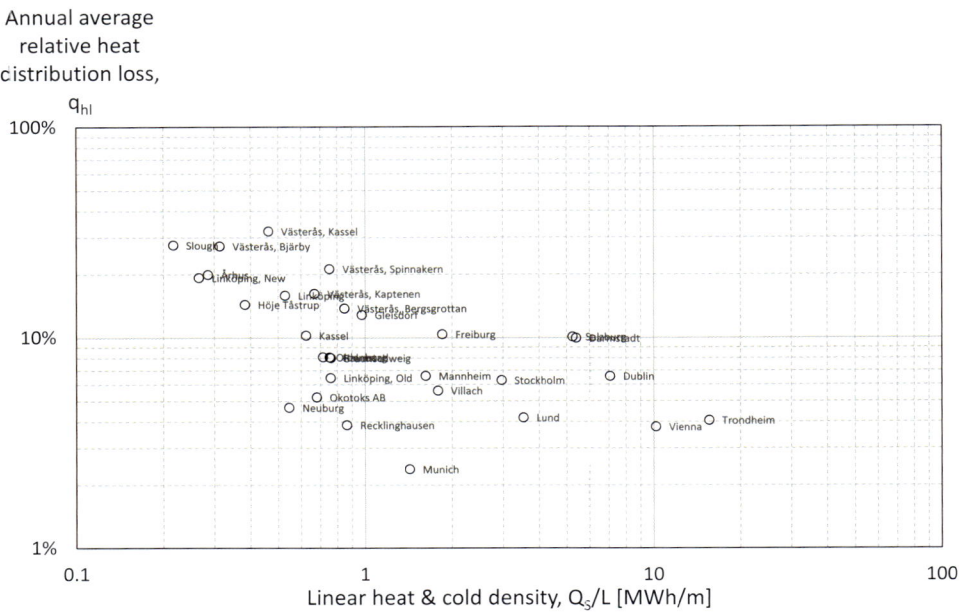

Figure 72. Relative annual heat distribution loss as a function of the linear heat density.

The average heat transfer coefficient for the studied cases was 0.98 ± 0.98 W/m²K, including one standard deviation. The minimum value (0.15 W/m²K) appeared in the Bamberg case, whereas the maximum value (4.9 W/m²K) was reached for the Salzburg case.

Values below the six manufacturer lines indicate underestimated losses, whereas the opposite is valid for values above the manufacturer lines. The figure also illustrates that the heat transfer coefficients are higher for pipes with small diameters. Both the case estimations and manufacturer estimations exhibit this pattern. Hence, it is more challenging to protect small-diameter pipes than wider pipes from heat loss.

6.9 Heat distribution costs

This final output parameter denotes the average distribution capital cost that dominates the annual heat distribution costs. The average distribution capital cost can be estimated using the annuity multiplied with average investment cost per metre divided by the linear density. The average capital investment cost per metre pipe depends linearly on the pipe diameter according to experience. Hence, this average investment cost can

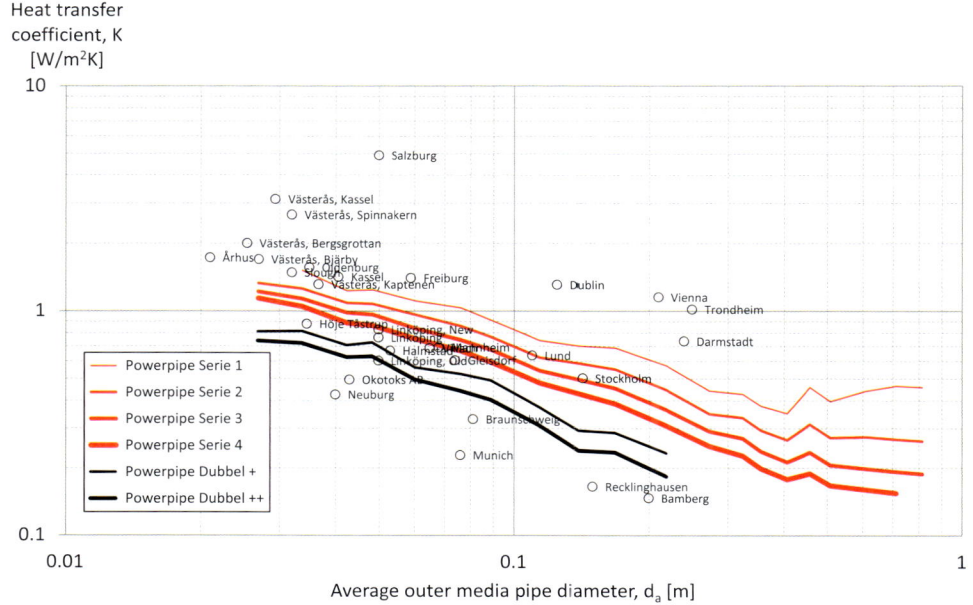

Figure 73. Annual average heat transfer coefficient as a function of the average pipe diameter.

be expressed by a linear function using one intercept C_1 and one slope C_2, as presented in Table 14.

The actual values of these two constants vary by street complexity, ground conditions, time, and country skills. Mature district heating countries have developed methods and experiences from long term learning that provide lower values for both constants. Two constants were recently estimated based on Swedish experiences, considering that the construction costs are more expensive in dense areas than sparse areas (Persson et al., 2019). The estimations were 212 euro/m for intercept C_1 and 4464 euro/m² for slope C_2. These two were also used in this analysis. It is a substantial simplification to apply these two estimations for cold networks, but it can be accepted for a first-order estimation. An annuity based on the real interest rate of 3% with a depreciation time of 30 years was applied in the previously mentioned study to obtain general values. The same annuity was chosen for this analysis.

In Figure 74, the estimated capital distribution costs are presented concerning linear densities. The capital distribution cost is strongly correlated to the combined linear heat and cold densities. For the 37 studied cases, the average marginal capital distribution cost was 28.3 ± 19.8 euro/MWh, including one standard deviation. This estimation is high compared to typical district heating systems, since the analysed group of networks has a high proportion of low heat density areas. The maximum value (83.4 euro/MWh) for the Slough case was 56 times higher than the minimum value (1.5 euro/MWh) for the Zürich FGZ case. Three conclusions are identified from Figure 74.

First, it is essential to install new networks with not so high capital distribution costs. According to Werner (2016), the average district heating price in Europe was 65 euro/MWh during 2013, which indicates a current (2020) probable price level of 70 to 80 euro/MWh. If capital distribution costs become too high in new systems, only limited space is left for other costs within the district heating system. It is questionable whether the networks in the upper left corner in Figure 74 can be competitive compared to individual heat pumps in the future.

Second, many of the low-temperature networks in Figure 74 have relatively low linear densities below 1 MWh/m because the connected buildings exhibit good thermal performance with low specific demands. This outcome is the new reality for European district heating providers when connecting new buildings. These efficient buildings can only be connected if they are in dense areas. It is questionable whether it would be possible to connect efficient detached single-family houses in the future.

Third, heat supplied from low-temperature sources can probably be acquired for lower cost compared to the supply cost in the current district heating systems according to Chapter 2. But this LTDH benefit can only

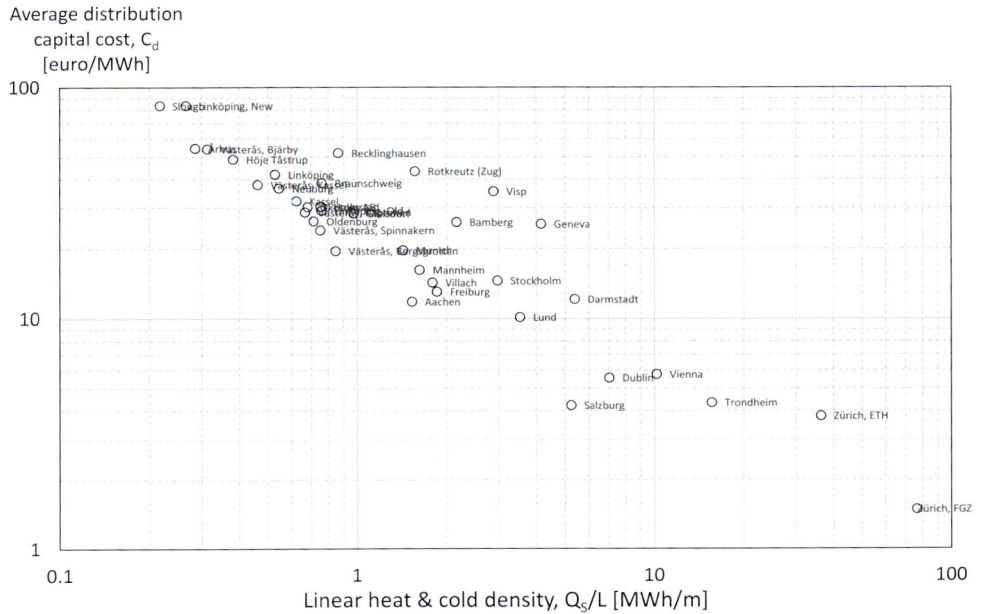

Figure 74. *Average heat distribution cost as a function of the combined linear heat and cold density.*

provide a limited cost space for areas with low linear densities having high specific capital costs.

6.10 Major conclusions concerning competitiveness

From an overall perspective, traits in LTDH business models can be complementary to the conventional district heating model. The selling point of a combination of the two in the context of the existing district heating system or as a stand-alone solution in greenfield investments is that local resources are used, minimising the carbon footprint. In an era of increased digitalisation, it is valuable to engage in dialogue and long term relationships, an upside for the LTDH prosumer relationship.

The market maturity of LTDH is low, and as such, an emphasis is placed on ensuring functional, technical solutions rather than a simultaneous development of the business case. For future installations, tandem development is recommended.

Retaining heat distribution costs in district heating systems at feasible levels is vital to maintain competitiveness. The most significant component of the heat distribution cost is the specific capital cost that is higher in low heat density areas. Second is the cost of the heat distribution loss. But only the latter cost can be considerably reduced by low-temperature heat distribution, since only the ability to use plastic pipes can reduce the capital cost for LTDH.

Furthermore, LTDH should be able to be supplied (input at heating plants) with heat from low-temperature heat sources, which are expected to yield a lower heat generation cost. In low heat density areas, lower heat generation costs and lower heat distribution losses obtained by LTDH cannot completely compensate higher specific capital costs. Hence, it is impossible to increase the total competitiveness of district heating with LTDH in low heat density areas.

6.11 Literature references in Chapter 6

Aronsson B., & Hellmer, S. (2011). Existing legislative support assessments for DHC: Ecoheat4EU.

Averfalk, H., Ingvarsson, P., Persson, U., Gong, M., & Werner, S. (2017). Large heat pumps in Swedish district heating systems. Renewable and Sustainable Energy Reviews, 79, 1275-1284. doi:https://doi.org/10.1016/j.rser.2017.05.135

Averfalk, H., & Werner, S. (2018). Novel low temperature heat distribution technology. Energy, 145, 526-539. doi:https://doi.org/10.1016/j.energy.2017.12.157

Cyert, R. M., & March, J. G. (1963). A behavioral theory of the firm (Vol. 2): Englewood Cliffs, NJ.

Dalla Rosa, A., Li, H., Svendsen, S., Werner, S., Persson, U., Ruehling, K., . . . Bevilacqua, C. (2014). Annex X Final report | Toward 4th Generation District Heating: Experience and Potential of Low-Temperature District Heating. IEA DHC|CHP. Retrieved from

Edwards W. (1954). The theory of decision making. Psychological bulletin, 51(4), 380.

Frederiksen, S., & Werner, S. (2013). District Heating and Cooling. Lund: Studentlitteratur AB.

Knight, F. H. (1921). Risk, uncertainty and profit (Vol. 31): Houghton Mifflin.

Leoni, P., Geyer, R., & Schmidt, R.-R. (2020). Developing innovative business models for reducing return temperatures in district heating systems: Approach and first results. Energy, 195, 116963.

Leonte, D. (2019). Market and stakeholder analysis, Delivarable 2.1 in EU project Reuseheat (H2020, 767429). Retrieved from https://www.reuseheat.eu/wp-content/uploads/2019/02/D1.4-Accessible-urban-waste-heat.pdf:

Lygnerud, K. (2019). Business Model Changes in District Heating: The Impact of the Technology Shift from the Third to the Fourth Generation. Energies, 12(9), 1778.

Lygnerud, K., Wheatcroft, E., & Wynn, H. (2019). Contracts, Business Models and Barriers to Investing in Low Temperature District Heating Projects. Applied Sciences, 9(15). doi:10.3390/app9153142

Myers, S. C. (1984). Capital structure puzzle: National Bureau of Economic Research Cambridge, Mass., USA.

Osterwalder, A., & Pigneur, Y. (2010). Business model generation: a handbook for visionaries, game changers, and challengers: John Wiley & Sons.

Persson, U., Averfalk, H., Nielsen, S., & Moreno, D. (2020). Accessible urban waste heat (Revised version): Deliverable report D1. 4.

Persson, U., & Werner, S. (2010). Effective Width : The Relative Demand for District Heating Pipe Lengths in City Areas. Paper presented at the 12th International Symposium on District Heating and Cooling; Tallinn, Estonia from September 5th to September 7th, 2010, Tallinn. http://urn.kb.se/resolve?urn=urn:nbn:se:hh:diva-6014

Persson, U., & Werner, S. (2011). Heat distribution and the future competitiveness of district heating. Energy, 568-576.

Persson, U., Wiechers, E., Möller, B., & Werner, S. (2019). Heat Roadmap Europe: Heat distribution costs. Energy, 176, 604-622. doi:https://doi.org/10.1016/j.energy.2019.03.189

Petersson, S., & Dahlberg Larsson, C. (2013). Samband mellan flödespremie och returtemperatur. Fjärrsyn report, 25.

Porter, M. E. (1990). The competitive advantage of nations. Competitive Intelligence Review, 1(1), 14-14.

Werner, S. (2016). European district heating price series: Energiforsk. Report 2016:316.

Wheatcroft, E., Wynn, H., Lygnerud, K., Bonvicini, G., & Leonte, D. (2020). The role of low temperature waste heat recovery in achieving 2050 goals: a policy positioning paper. Energies, 13(8), 2107.

7 PRACTICAL IMPLEMENTATION OF LOW-TEMPERATURE DISTRICT HEATING

Author: Dietrich Schmidt, Fraunhofer

7.1 Introduction

The introduction and application of new concepts and technologies such as LTDH often face concerns on being actually feasible and working in a reliable way. The previous chapters discussed measures to be taken on building and system levels and showed the various technical and economic benefits of LTDH. The aim of this chapter is to highlight, that many innovative LTDH systems are already implemented and operated successfully. Therefore, various demonstration projects and activities have been collected to

- to identify suitable case studies and present them
- to collect data for a closer evaluation and assessment of selected cases to compare the potential of LTDH with conventional solutions
- to calculate KPIs for the studied cases

By doing so, real evidence is given, that LTDH systems considered as the newest generation of district heating are a proven, reliable, and market-ready technology.

This collection of demonstration cases aggregates validated knowledge on the experiences and findings from different projects. To foster the practical implementation of LTDH, it is important that already existing know-how is transferred to the district heating market and all stakeholders, respectively. Moreover, these showcases of innovative and brave frontrunners are important to give practitioners better ideas and examples about the potentials and the paths to implement LTDH systems.

The demonstrators considered in this guidebook include various network configurations and different boundary conditions, such as already realised low temperature community energy system concepts as well as planned or designed systems. Furthermore, projects showing an innovative use or operation of buildings, advanced technologies, and the interaction between components within a system are included. The evaluation results of the collected examples of the newest generation of innovative district heating and advanced building technologies and their competitiveness as well as relevant factors for market implementation are included into the previous Chapters 3 to 6.

In total, six different classes of demonstrators have been identified:

 Realised demonstration project on existing or conversion areas with an existing heating network

 Realised demonstration project on existing or conversion areas with a new heating network

 Realised demonstration projects on new constructed areas with a new heating network

 Realised demonstration projects on the single building scale

 Simulation and design studies on areas

 Demonstrators on a laboratory scale

The demonstrators included in this guidebook are analysed regarding which elements of new knowledge they can generate. For each demonstrator a specific innovation is in focus. The innovation is validated from a technical point of view. Fifteen selected demonstration activities within Europe were analysed in detail, each reported within a single section in this chapter. The case examples described are from Austria, Denmark, Germany, Sweden, Switzerland, and The Netherlands. Further 25 cases are listed in Section 10.2, also including cases from three other countries. Additionally to that, a gross list with approx. 160 realized low temperature initiatives all over the world were collected (see Section 10.3). Figure 75 shows all collected demonstration cases within Europe. This map impressively proves the progress of the LTDH technology.

★ Cases analysed in detail and presented in Chapter 7
● Cases analysed in detail in this project, see Section 10.2
○ Cases described in the gross list of low-temperature initiatives, see Section 10.3

Figure 75. Location of the regarded demonstrator cases.
© Fraunhofer IEE, own representation. Map taken from Eurostat: https://ec.europa.eu/eurostat.

7.2 Successful implementation stories

In Table 16, fifteen cases have been selected for a presentation to display the variety for the implementation of LTDH systems based on different boundary conditions. For the description of cases, each of the following sections includes a general overview of the studied system, technical description of the technologies used and the main results and outcomes. In addition, many also include information of the related project and the involved stakeholders.

7.3 Transition of a small-scale system - City of Gleisdorf, Austria

Gleisdorf (10,000 inhabitants) is a small but continuously growing city with good infrastructure and high quality of life in the south east of Austria near Graz. Since decades the municipality of Gleisdorf fosters renewable energy and the cities policy guidelines in the early 1990s already included a strong commitment to an "environmentally friendly energy policy", to "demand and promote the use of all renewable forms of energy and energy saving measures" and to "cooperate with research institutes to increase knowledge, realize projects and support a trans-national know-how exchange". Since those days, various solar projects and other related activities such as energy efficient buildings and a GIS-based mapping of energy related data as an early approach for spatial energy planning were carried out together with a working group of solar pioneers which later became AEE – Institute for Sustainable Technologies.

As a logical consequence the local multi-utility company "Stadtwerke Gleisdorf" started in 2009 with the construction of a biomass and solar thermal based district heating system as a renewable alternative to the predominant natural gas supply until then. In the course of ongoing extensions, several larger consumers and existing microgrids, which were initially supplied by natural gas boilers, have been connected. Therewith related, further decentral heat production units with existing natural gas boilers for peak load coverage and distributed solar thermal systems were integrated in order to utilise the existing facilities best possible. This led to a rather complex energy system compared to its size consisting of 2 separate networks with in total 6 distributed heat production plants containing four biomass boilers, five gas boilers, six thermal solar plants, and various heat storage units at production and consumer sites. From the very beginning, extensive measurement equipment for all relevant plant components and most consumers were considered to enable a comprehensive monitoring. The general system configuration allows a highly flexible operation strategy for production units and the network, a high capacity utilisation of production units and investments respectively and a highly reliable heat supply. Nevertheless, it has some challenges due to the high system complexity (e.g. automatic and reliable control strategy without continuously manned control room).

However, the project impressively proves that renewable district heating is highly attractive and thus, is continuously growing (15% increase of connected load in last 3 years) even though the existing natural gas grid and still missing carbon pricing schemes in Austria cause high economic pressure.

As a consequence, the renewable production capacity of the system is already at its limit while another 30% increase of connected load is already upcoming. Although, the production units will be still capable to provide the required load capacity and energy demand, the share of natural gas will significantly increase from

Table 16. The fifteen selected and described demonstrators for the implementation of LTDH systems.

City and Area	Country	Demo class	Network config.+	Temp. level++	Heat Supply	System in operation	Network size+++
Gleisdorf Transition strategy	Austria		Multilevel	Warm (80/50 °C)	Biomass & gas boilers, solar collectors	Yes	Medium (6.4 km)
Wüstenrot Weihenbronn	Germany		Classic	Warm (70/40 °C)	Biomass boiler, solar collectors	Yes	Micro (0.37 km)
Darmstadt Lichtwiese*	Germany		Classic	Warm (88/58 °C)	District heating, district cooling with absorption chiller	Yes	Medium (4.2 km)
Heerlen Parkstad Limburg	The Netherlands		CHC	Cold (28/16 °C)	Mine water	Yes	Big (40 km)
Zürich FGZ	Switzerland		CHC	Cold (< 25 °C)	Data centre	Yes	Small (2.1 km)
Salzburg Lehen	Austria		Classic	Warm (65/40 °C)	Solar collectors, heat pump, district heating	Yes	Small (0.68 km)
Graz Reininghaus	Austria		Classic	Warm (69/43 °C)	Industrial excess heat, heat pump	Yes	Medium (6.0 km)
Braunschweig Rautheim	Germany		Classic	Warm (70/40 °C)	Data centre	Yes	Small (2.75 km)
Lund Brunnshög	Sweden		Classic	Warm (65/35 °C)	Excess heat from research facilities	Yes	Medium (6.5 km)
Kassel Feldlager	Germany		Ultra-low	Cold (40/30 °C)	ATES, GSHP and solar collectors	No	Medium (5.7 km)
Bamberg Lagarde	Germany		Multilevel	Cold & Warm (1–2/80–50 °C)	GSHP, sewage, CHP, district heating	No	Small (1.2 km)
Bjerringbro Tyttebærvej	Denmark		Ultra-low	Warm (50/30 °C)	Industrial excess heat, ATES, heat pump	Yes	n.a.
Viborg Multi-family building	Denmark		Classic	Warm (68/40 °C)	District heating	Yes	Big (345 km)
Frederiksberg Multi-family building	Denmark		Classic	Warm (80/48 °C)	District heating	Yes	Big (182 km)
Kassel District LAB	Germany		Multilevel	Cold & warm (5–130/ varying °C)	Heat pump, boiler	No	Micro (0.39 km)

*The project "Darmstadt Lichtwiese" is described in detail in Chapter 5.   +The system configurations are described in detail and classified in Section 10.1.
++Typical/mean supply and return temperatures   +++Trench length

an average of 15% to over 20%. Since the district heating system of Gleisdorf claims to be renewable and high fossil fuel shares prohibit public funding, "Stadtwerke Gleisdorf" supported by AEE INTEC developed an ambitious plan to further develop the system, to lower the system temperatures and integrate new low temperature heat sources. It is a perfect example of the transition to 4th generation of district heating in small scale and the therewith related challenges and requirements regarding optimisation of system temperatures, intelligent operation and control of the system, integration and management of multiple heat sources of different kind and size, management of distributed storages, system hydraulics and short and mid-term enlargement scenarios.

7.4 Housing estate Weihenbronn in Wüstenrot, Germany

The district of Weihenbronn in Wüstenrot consists of a ring road connecting a small housing estate, the city hall building and the fire brigade building complex.

The unusual situation of having a central administrative office outside the city centre can be explained by the Baden-Württemberg municipal reform at the beginning of the 1970s. At that time, the municipalities of Finsterrot, Maienfels, Neuhütten, Neulautern and Wüstenrot merged and then moved the seat of the town hall and fire brigade to the geographical centre of the new municipality.

At the beginning of the project, a small heating network was available for the town hall and the buildings of the fire brigade, the German Red Cross building, and the building yard. The oil-fired system was to be renewed. The project partner UBP-consulting GmbH & Co. KG took over the project planning of a network extension in the nearby residential area, combined with the construction of a new biomass heating plant (Figure 76).

The network is designed for the large municipal customers with a heat demand of 515 MWh per year and the residential buildings with a total heat capacity of 80 MWh per year. The conversion from oil-fired heating to the biomass-fired district heating network is associated with an annual avoidance of 190 t of carbon dioxide emissions. Furthermore, there is capacity for additional connections.

Figure 76. Heating system with storage tank (Source: UBP-consulting GmbH & Co. KG community Wüstenrot C. Wiederholl)

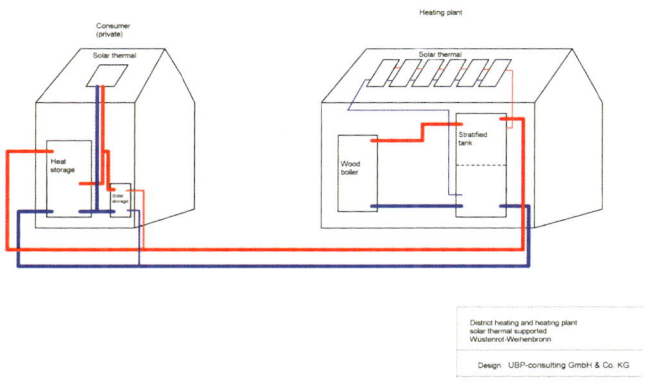

Figure 77. Integration of the solar thermal system into the grid (Source: UBP-consulting GmbH & Co.KG)

The new heating network was put into operation in autumn 2015. The heating system consists of two biomass boilers (120 kW each) installed in a container and a solar thermal system with a surface area of 10 m² installed on the roof (Figure 76 & Figure 77). For peak load and emergency supply a buffer storage tank is placed next to the boiler room.

7.5 Mijnwater project in Heerlen, The Netherlands

The municipality of Heerlen started this project in 2005 to investigate whether anything could be done with the groundwater in the flooded mine galleries. The mine water initiative in Heerlen is a geothermal project originating from the European Interreg IIIB NWE programme and the 6th Framework Programme project EC-REMINING-lowex.

The Mijnwater project was built as a 4th generation District Heating and Cooling network. In the winter, warm water of 28 °C was fed from the mine into the grid to deliver warmth, while in the summer cool water of 16 °C from a shallower cool source was distributed. This was a '4th Generation' grid in the sense that it used a low temperature heat source (or high temperature cooling), and also that it distributed the heat or the cold from a central point to the customers. This grid started by serving one large office building (national statistics bureau CBS) and a social housing project in Heerlen. Several utility buildings were connected, but the first installations slowly exhausted the geothermal source, limiting its scalability. In 2012, the upgraded and new system tried to counter the limitations by integrating the storage of heat from buildings for other moments in time or locations. Since then, the grid is able to exchange heat and cold between all customers, simultaneously, while the mine water system is used to store heat and cold. Mijnwater became a so-called cold network based on the combined heating and cooling (CHC) configuration with the following five ambitions:

1. Closing the energy loop : An optimized system allowing exchange of heat and cold between end users. To prevent waste, energy exchange occurs first at the scale of the building, then within the neighbourhood and finally at the city level.
2. Using low-graded sustainable sources: Energy sources can be classified according to their application opportunities. The high graded sources have the potential to be used twice or more, like first serving industrial processes and the waste flow for heating buildings. In CHC, the supply is matched with the requested quality level of the demand.
3. Decentralized & demand-driven energy supply: Circulating energy within the system only when and where needed, as close as possible to the end-user
4. An integral approach of energy flows at the city scale: Connecting heating and cooling to other energy flows (power grid, hydrogen conversion, solar plants, etc.) within the city, to avoid energy waste across sectors and reduce peak loads.
5. Local sources as a priority: Avoiding big investments and energy losses in transport, while stimulating the local economy.

In 2020, Mijnwater supplied sustainable heat and cold to more than 400 dwellings and 250,000 m² of commercial buildings. It gives a major contribution to the sustainability of the built environment in Heerlen and Parkstad-Limburg more. The project is positioning the city of Heerlen as an innovative green tech region in the field of thermal smart grids. The goal is to connect 30,000 homes and offices in Parkstad by 2030.

Mijnwater is using a cluster approach to transform the concept from a mine water pilot project into a modern, intelligent, and sustainable hybrid energy infrastructure. The geographically dispersed local cluster grids are supported by a mix of local sustainable resources, which are supplied by the mine water grid (backbone) and mine water reservoir. Energy exchange will be realized between buildings by means of local cluster grids and between the cluster grids through the existing mine water grid.

Buildings are no longer just an energy consumer but also an energy supplier. A building that extracts hot water (e.g. 27 °C) for heating from the hot pipe of the cluster grid returns cold water back to the cold pipe of the cluster grid (< 15 °C). Other buildings connected to the grid for cooling can instantly use this cold water. Heating and cooling of the buildings can occur passively and/or actively by using heat pumps. This depends on the available temperatures in the cluster grid (cold 8–20 °C; heat 27–50 °C) and the requested release temperatures of the building (cold 5–18 °C; heat 30–50 °C). At the end user location, heat pumps adjust the supply temperature to the necessary temperatures for heating or cooling. A special booster heat pump produces hot tap water at delivery.

Solar collectors and other heat generators can deliver additional heating and cooling, like a bio-CHP, which can raise the supply temperature for heating up to 50–55 °C. To achieve high exergy efficiencies by maximizing passive (re-) use of heat and cold and by raising the heat pump efficiencies up to a COP of 7(+), a boiler house is designed.

The production wells (HH1 and HLN1) supply the shortage of heat and cold to the mine water backbone. The surplus of heat and cold will be stored in the mine water reservoir through the injection wells (HH2 and HLN2). The current return/injection well (HLN3) will be out of order and only be used in case of exceptional situations.

The capacity of the mine water system is finite, if the heat-, cold extraction and infiltration is not balanced on a yearly basis. To ensure that the extracted hot (27–50 °C) or cold water (8–20 °C) from the cluster grid is cooled down (< 15 °C) or heated up (> 29 °C) sufficiently before injection in the mines, a temperature condition is included in the contract with the end users.

For realizing the objectives of the Sustainable Structure Plan of Heerlen a combination of mine water with other renewable energy sources such as biomass and/or solar energy and waste heat is necessary. All these energy sources are locally situated and will be connected to the nearest cluster grid to supply their heat and cold to the corresponding cluster and through the mine water backbone to other clusters.

The existing mine water return pipe will be used for additional supply and disposal of hot or cold mine water. At the cluster grid a cluster installation with booster pumps for energy exchange between the mine water

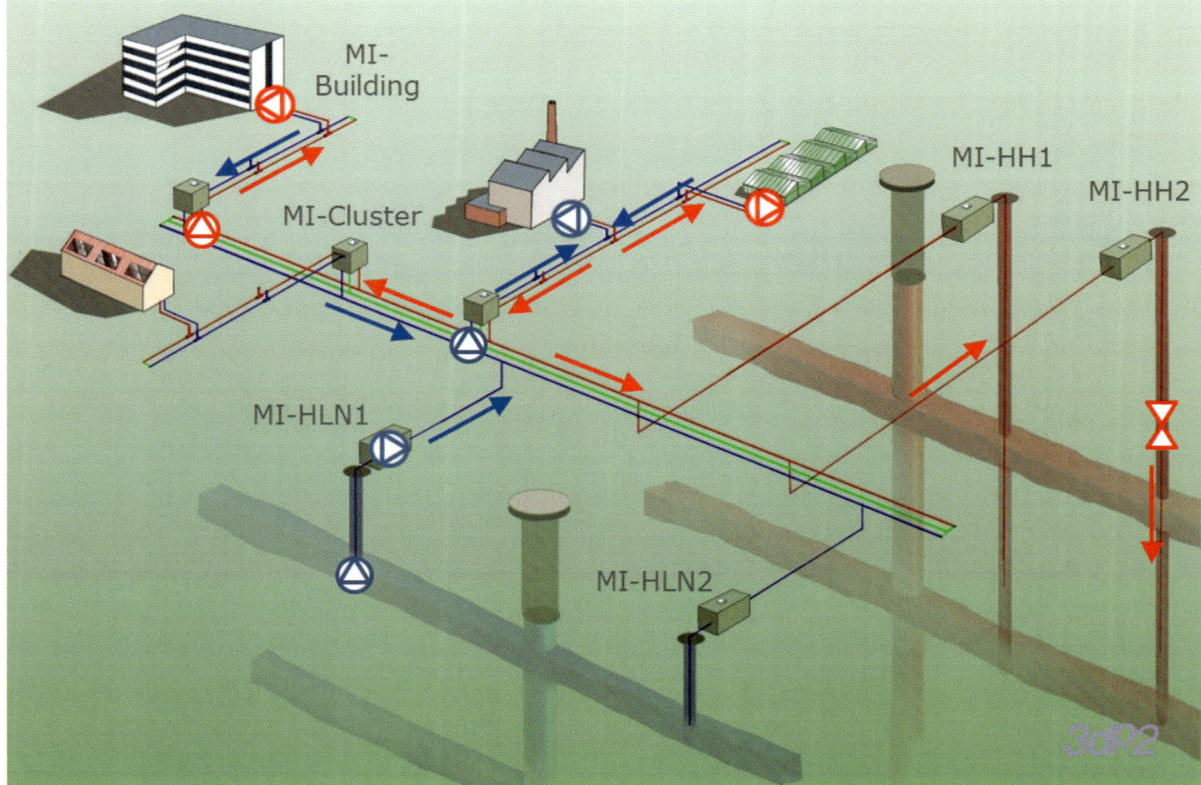

Figure 78. Scheme of Mijnwater combined heating and cooling configuration cluster concept

and cluster grid are installed. Sophisticated injections valves are applied at the hot and cold injection wells and in the near future all wells become bidirectional for further capacity enlargement, back-up and smart production and injection of mine water.

The Minewater system is fully automatic and demand driven with three levels of control. All buildings (first level) are connected to a cluster network (second level). Several clusters are connected to the mine water backbone and reservoir (third level). At each level (building, cluster, mine water) there is a net heat or cold demand. The buildings determine the demand of the cluster. The cluster provides what the buildings demand. The clusters determine the demand of the mine water backbone. The mine water backbone and mine water wells provide what the clusters demand.

Exchange at the interface between the levels takes place with autonomous substations (MI = Minewater Installation). Each level works with another independent process control parameter. To show how it works a typical process situation is shown in the artist impression of Figure 78.

7.6 Cold district heating at FGZ Zürich, Switzerland

'Familienheim-Genossenschaft Zürich' (FGZ) is a non-profit housing cooperative owning and operating more than 2,200 mainly residential buildings (185,000 m² heated floor area) for about 5,300 inhabitants forming a city quarter on its own within Zürich with a heat demand of 35 GWh/a for space heating and domestic hot water.

FGZ decided on a master plan regarding the further development and stepwise refurbishment of the FGZ district and its buildings to reduce the heat demand to 13 GWh/a in 2050. Therewith complementary, a new heat supply concept was developed by experts from the Zürich based company 'anex Ingenieure AG' for energetic crosslinking of low temperature heat sources, borehole thermal energy storages (BTES) and central heating plants with heat pumps using a cold district heating system (CDH) with a temperature level < 28 °C. This replaces the priory used oil and gas boilers.

The construction started in 2011 and the CDH was commissioned in 2014 and has been successfully in operation since then. An enlargement project to connect

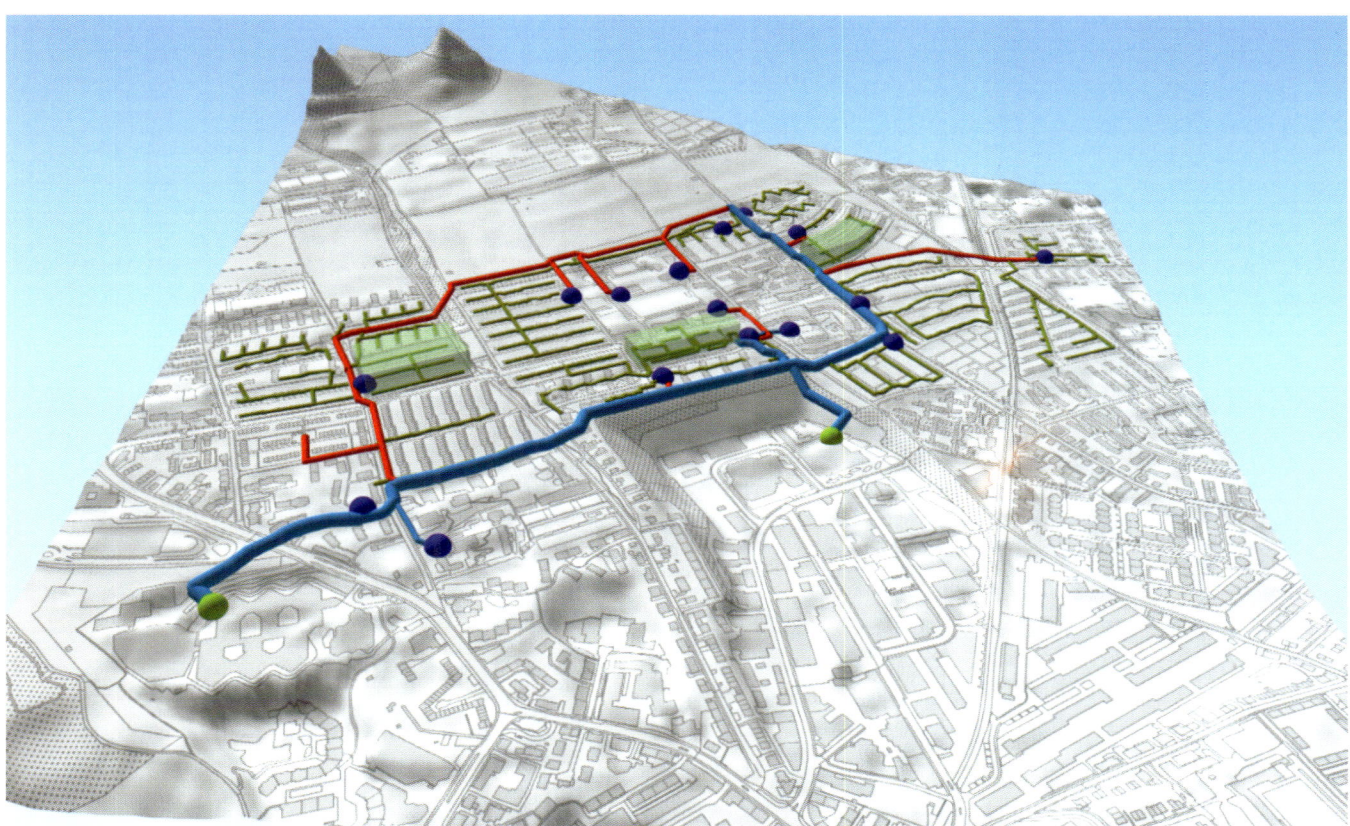

Figure 79. Layout of the FGZ cold district heating system including heat sources, borehole storages, heat pump locations and secondary distribution network sources: AEE INTEC, anex Ingenieure AG, opendata.swiss

further central heating plants, borehole storages and a second waste heat supplier was finished in 2019. In future, the whole city quarter will be supplied by the CDH.

The CDH grid consists of a flow- and a return line made of uninsulated pipes and uses water as heat transfer medium. Currently it has a linear topology (Figure 79, blue line) but will be a ring with linear connections to heat sources, BTES and central heating plants after the final construction stage (Figure 79, red line). There is no central pumping and predefined flow direction respectively as in conventional district heating systems. Individual pumps in each central heating plant and at feed-in points generate the individually required mass flow. The BTES within the system are used to compensate short term as well as seasonal mismatches of thermal supply.

Each central heating plant connects the ULTDH with a district heating sub-grid using high efficient large scale heat pumps to provide the required temperature level of each sub-grid while using the CDH as heat source (approx. 80% of the heat is provided with 68 °C, 20% with 40 °C). The annual COP ranges between 4 and 5 (including pumping).

The temperature level in the CDH grid ranges between 4 and 28 °C and is mainly depending on the charging status of the BTES and the time of the year respectively. The heat is provided by the cooling load of two large data centres. The CDH allows free-cooling of these data centres and thus reducing their cooling efforts.

General plant configuration:
- Ultra-low temperature district heating grid (primary grid)
- Two low temperature heat sources (data centres)
  - total heat capacity of approx. 4.5 MW
- 2 BTES with 332 boreholes, each with 250 m depth (450 boreholes planned for 2050)
- 6 central heating plants (10 in 2050) connecting the CDH with WDH sub-grids for heat distribution to buildings (existing secondary district heating grids)
  - Heat pumps with total capacity 4.8 MW (8.8 MW in 2050)
  - Total heat capacity incl. peak load coverage 7.2 MW (12 MW in 2050)

7.7 Stadtwerk Lehen in Salzburg, Austria

New low-energy apartments, a kindergarten and a student dormitory were built on an area of approx. 43,000 m² in a new city Quarter in Salzburg/Lehen. It is named 'Stadtwerk Lehen', is an energy-saving construction and is the core project of Salzburg's contribution to the EU project 'Green Solar Cities' in the Concerto programme. It entails heat supply for low-temperature network with solar system and storage resulting in a win-win.

The heat is stored from a total of 2,047 m² (gross collector area) of thermal collectors on the roofs in the central 200 m³ storage. Here, a heat pump uses the temperature difference in the buffer storage and additionally cools the lower volume of the storage. The heat pump increases the yield of the solar system by approximately 15%. The heat is distributed to the users from the storage. The annual average supply temperature is 65°C and the return temperature ranges from 35 to 45°C. An additional 20.2 kWp PV system on the roof supplies electricity for general systems (ventilation, heat pump, underground car park lighting), a further 30 kWp were installed by the utility Salzburg AG in the immediate vicinity.

In each apartment, there is a substation where the domestic hot water is produced locally as required (hygienically the best solution) and the heat requirement for heating and domestic hot water is counted. A LED visualisation on the storage informs all passers-by how much thermal energy and how much electricity is currently produced by the sun and how much of the demand has been supplied in the last 24 hours (Figure 80). Due to the low-energy construction of the buildings, the total energy requirement was approx. 1.5 GWh/a, approx. 30% lower compared to the usual construction standard. Furthermore, 68% of the primary energy and 76% carbon dioxide can be saved compared to a conventional oil-heated residential complex.

After commissioning in November 2011 an energy monitoring started with special emphasis on informing residents. Stadtwerk Lehen was selected as a test area for smart meters and interested residents can obtain information on energy consumption online on an ongoing basis. This should have a positive influence on user behaviour and the satisfaction of the residents. Monitoring supports the operator (Salzburg AG) in system optimisation and makes it transparent for the user that a

Figure 80. An LED visualisation on the buffer tank informs all passers-by how much thermal energy and how much electricity is currently produced by the sun and how much of the demand has been supplied in the last 24 hours (© SIR Strassl).

high proportion of renewable energy is used in this project. The monitoring shows, that especially due to the low network temperatures and the heat pump the solar yield is higher than calculated. But the energy demand of the buildings is much higher than calculated as well, due to higher room temperatures than designed for.

7.8 Smart City Reininghaus in Graz, Austria

The „roadmap" for heat supply strategy for the city of Graz from renewable sources envisages that, based on a share of around 25% in 2017, a share of 50% can realistically be achieved within the next 10 years. Before 2050 it should be possible to generate all the district heating with renewable resources - provided the appropriate framework conditions are in place.

Triggered by the shutdown of the heat supply from the coal-fired Mellach CHP plants by 2020, innovative solutions for a transition of the district heating system were developed and some cases already implemented. In pilot projects, several decentralised heat generators and low-temperature systems are used to support the integration of alternative energies (e.g. heat pumps, solar thermal, waste heat, and biomass).

The energy model Graz Reininghaus is a good example of a modern concept for heat supply in new 'Smart-City' construction areas. It is an urban development area in the west of Graz with an area of approx. 54 ha and up to 12,000 inhabitants in final development. The former area of the Reininghaus brewery is the largest development area in the city of Graz, offers the opportunity to develop an urban, dense and energy-optimized district (1.8 km from the historic old town). In the course of the development of the Reininghaus district, an innovative energy model is developed in cooperation between Energie Graz and Marienhütte. The energy utility Energie Graz initiated and developed the energy model Graz Reininghaus. Furthermore, they are the investor and operator of the plants. The cornerstones are:

- low-emission heat supply
- low-temperature heat network (appr. 69 °C) with the use of industrial waste heat sources from the steelworks Marienhütte
- two highly efficient large-scale heat pumps (each with up to 5.75 MW thermal) use the industrial waste heat and feed the district heating subnetwork Reininghaus at 69 °C and to the district heating network Graz at up to 95 °C
- annual amount of heat generation: approx. 40–45 GWh/a (equals to 4% of the district heating demand of the whole city of Graz)
- modular heat storages, volume up to 1,600 m^3 as daily and weekly storage (so called "Power Tower")
- photovoltaic system with (85 kWp) built on the façade of the "Power Tower"

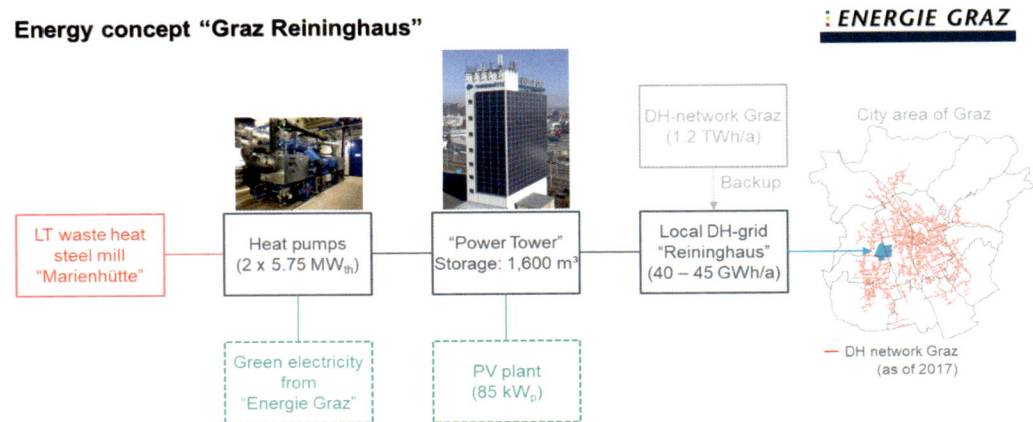

Figure 81. Supply of the new urban district "Graz Reininghaus" with an innovative heat supply concept. Adapted from (Unger, 2018).

7.9 Excess heat recovery from data centre in Braunschweig, Germany

Veolia's subsidiary BS|ENERGY is a local energy (electricity, heat, and gas) provider in Braunschweig, Germany. It operates an existing district heating network that spans the whole city. It is powered by high-efficiency coal and gas cogeneration plants.

A local property developer requested district heating during the early planning phase of a new residential area. With the simultaneous construction of a new data centre in the adjacent parcel, Veolia identified this as an opportunity to develop an innovative network that would utilise the waste heat from the cooling system of the data centre. Extracting heat from the data centre reduces the need to cool the data centre and associated energy consumption accordingly.

To supply a newly built energy efficient housing area consisting of 600 housing units with a net floor of 48,000 m^2, the low temperature network will be used. The peak heat demand is estimated around 1.8 MW, and the base load will be covered by the waste heat recovery. The low temperature of the heat source, 25 °C, requires a heat pump to increase the source temperature to a satisfactory level for the network. At the same time, keeping the temperature level of the DHN supply as low as possible is desired for high efficiency. A connection to the existing high temperature DHN of BS|ENERGY will also be provided, enabling flexibility and demand peak shaving.

Data centre operator transfers its waste heat to a cooling water cycle. The cooling water with a temperature of 25 °C is forwarded outside of the data centre to the energy station. The heat pump cools the cooling wa-

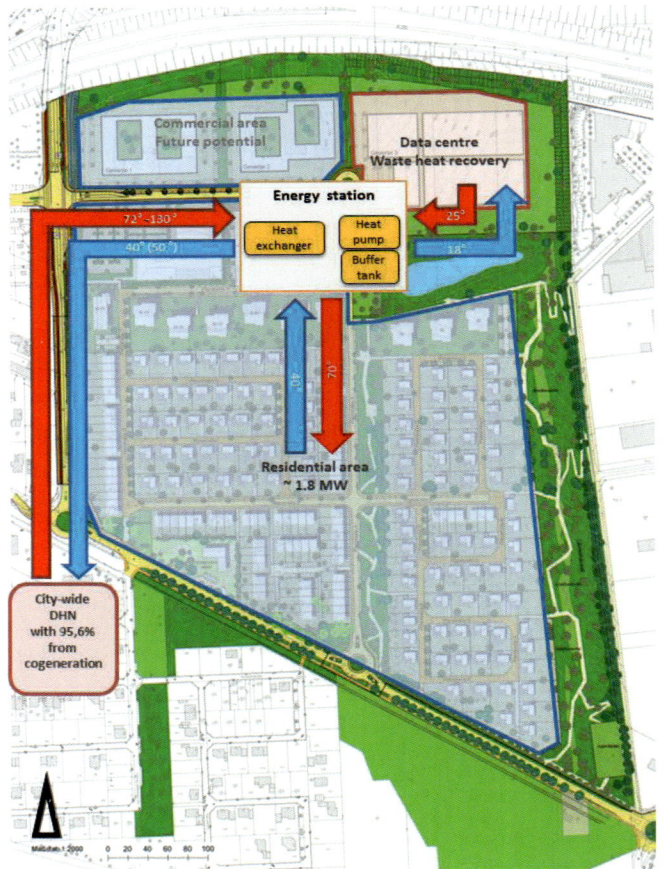

Figure 82. Excess heat recovery from data centre in Braunschweig/Germany.

ter to 18 °C and uses the removed heat to raise the return temperature in the network to the required supply temperature of 70 °C. The cooling water is now able to take up the excess heat again. Connection to the existing high temperature network enables demand peak shaving. The project received the 6th Global District Energy Climate Award in 2019 for the vision of a new, highly energy-efficient district energy system (Euroheat, 2019).

Concerning pricing scheme, BS|ENERGY does not apply any diversification in terms of tariff on their customer base. No differentiation was applied to customers supplied from the traditional high-temperature district heating network. Pricing scheme is available at the BS|ENERGY homepage. In addition, a fibre optic internet access infrastructure is part of the package for all clients.

7.10 Excess heat from research facilities in Brunnshög in Lund, Sweden

This planned low-temperature heat distribution network is in Lund, Sweden. The city is dominated by Lund University and will expand with a new major residential area called Brunnshög, located northeast of Lund. This description will present the expansion until 2035. This initial phase will consist of 110 hectare land area and a building space of 0.64 million m².

Within this Brunnshög area, two major research facilities have been located. The first one is the MAX IV Laboratory that is a Swedish national laboratory providing scientists with X-rays for research. The other research facility is the European Spallation Source (ESS) that is a multi-disciplinary research facility with 13 European nations as members.

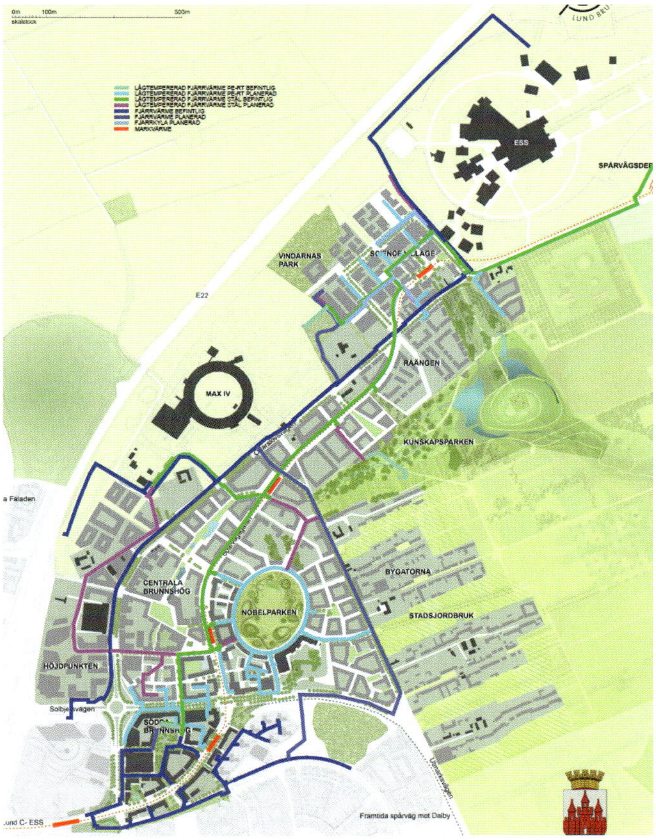

Both research facilities will generate large quantities of excess heat. At full activity, the annual heat volumes can be estimated to about 200-250 GWh per year at various low temperatures. The aim with the low-temperature heat distribution network is to recycle this excess heat and use it for the new Brunnshög residential district. However, the expected heat demand from these buildings will only be 23 GWh in 2035. Hence, the future challenge will be to use further excess heat in the traditional Lund district heating system.

The Brunnshög network is a part of the 2013 Brunnshög agreement between the Municipality of Lund and the local infrastructure providers. Kraftringen is the municipal local energy provider in the Lund area that also own and operate this low-temperature heat distribution network. The first heat was delivered from this low-temperature network in September 2019. The heat demand for all new buildings in the Brunnshög will be at the NZEB level. The expected specific heat demand for planning the heat distribution network is 36 kWh/m².

The distribution network is designed for using a supply temperature of 65 °C and a return temperature of 35 °C. It will be a classic network configuration with two parallel pipes and one substation in each building. The network layout is presented Figure 83. The chosen supply temperature is the minimum supply temperature that can fulfil the Swedish legislation concerning the *Legionella* risk in hot water circulation system in the connected buildings.

The total trench length in the Brunnshög area will be about 6.5 km until 2035, and plastic pipes will be used for 2.8 km. These plastic pipes will be one innovation developed together with the Logstor pipe manufacturer, since these pipes will have a design pressure of 10 bar.

The first major lesson learnt is that it is not possible to use lower heat distribution temperatures than 65/35 °C when the classic network configuration is used with one substation for each multi-family building. Further lessons learnt will be communicated during 2021 through the COOL DH project that Kraftringen participates in order to support this pilot project.

Figure 83. The planned Brunnshög low-temperature heat distribution network. © Kraftringen Energi AB.

7.11 Geosolar District Heating in "Feldlager" in Kassel, Germany

The planning area „Zum Feldlager" in Kassel is surrounded by existing buildings of the district and is located in an urban ventilation path. For that reason, combustion of oil or wood (fine dust emissions) should be avoided. Due to the location of the area a connection to the existing district heating network of Kassel is not feasible because of logistical and economic reasons. Instead, a local district scheme is implemented. The concept involves principally the use of renewable energy sources (RES) such as geothermal and solar energy for LTDH supply. The new housing estate will be characterized by a very compact and south oriented construction; 1-2 storey detached and semi-detached houses in the north, two-storey terraced houses in the centre and large three-storey apartment buildings in the south. All buildings have a specific heat demand of 45 kWh/m²·a and a specific domestic hot water (DHW) demand of 730 kWh/person·a. Thus, the demand is below the maximum energy demand for new buildings (<50 kWh/m²a) according to the valid German energy saving ordinance EnEV 2014.

Unlike typical new (smaller) district heating systems in Germany the district heating in the geosolar concept shall have only 40 °C supply temperature. This temperature level is, however, enough to provide the space heating of buildings via surface heating systems. The LTDH network is fed by the central heat pump and the electric boiler during the heating period of 7 months. The heat pump uses the ground via borehole heat exchangers (92 boreholes, 120 m depth) as a heat source. The geosolar heat supply concept combines central and decentralized heat sources (see Figure 84 below).

Domestic hot water preparation requires higher temperatures of minimum 45 °C for single-family houses and 60 °C for multi-family houses. It is mainly provided by distributed solar thermal systems in every building with an electrical back-up. Due to the rather high electricity prices in Germany, it is advantageous to preheat the hot water storage during winter using the district heating supply.

Simulation studies showed the collector area and preheating by district heating in winter will reduce the auxiliary energy demand to 17% of the total energy demand for domestic hot water. Decentralized solar domestic hot water systems enable low supply temperature and seasonal operation of the district heating. Both factors lead to very low heat transport losses of 2.5 % of annual heat generation. Such low heat losses are crucial for the economic feasibility of the district heating in areas with very low linear heat density (e.g. 500–1000 kWh/a·m).

The results show that the geosolar heat supply concept leads to 61% lower primary energy demand and to 64% lower CO_2 emissions comparing to the reference system (gas condensing boiler with solar DHW).

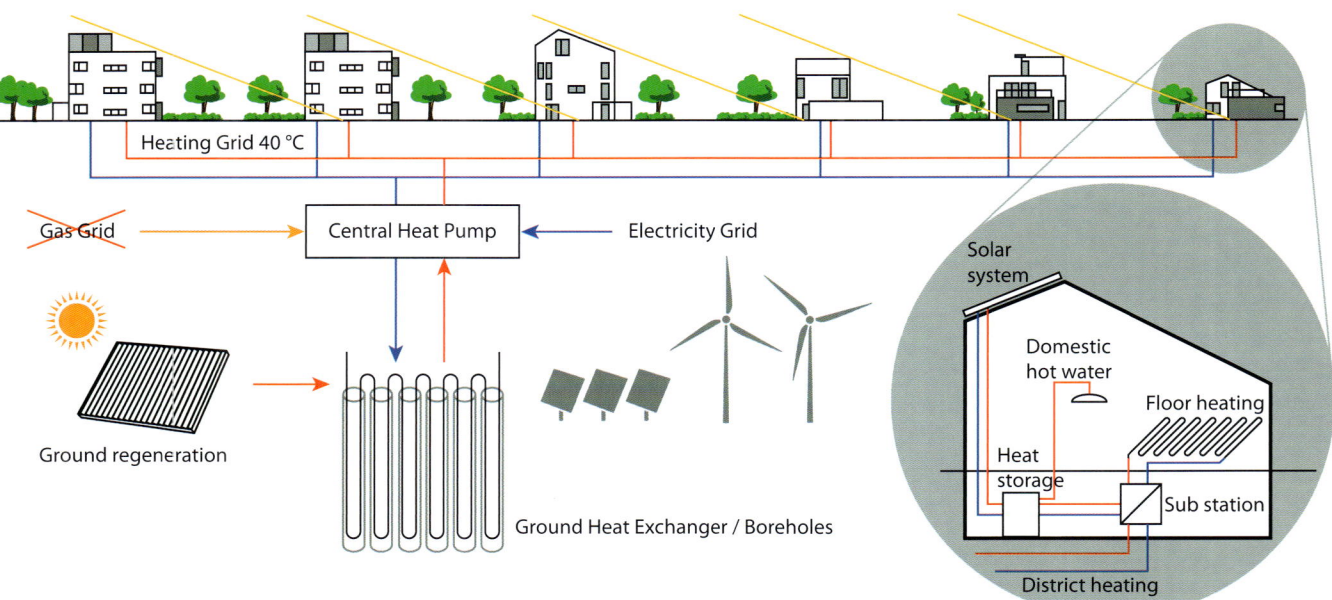

Figure 84. General description of hydraulic of centralized geothermal powered heat pump for district heating supply (Schmidt and Kallert 2017)

Furthermore, the total annual costs of both systems (include capital expenditures, operation, and maintenance costs) were calculated using current prices and interest rates. The total costs for the geosolar heat supply concept were estimated to be about the same or even lower than the reference system (Schmidt et al., 2017 / Schmidt and Kallert, 2017).

Finally, it was decided by the city council and the local utility company not to realise the suggested project concept. The main reason was that it would take too long time to get a drilling permission for the ground heat exchanger. Instead, a standard district heating system (supply temperature of about 80–90 °C) operated with a gas-fired CHP unit will be realised and no innovative and future concept will be built. The chance to realise a good and carbon emission saving concept has not been taken, which is a tragedy for the implementation of the necessary energy transition.

7.12 Conversion area 'Lagarde' in Bamberg, Germany

Since the withdrawal of the US army in 2014, the city of Bamberg has planned to develop and integrate the former military area (Lagarde Campus), which consists of existing and new building stocks, into the existing city structure. Above all, the urban and architectural identity should be preserved during the planning process, at the same time the energy system should be reinterpreted in a contemporary way. Thereby, the energy supply of the Lagarde Campus should be primarily based on renewable energies, which generate heat and power in the quarter itself, in order to contribute to the local climate goal, within the German national framework of the funding programme 'Wärmenetzsysteme 4.0' (Kaiser and Loskarn, 2019 / Kaiser and Loskarn, 2019a).

The purpose of the feasibility study was to develop a cold district heating system (CDH), which supplies low temperature heat from near-surface geothermal collectors to decentralized heat pumps, located in the connected buildings, based on a thermal-hydraulic analysis with the simulation environment TRNSYS.

The main heat supply of the cold district heating network is supplied by a combination of near-surface geothermal collectors and different heat sources, such as sewage (about 12 °C) and drinking water networks (8–14 °C). Due to the limited area of undisturbed ground, geothermal collectors are buried horizontally under buildings, 3.6 meters deep in the ground. Regarding the heat supply with near-surface geothermal collectors, the distribution networks are designed with a supply temperature of 1 °C and a return temperature of -2 °C in the winter periods. Because of the low temperatures, the system is intended to be operated with a brine. For the additional heat gain through the distribution networks, plastic pipes without thermal insulation are used for the CDH. Besides the CDH, PV-modules are implemented to cover the local electricity demand of household devices and partly that of the decentralized heat pumps.

The results of the simulation study show that the network temperature is significantly affected by the ground temperature during the operation of the CDH. Thereby, the operation of cold distribution networks with a supply temperature of around 1 °C in the winter periods and ca. 10 °C during the summer months allows an annual heat gain of ca. 19 MWh/a through distributi-

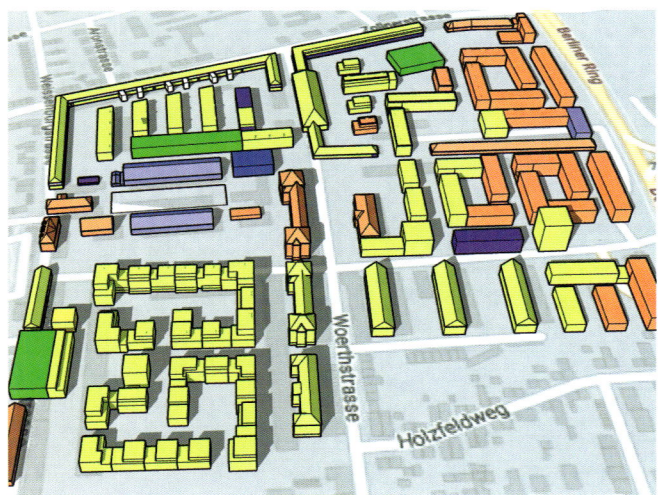

Figure 85. The Lagarde Campus in Bamberg / Germany © Fraunhofer IEE

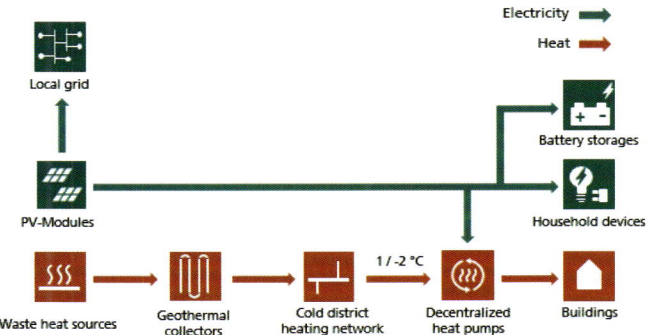

Figure 86. Scheme of the energy system concept in Lagarde Campus Bamberg / Germany © Fraunhofer IEE

on pipes, which corresponds to around 1% of the total heat demand (ca. 1,820 MWh/a).

However, the simulation results of near-surface geothermal collectors indicated that the limited amount of heat sources from sewage and drinking water networks could lead to the icing over of ground collectors in freezing weather. For this reason, further integration of heat sources such as ice storages, PVT-Modules etc. are investigated in the second phase of the study.

7.13 Ultra-low-temperature district heating in Bjerringbro, Denmark

In the town of Bjerringbro in Denmark, surplus heat from an industrial plant in the city centre is utilised in the local district heating network. The surplus heat is available from a ground water cooling system, from where heat is extracted and boosted in a heat pump before it is fed into the system. To obtain a high efficiency of the heat pump, a pilot study was carried out, to test if it would be possible to utilize supply temperatures as low as 47 °C in the district heating system. The test was carried out in an area with 21 single-family houses constructed around 1980. A picture of the houses is seen in Figure 87.

The district heating network in Bjerringbro is normally operated at annual average supply and return temperatures of around 70 °C/40 °C. When the supply temperatures were lowered, the single-family houses were therefore equipped with one of five different domestic hot water substations including electric boosting: a small booster heat pump, a domestic hot water tank with immersion heater, a district heating storage tank with an immersion heater, or a combination of a heat exchanger and a local direct electric booster at the tap. Based on measurements and theoretical evaluation of the local units, it was found that the most feasible substation layout for this type of network was the installations of a heat exchanger combined with a direct electric booster at the tap (Yang et al., 2016). The main reason for this was that the installations had much lower heat losses and needed a much smaller supply of electricity to ensure comfort and safety requirements for domestic hot water. Furthermore, the installation is both cheaper and less space consuming than the alternatives.

Apart from the installation of domestic hot water boosters, no changes were made to the existing single-family houses or space heating installations. None of the

Figure 87. Single-family houses in Bjerringbro that were supplied with cold district heating

occupants complained about problems maintaining thermal comfort in the houses, despite of the fact, that the existing radiator systems where supplied with supply and return temperatures with an annual average in the range of 45 °C/ 30 °C (Østergaard & Svendsen, 2017).

During the test, the central heat pump in the district heating system obtained COP-values in the range of 3.35–5.04 (Yang & Svendsen, 2018). Due to the low heat density of the area, it was found that the district heating operation with lower temperatures would be favorable for the efficiency of the distribution network and would also be economically viable due to availability of cheap surplus heat.

7.14 Temperature reductions in existing apartment building in Viborg, Denmark

The temperatures in the district heating network in Viborg, Denmark are gradually reduced these years, as the district heating operators plan to make use of large heat pumps in the heat supply. Current annual average district heating supply and return temperatures are therefore as low as 68 °C/40 °C. In order to reduce the district heating temperatures, a large effort has been put into reducing supply and return temperatures in the buildings in the network. One example of this is a test that was carried out in an existing apartment building from the 1970s containing 33 apartments. The building is seen in Figure 88.

The radiator system in the building is connected to the district heating network by a central heat exchanger located in the basement of the building. Inspection of the apartment building showed that the radiator system was not equipped with proper equipment for hydronic balancing, and that the behaviour of occupants in the building had a negative impact on the heating system temperatures (Benakopoulos et al., 2019). As suggested in Chapter 3, the supply temperature of the heating system was therefore optimized according to obtain a low supply temperature and a high flow in the heating system. This strategy minimizes the negative impacts and malfunctions in the control of individual heating elements. The result was that the annual average supply and return temperatures were reduced from 46 °C/38 °C to 40 °C/37 °C.

Domestic hot water is prepared in a central heat exchanger in the basement of the building and circulated to the hot water taps in all apartments. To reduce the risk of *Legionella* bacteria, a disinfection system based on electrolysis of saltwater was installed in the building (Danish Clean Water, 2020). The installation is seen in Figure 88. The disinfection system reduces the risk of *Legionella* growth in the domestic hot water system, and hence it may be possible to reduce supply and return temperatures in the circulation system from 55 °C/51 °C to approximately 47 °C/44 °C. If this is possible, the heat loss from the domestic hot water system can be reduced by approximately 38% and it may also be possible to reduce district heating supply and return temperatures (Benakopoulos et al., 2021). An ongoing research project will test the thesis that the disinfection system can deliver *Legionella* safe domestic hot water even if circulation temperatures are lowered, which would be a prerequisite to apply the strategy in the future.

7.15 Automatic return temperature limitation in Frederiksberg, Denmark

In a demonstration project carried out in Frederiksberg in Denmark, a new smart radiator thermostat was tested to investigate a new way of providing robustness towards excessive water flows in existing space heating system, and thereby ensuring low district heating return temperatures.

The new thermostat is a modification of current electronic Danfoss Eco thermostats. However, the new version includes a return temperature sensor to be attached to the return temperature pipe of the radiator, and a control algorithm that can reduce the flow through the radiator valve if the return temperature exceeds a certain threshold. To test the functionality of the new thermostat, prototypes were installed on radiators in two different apartment buildings with a floor area of 2,600 m²–2,900 m². Figure 89 shows one of the test buildings along with a picture of the thermostat prototype.

The tests of the thermostat demonstrated that the thermostat was capable of limiting radiator return temperatures according to the control algorithm. In both test buildings, space heating return temperatures below the set limit of 35 °C were thereby obtained at the end of the test period.

Figure 88. Picture of the building in Viborg where the tests were carried out (left) and the installed sterilization system (right).

Figure 89. Picture of one of the test buildings and the thermostat prototype with the return temperature sensor.

In the first building site, the space heating system was successfully operated with a large cooling in the space heating system, obtaining heating system temperatures of approximately 60 °C supply and 32 °C return at outdoor temperatures around 5 °C. During the test, the cooling of the district heating water was approximately 5 °C higher during the winter period, than what it had been previous years.

At the second test site it was however seen that the new thermostat was not completely robust towards occupant interference and errors in heating system operation. Excessive flows through some radiators and pump pressures much above what was recommended, were found to cause space heating return temperatures to be somewhat above 40 °C during part of the prototype test. When the pump pressure in the space heating system was modified to the recommended level and the flow through problematic string balancing valves were limited, it was demonstrated that the new return temperature thermostat limited the space heating return temperature to 35 °C in the building. The results thereby indicate that the new thermostat can be a useful alternative to pre-set radiator valves, and help provide balancing in the space heating system and limit space heating return temperatures to 35 °C.

7.16 District LAB Experimental Facility in Kassel, Germany

Especially LTDH is recognized as a key technology for the (cost-) efficient integration of renewable energy and waste heat sources in our energy systems. A further development of LTDH systems is needed for a decarbonisation of the heating sector. On the other hand, district heating is also put into question. To face these challenges research on innovative district heating concepts integrating decentral feed-in of renewable energy is needed.

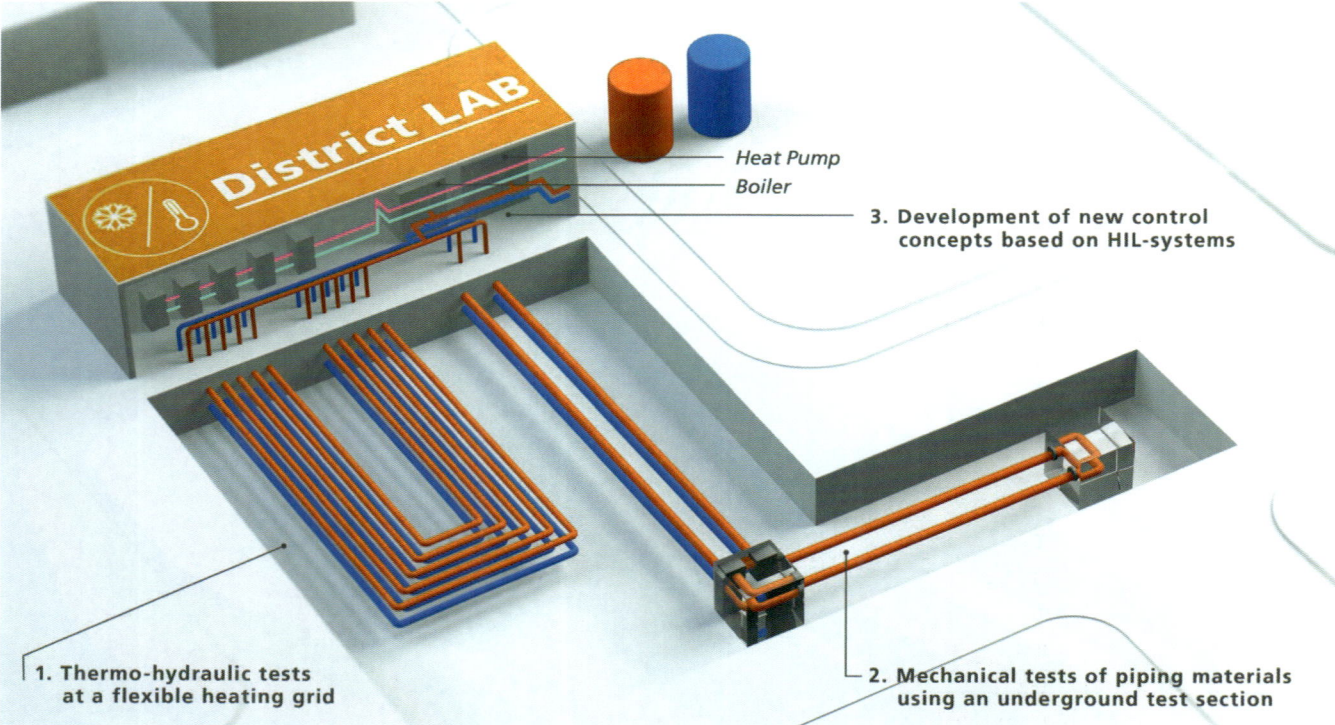

Figure 90. The system layout from the District LAB test facility for innovative district heating systems on a community scale, © Fraunhofer IEE

The Fraunhofer IEE in Kassel has started to set up a new research and experimental facility as a test and development platform for innovative district heating systems in close cooperation with industry partners. This facility consists of a real district heating grid in a lab-scale and expands the possibilities of the already existing facilities. This experimental environment is more flexible than experiments compared to tests in real operating district heating grid (demonstration) since no public utility mandate is connected to the District-LAB facility. Tests with varying temperatures and pressures can be conducted; new components and operating modes can be tested. The real behaviour of components and systems can be tested in a real life environment and simulation models can be validated in this facility (Schmidt, 2019).

First project ideas are within three different fields as the tests of new operational strategies (e.g. dynamic and changing boundaries for feed-in and utilization, grid operation with new temperature regimes, dynamic pressure and temperature changes) or tests of components (as piping systems, heat exchanger / sub-stations, pumps, control elements) or within the field of the development of simulation tools and validation (e.g. static hydraulic simulations incl. heat losses, dynamic simulations of control strategies and pressure changes).

As shown in the Figure 90, the District LAB has three main fields of application:

1. Thermo-hydraulic tests at a flexible heating grid: As part of the District LAB, a flexible heating grid in a lab scale provides the investigation of different district heating scenarios. Based on a flexible design of distribution pipes with supply and return pipes as well as heat sources and sinks, a wide range of temperature levels (5...130 °C) can be implemented.
2. Mechanical tests of piping materials using an underground test section: Due to the increasing implementation of volatile energy sources, district heating systems are faced with new and fluctuating operating conditions. To measure the impact of these and other conditions on the piping system, the District LAB includes a 40-meter underground test section for various investigations regarding the mechanical piping behaviour.
3. Development of new control concepts based on Hardware-in-the-Loop systems: Based on an extensive measurement and control system as well as a digital management system, the experiments will be parameterised and monitored. Due to Hardware-in-the-Loop (HiL) systems, defined loads, temperatures, and pressure levels can be applied to the test network enabling the development and evaluation of control concepts and energy management systems.

7.17 Summary and lessons learnt from the case studies

Core objective for the description of case studies was to identify and collect innovative demonstration concepts as examples of success stories for communities interested in developing LTDH systems. Demonstrated cases include use of advanced technologies and interaction between different components within the systems. Based on these experiences, principles and lessons learned in designing these systems are given. There were a total of 40 case studies from Austria, Denmark, Germany, Ireland, Norway, Switzerland, Sweden, The Netherlands and United Kingdom (see Section 10.2). The district heating systems were of very different sizes, from miniature to city wide systems. Network lengths were from approximately 370 m to more than 340 km. The connected buildings were detached, terraced and block houses, and many low energy or passive houses. Sources of heat were solar collectors, heat pumps, CHP plants, excess heat from industry or the systems were connected to a larger network close by with heat exchangers. The temperature levels recorded were typical for cold and low-temperature systems, varying from 1 to 88 °C in supply and -2 to 58 °C in return. Savings and increased efficiencies were observed in every case studied. Table 16 is summarising the cases by giving some key parameters as distribution temperatures, system configuration, heat supply technologies and trench length. A short description of each analysed project concept can be found in this chapter. The following Table 17 is giving an overview on the main lessons learnt from the different cases.

Table 17. Overview of demo sites and lessons learnt collected for this implementation guidebook

City and Area	Country	Summary
Gleisdorf Transition strategy	Austria	It is a perfect example of the transition to 4th generation of district heating in small scale and the therewith-related challenges and requirements regarding optimisation of system temperatures, intelligent operation and control of the system, integration and management of multiple heat sources of different kind and size, management of distributed storages, system hydraulics and short and mid-term enlargement scenarios.
Wüstenrot Weihenbronn	Germany	Preliminary analysis of the heating distribution systems in the buildings to reduce the supply temperature and to implement measures to increase the spread between supply and return temperature. Identification and support in the implementation of optimisation measures in the heating distribution system. Planning of a heating network with a lower flow temperature of around 55–60 °C is possible if the domestic hot water preparation and distribution is also adapted (flat stations). However, this would be too much work for existing buildings. Use of innovative and cost-effective materials for the realisation of the heating network (duo-plastic pipes). Design of the solar thermal plant for the entire grid coverage in summer operation (no heat demand in communal buildings) Avoidance of inefficient operating conditions in the wood chip boiler and of plant losses.
Darmstadt Lichtwiese	Germany	High network temperatures result from issues in a few critical buildings of the campus, while most buildings operate at low temperatures already. This makes it possible to decrease network temperatures significantly, fixing the most critical reasons for increased temperatures. Furthermore, comprehensive monitoring on both the primary and secondary sides of the district heating substations is crucial to identify which measures need to be prioritized.
Heerlen Parkstad Limburg	The Netherlands	It is a great example of a realised heating and cooling grid based on renewable sources. To become an attractive choice for potential customers, Mijnwater accomplishes a price reduction of about 10% compared to conventional solutions. Moreover, the company is owned by the municipality, so if a profit is made this will be returned to the community. The challenge is to deliver quality at this price and grow from a core in Heerlen to a heat/cooling grid covering all of the agglomeration of Parkstad Limburg. New types of buildings are to be added, increasing the fraction of apartment blocks, but also row-housing and separated dwellings, with a larger fraction of smaller clients.
Zürich FGZ	Switzerland	One of the first ULTDH systems impressively proving its feasibility. It is possible to couple and utilize conventional HT-DH with ULTDH to supply older buildings. The rather simple hydraulic and control concept is successful and reliable. Long term monitoring proof high heat pump efficiency. Suitable pipe dimensioning lead to reasonable pumping power demand. No significant higher space demand for piping since no insulation is required.
Salzburg Lehen	Austria	For this case, the energy demand of the buildings is much higher than calculated. A good monitoring supports the operator (Salzburg AG) in the optimization of the system and makes it transparent for the user that a high proportion of renewable energy is used in this project.
Graz Reininghaus	Austria	The major lessons learnt is, that LTDH networks facilitates low-emission / decarbonised heat supply. Low-temperature systems allows the exploitation of low-grade heat sources within an urban and locally context. For developing innovative supply concepts, strong partnerships beyond the existing ones are key – for initiating, planning, building, and operating – to get full trust and acceptance.
Braunschweig Rautheim	Germany	This case is showing an award winning vision of a new, highly energy-efficient district energy system based on waste heat from a data-centre. For this case the heat pump choice has been determined by cooling liquid (natural refrigerant only), LTDH temperatures (40 °C–70 °C), and nominal heat output.
Lund Brunnshög	Sweden	The first major lesson learnt is that it is not possible to use lower heat distribution temperatures than 65/35 °C when the classic network configuration is used. Further lessons learnt will be communicated during 2021 through the COOL DH project.
Kassel Feldlager	Germany	This extensive study shows in particular that renewable energy sources offer great potential for sustainable and efficient heat supply of buildings at similar cost as conventional systems. Although the project gave convincing results, it has not come to a realisation.
Bamberg Lagarde	Germany	The main findings from this project with partly existing buildings are that it is possible to get a cover rate of more than 50% of heat from renewable sources from a techno economical viewpoint and that the regulatory framework in Germany is blocking cross sectoral energy concepts.

City and Area	Country	Summary
Bjerringbro Tyttebærvej	Denmark	It can be possible to provide space heating for existing single-family houses with district heating supply and return temperatures of 45 °C/ 30 °C without performing any renovations.
		The combination of a central heat exchanger and a decentral direct electric heater is an efficient solution to provide domestic hot water in single-family houses heated with district heating with supply temperatures around 45 °C. Ultra-low temperature district heating networks can be an efficient solution in low heat density areas where cheap surplus heat is available at low temperatures.
Viborg Multi-family building	Denmark	It is possible to reduce district heating temperatures in existing buildings even with a small effort and applying solutions with a short pay-back time. Installation of a water sterilization system can provide reductions in the necessary district heating supply and return temperatures as well as reduced risk of *Legionella* contamination and reduced energy consumption due to reduced heat losses inside the building.
Frederiksberg Multi-family building	Denmark	New electronic thermostats with smart return temperature reduction can help limit heating system return temperatures and provide automated continuous optimization of heating system water flows. New electronic thermostats are currently not completely robust towards occupant interference and heating system errors, that may therefore still cause excessive water flows in heating systems if issues are not identified and corrected.
Kassel District LAB	Germany	The District LAB test facility for innovative district heating technologies is unique and needed to tackle the challenges with future district heating networks based on fluctuating renewable energy sources.

7.18 Major conclusions from the case studies

The above shown implementation cases are providing evidences that LTDH is a proven and market ready heat supply technology and works under various boundary conditions. As shown, there are many possible and different strategies for the heat supply, as well as a larger number of possible system designs to successfully realise these systems.

Especially from the displayed case studies in this chapter the following main conclusions can be drawn:

- From a technical point of view the large variety of system configurations in the cases, based on the local boundary conditions, show the flexibility in the implementation and realisation of LTDH systems. For the operation a good monitoring and management system is securing the success of the project. When it comes to the integration of multiple heat sources and more complicated systems, a wider digitalisation of the processes is needed.
- The regulatory boundary conditions are not always and not in all regarded countries beneficial. So, e.g. for the integration of geothermal heat a long (or too long) process of approving needs to be considered, which can potentially derail the realisation. Furthermore, real cross sectoral energy systems are not foreseen in today's rules, which makes a realisation very complicated.
- The cases clearly show that high connection rate and a good support from the customer could be gained, when the system is owned by the municipality or by a cooperative.
- For a business point of view also the above mentioned ownership issue is of importance. Interest rates might be lower and long payback times are manageable. Some cases show, that a transition to LTDH systems is economical feasible, some cases point out that the price level for the heat supply could also be up to 10% lower compared to a conventional solution even so future damage cost from global warming is not included yet.

In summary, the above shown cases give the evidence that LTDH is a proven and market ready heat supply technology and works under various boundary conditions. Furthermore, experiences from the cases show a good support for a needed implementation of digitalisation measures to secure a good operation under the new boundary conditions, such as integration of fluctuating renewable or waste heat sources or changed network (bidirectional) operation.

7.19 Literature references in Chapter 7

Benakopoulos, T., Salenbien, R., Vanhoudt, D., Tunzi, M., & Svendsen, S. (2021). Low return temperature from domestic hot-water system based on instantaneous heat exchanger with chemical-based disinfection solution. Energy, 215, 119211. https://doi.org/10.1016/ .energy.2020.119211

Benakopoulos, T., Salenbien, R., Vanhoudt, D., & Svendsen, S. (2019).
Improved Control of Radiator Heating Systems with thermostatic radiator valves without pre-setting function. Energies, 12(3215). https://doi.org/10.3390/en12173215

Dammel, F., Oltmanns, J. & Stephan, P. (2019). Nutzung der Abwärme eines Hochleistungsrechners am Campus Lichtwiese der TU Darmstadt. In: EuroHeat&Power, Brussels, Belgium, 2019.

Danish Clean Water. (2020). Clean and safe solutions to control and prevent bacteria. Retrieved December 8, 2020, from https://danishcleanwater.com/

Euroheat (2019). 6th Global District Energy Climate Awards revealed | Looking out for excellence! Retrieved from https://www.euroheat.org/news/6th-global-district-energy-climate-awards-revealed/

Heatpumpingtechnologies (2018). Waste Heat Recovery at the Steel and Rolling Mill Marienhütte GmbH, Graz, Austria. Retrieved from https://heatpumpingtechnologies.org/annex47/wp-content/uploads/sites/54/2018/12/marienhuett.pdf

Hiddes, L. et al. (2014). The Netherlands Smart Energy Region 2014– WG1 publication; WG1 publication Smart Energy Regions: The Netherlands Smart Energy Region 2014.pdf, ISBN-978-1-899895-14-4, pages 169-180. http://www.smarter.eu/content/wg1publicationsmartenergyregionsdownloadsinglesections

Jurns, J.M., Bäck, H. & Gierow, M. (2014). Waste Heat Recovery from the European Spallation Source
Cryogenic Helium Plants – Implications for System Design. In: AIP Conference Proceedings 1573, 647 (2014). doi:10.1063/1.4860763

Kaiser, J. & Loskarn, S. (2019). Entwicklung eines Strom- und Wärmekonzeptes auf Quartiersebene unter Berücksichtigung von Sektorenkopplung für den Lagarde Campus Bamberg. Presentation at Berliner Energietage 2019, Berlin, Germany, 20 Mai 2019.

Kaiser, J. & Loskarn, S. (2019a). Der Lagarde-Campus Bamberg – Integration eines gekoppelten Strom- und Wärmesystems in die bestehende Netzstruktur. Presentation at VKU Netzforum 2019, Berlin, Germany, 14 November 2019

Mauthner, F. & Joly, M. (2017). Technology and Demonstrators. Technical Report Subtask B – Part B3 (Case studies) and Technical Report Subtask C – Part C2 (Analysis of built best practice examples and conceptual feasibility studies). Gleisdorf and Ecublens: IEA SHC Task 52 – Solar Heat and Energy Economics in Urban Environments. AEE Intec, Gleisdorf, Austria. http://task52.iea-shc.org/Data/Sites/1/publications/IEA-SHC%20Task%2052%20STC2-Best%20practice%20summary%20report_2017-08-31_FINAL.pdf

Oltmanns, J., Dammel, F. & Stephan, P. (2017). Modeling the heat supply system of TU Darmstadt Campus Lichtwiese. In: Proceedings of ECOS 2017 – The 30th Int. Conf. on Efficiency, Cost, Optimization, Simulation and Environmental Impact of Energy Systems, July 2-6, 2017, San Diego, USA.

Oltmanns, J., Freystein, M., Dammel, F. & Stephan, P. (2018). Improving the operation of a district heating and a district cooling network. In: Energy Procedia 149, pp. 539-548, 2018.

Oltmanns, J., Sauerwein, D., Dammel, F., Stephan, P. & Kuhn, C. (2020). Potential for waste heat utilization of hot-water-cooled data centers: A case study. In: Energy Science and Engineering 8.5, pp. 1793-1810, 2020.

Pietruschka, D. (2016). Vision 2020. Die Plusenergiegemeinde Wüstenrot. Fraunhofer IRB Verlag, Stuttgart, Germany. ISBN 978-3-8167-9545-2

Schmidt, D. (2019). New Experimental Facility for Innovative District Heating Systems on a Community Level – District LAB. 1st Nordic Conference on Zero Emission and Plus Energy Buildings (Poster); 6-7 November 2019; Trondheim, Norway.

Schmidt, D. et al. (2017). Development of an innovative heat supply concept for a new housing area
In: Clima 2016, 12th REHVA World Congress. Proceedings. Vol. 3., 10 pages. http://publica.fraunhofer.de/documents/N-402348.html

Schmidt, D. & Kallert, A. (ed.) (2017). Future Low Temperature District Heating Design Guidebook
Final Report of DHC Annex TS1. ISBN 3-899999-070-6, AGFW Project Company, Frankfurt am Main, Germany

SIR (2014). Wohnbauforschungsprojekt Stadtwerk Lehen. Salzburg: SIR - Salzburger Institut für Raumordung und Wohnen, Salzburg, Austria https://www.salzburg.gv.at/bauenwohnen_/Documents/wohnbauforschungsprojekt_stadt_werk_lehen_publizierbarer_endbericht_2014.pdf

Torío, H. & Schmidt, D. (ed.) (2011). IEA ECBCS Annex 49 - Exergy Assessment Guidebook for the Built Environment, Fraunhofer Verlag, Stuttgart 2011. https://www.bookshop.fraunhofer.de/buch/exergy-assessment-guidebook-for-the-built-environment/235786

Unger, H. (20. November 2018). Wärmepumpenprojekte Energie Graz. Graz: Grazer Energiegespräche.

Verhoeven, R., Willems, E., Harcouët-Menou, V., De Boever, E., Hiddes, L., Op'tVeld, P. & Demollin, E. (2014). Minewater 2.0 Project in Heerlen The Netherlands: Transformation of a Geothermal Mine Water Pilot Project into a Full Scale Hybrid Sustainable Energy Infrastructure for Heating and Cooling. In: Energy Procedia, Vol. 46, 2014, p. 58-67. Doi: 10.1016/j.egypro.2014.01.158

Wagener, P. (2017). IEA Annex 42 Heat Pumps in Smart Grids – Final report: Projects, start-ups in the field of Smart Grid https://heatpumpingtechnologies.org/publications/heat-pumps-in-smart-grids-final-report

Yang, X. & Svendsen, S. (2018). Ultra-low temperature district heating system with central heat pump and local boosters for low-heat-density area: Analyses on a real case in Denmark. In: Energy, Vol. 159, 2018, p. 243-251.

Yang, X., Li, H. & Svendsen, S. (2016). Evaluations of different domestic hot water preparing methods with ultra-low-temperature district heating. In: Energy, Vol. 109, 2016, p. 248-259.

Østergaard, D. S. & Svendsen, S. (2017). Space heating with ultra-low-temperature district heating - a case study of four single-family houses from the 1980s. In: Energy Procedia, Vol. 116, 2017, p. 226-235.

8 TRANSITION STRATEGIES

Author: Sven Werner, Halmstad University.

The essence of conversion, transformation and transition is that all humans are capable of profound change. Required changes in our communities should be initiated, communicated and implemented by visions, strategies and planning measures. Changes to our energy system must also be identified at all levels in our global community. Although changes are necessary and inevitable in all areas, they are often accompanied by concerns and doubts about the new. The purpose with this chapter is to show how transition strategies have addressed these concerns and doubts in some urban areas.

Regarding climate change mitigation, major global change initiatives include the formation of the Intergovernmental Panel on Climate Change in 1988 and United Nations Framework Convention on Climate Change in 1992. Despite being formed approximately 30 years ago, these global initiatives must still be supported by the actions of international, national, regional and local communities. The two latter levels are particularly important in the context of district heating systems since no international or national district heating system exist. These systems provide local heat supply from utilisation of available local options as added values. However, support from global and national communities are essential to obtaining momentum for decarbonised district heating systems.

The European vision for heating and cooling has been expressed by the Renewable Heating and Cooling (RHC) Platform as: '100 percent renewable energy-based heating and cooling is possible by 2050' (RHC, 2019b). This vision is based on the fact that many districts in Europe have reached – or are close to reaching – this vision already (RHC, 2019a).

Euroheat & Power, the European association for district heating and cooling, participated in developing the RHC vision for the European heat market. This long term vision has been supported by the following commitment from the European district heating industry in October 2019: 'We will relentlessly pursue the full decarbonisation of district heating and cooling networks in Europe before 2050' (Laufkötter, 2019).

Notably, similar commitments are also present in national contexts. In Sweden, the heating industries representing district heating, electricity, biomass and heat pumps have – together with major heat users – adopted a national roadmap for obtaining a fossil-free heat supply. The promise from these heating industries is to achieve this ambitious goal by 2030 (Energiföretagen, 2019).

In this overall decarbonisation context, the issue of LTDH should always be included since lower temperatures will make the transition more profitable (as earlier presented in Section 1.6 and Chapter 2). Hence, decarbonisation is the major issue and primary step for all local transition strategies. Moreover, having lower temperatures in heat distribution networks is a very important secondary step to ensuring more effective transitions.

In this chapter, some evidence concerning adopted visions, strategies and planning measures from certain urban areas are presented (Sections 8.1–8.4) when the heat distribution temperature issue has been properly identified within the decarbonisation context. These three steps are vital to implementing district heating and cooling systems based on renewable or recycled heat or cold in every urban area. In Chapter 5, an applied study concerning the Lichtwiese campus area at TU Darmstadt was provided. Additional university examples are briefly provided in Section 8.5 since some universities have been forerunners in their choice of LTDH. Finally, the major conclusions from this chapter are presented in Section 8.6.

8.1 Adopted transition initiatives from five urban areas in Europe

Five urban areas in Europe are used as examples of visions, strategies and planning measures within their overall transition strategies. These urban areas are presented in Table 18 with the number of urban inhabitants and the current heat deliveries from existing district heating systems. The sizes of these urban areas range from a small town in Austria (Gleisdorf) to a major city in Germany (Munich).

The current average heat demand within European buildings is approximately 5.5 MWh per capita and year. Some urban areas have heat delivery demands close to this average, which reveals that the market share for district heating is already high. However, some of the heat delivery data in Table 18 also includes delivery beyond buildings. Therefore, it should be possible to deliver more heat through district heating in these five urban areas.

The corresponding original information sources for the applied transition strategies (i.e., visions, strategies, and planning measures) are presented in Table 19.

Table 18. Overview of the five urban areas used as examples of visions, strategies, and planning measures for lower temperatures within decarbonisation strategies.

City or town	Country	Urban population, thousand inhabitants	Current heat deliveries from the existing district heating system, GWh/year	Current heat delivery per capita, MWh/year
Gleisdorf	Austria	10	7	0.7
Viborg	Denmark	40	210	5.3
Basel	Switzerland	170	830	4.9
Geneva	Switzerland	500	520	1.0
Munich	Germany	1400	4300	3.1

Table 19. Information sources used for inputs regarding visions, strategies and planning measures in the five urban areas.

City or town	Information sources
Gleisdorf	(Stadt Gleisdorf, 2012), (AEE INTEC, 2020)
Viborg	(Abildgaard, 2017)
Basel	(Regierungsrat des Kantons Basel-Stadt, 2017), (Küng, 2019)
Geneva	(Secrétariat du Grand Conseil, 2013), (Quiquerez, Lachal, Monnard, & Faessler, 2017), (Durandeux, 2019)
Munich	(Henke et al., 2015), (Frey & Miller, 2017), (Stadtwerke München, 2017), (Theis, 2019)

8.2 Visions

Visions are required to address the long term direction to follow. They should concentrate a common goal from a company's owners, board and management for their customers, staff and the general public regarding what they aim to achieve in the future. The visions adopted in the five urban areas are presented in Table 20.

The visions for the five urban areas are highly variable. For example, the Gleisdorf vision relates to its general future direction, while the Geneva vision relates to tangible goals concerning both expansion and moving to non-fossil heat sources by 2035. The Viborg vision is influenced by a recent change to their CHP plant that requires urgent action. The Basel vision focuses on heavy reductions in carbon dioxide emissions. The Munich vision is very simple, stating that the district heating system should be completely renewable by 2040.

Table 20. Visions adopted in the five urban areas concerning their district heating systems.

City or town	Visions adopted
Gleisdorf	The municipality of Gleisdorf advocates renewable energy and the city policy guidelines from the early 1990s already included a strong commitment to an "environmentally friendly energy policy" to "demand and promote the use of all renewable forms of energy and energy-saving measures" and to "cooperate with research institutes to increase knowledge, realise projects and support a transnational skill exchange".
Viborg	As soon as possible, • Reduce heat supply costs compared to current cost level for the combined cycle CHP plant using natural gas since this plant lost governmental support at the end of 2018.
Basel	In 2035, • The proportion of non-renewable input should not exceed 30%. • The carbon dioxide emissions per inhabitant should be 2.3 ton or less for the Basel region. Both visions correspond to reductions of approximately 50% compared to the 2010 situation.
Geneva	In 2035, • 40% of heat demand should be supplied by district heating. • 80% should be supplied from non-fossil sources. • The length of the district heating networks should double. • Heat sales should triple. These visions should be adopted for the Geneva district heating systems to support the Geneva region energy strategy from 2013.
Munich	In 2040, • The heat supply in the Munich district heating system should be completely renewable.

8.3 Strategies

Strategies shall express what to do in both long and short term. The strategies identified for vision implementation in the five urban areas are presented in Table 21. The content in this table considers only the context for developing the district heating systems. A vital condition for these strategies is the lower expected future heat demand for buildings. For example, the total heat demand in Geneva is expected to be 37% lower in 2035 when compared to 2000 (Secrétariat du Grand Conseil, 2013). Similar expectations also exist in other urban areas.

In all strategies, the reduction of heat distribution temperatures is acknowledged as an important part of the overall strategy. The Basel strategy identifies the need for less gas delivery to reach the emission reduction goal for 2035. The Munich strategy involves the use of deep geothermal heat to achieve the 2040 goal of a completely renewable heat supply. Geothermal heat is also part of the Geneva strategy. Although deep heat mining for a geothermal CHP plant could be an opportunity in Basel, it became highly controversial following an unexpected earthquake in December 2006 caused by an earlier project. The expansion of heat distribution networks is included in the strategies of Gleisdorf, Basel, Geneva and Munich. The short term strategy in Viborg is to replace its fossil CHP plant with a large heat pump. This replacement becomes more profitable if lower temperatures can be obtained in the heat distribution network.

Table 21. Strategies identified for district heating in the five urban areas.

City or town	Strategies adopted
Gleisdorf	• Expansion of the district heating grid. • Transition from natural gas to renewables with preference for non-combustible renewables. • Lower heat distribution temperatures. • Heat storage for mitigating daily peak heat loads. • Sector coupling with the sewage water plant. • Intelligent control of the district heating system.
Viborg	• Heat supply from heat pumps based on ambient heat or heat recovered from the Apple data centre in Foulum, located 10 km east of Viborg. • Lower heat distribution temperatures based on the 2002 motivation tariff and the 2011 temperature strategy.
Basel	• Expansion of the district heating grids. • Lower heat distribution temperatures. • Implementation of biomass and geothermal CHP. • Heat storage for mitigating daily peak heat loads. • Reduction of gas delivery for heating.
Geneva	• Expansion of the district heating grids. • Lower heat distribution temperatures. • The heat supply should be based on the geothermal potential, ambient heat in water and recovery of excess heat. • Heat storage for mitigating daily peak heat loads.
Munich	• Expansion of the district heating grid. • Lower heat distribution temperatures. • The main resource for fossil-free district heating should be the geothermal heat sources in the Munich region.

8.4 Planning measures

Planning measures shall address both on long and short term how to do the change. They are required to fulfil the adopted visions and the implementation of the identified strategies. Otherwise, the goals associated with the adopted visions will not be fulfilled. The planning measures taken in the five urban areas concerning lower temperatures are summarised in Table 22.

Local energy companies are familiar with current district heating technology. If they must use new technology, they must also leave their current comfort zone and learn the conditions for this new technology. To overcome this barrier, one common denominator among planning measures is that research organisations have been identified as suitable partners in Gleisdorf, Geneva and Munich.

One interesting strategy is the 2035 closure deadline for gas distribution in Basel. In Viborg, a programme for lower customer heat demand has been initiated to support the goal of lower distribution temperatures. Viborg has also a programme for the reduction of unintended circulation flows, which also contributes to lower distribution temperatures.

Table 22. Initiated planning measures concerning district heating in the five urban areas.

City or town	Planning measures initiated
Gleisdorf	• Participation in the national Thermaflex flagship project concerning a demonstrator of fully renewable heat supply with smart control (virtual heat supply). • Accompanying spatial energy planning identified four areas for extension of the heat distribution grid. • The sewage water plant will be modernised, thus providing higher biogas yields and the possibility to install a heat pump using purified sewage water as heat source and biogas for booster heating. • Systematic and continuous optimisation and reduction of system temperatures.
Viborg	• Action plan for lower supply temperatures. • Reduction of customer heat demands. • Reduction of unintended circulation flows in the network. • Energi Viborg Kraftvarme is planning an air-based heat pump, which should be ready for operation by 2022.
Basel	• Reduction of the maximum supply temperature in the heat distribution network from 170 °C to 120 °C by 2025. • Closure of the gas distribution grid in the urban parts of Basel by 2035.
Geneva	• Participation in the international ERA-NET Geotermica research programme through the HEATSTORE project. • Cooperation with research organisations within GEothermie 2020 to exploit the geothermal potential in the Geneva region. • Identification of areas suitable for new district heating grids.
Munich	• Participation in the national EnEff: Wärme research programme within the LowEx-Systeme project to learn how to achieve lower heat distribution temperatures. • Cooperation with research organisations to exploit geothermal potential in the Munich region. • Introduction of low-temperature heat distribution in the Freiham network.

8.5 University campus systems as forerunners

District heating and cooling systems are often used within university campus areas. Typically, these systems are operated by the facility management departments within the university administrations. Energy researchers at some of these universities have taken the initiative to redesign these systems with LTDH. Hence, university campus areas have become testing areas for new district heating and cooling technologies emerging from these university transition initiatives. These early transition areas are also suitable living labs for university students that are expected to participate in the energy transition upon graduation.

In this section, some early university adopters that have introduced or plan to introduce lower temperatures in their district heating networks are briefly presented. These examples show that the applied study for the Lichtwiese campus at TU Darmstadt (presented in Chapter 5) is not an exceptional case.

In Zürich, Switzerland, the heating and cooling transition began at the ETH Hönggerberg campus area in 2013 and will be finalised in 2026 (Häusermann & Mast, 2017). Excess heat from the cooling supply during the summer is stored in low-temperature borehole heat storage for input to the heat supply during the winter. This recycled low-temperature heat is distributed in a cold network and upgraded in each building with heat pumps.

In Zwickau, Germany, the Westsächsische Hochschule will use a cold heat distribution network to distribute low-temperature heat from mine water that has filled an abandoned coal mine (Tiefe Geothermie, 2018).

In Paris, France, Saclay University will use a low-temperature geothermal heat source for a cold network to supply both heating and cooling (EPAPS, 2019). The intention is to connect buildings with a total area of 2.1 million m^2.

In Delft, The Netherlands, the Technical University has started the transition from a high-temperature district heating network to using a warm heat distribution network with considerably lower heat distribution temperatures (Mlecnik, Hellinga, & Stoelinga, 2018). This transition will involve a geothermal heat source and will be combined with a deep renovation programme that will include smart control of the heat use.

In Bergen, Norway, the Bergen University introduced a cold network in 1994 that supplied university buildings in the city centre (Stene & Eggen, 1995). A seawater heat pump plant provides heat during the winter season. Some connected buildings are also cooled by the cold network during the summer.

In London, the United Kingdom, South Bank University (LSBU) will use a cold network to balance heating and cooling demands (Song, Wang, Gillich, Ford, & Hewitt, 2019). The network involves the efficient use of heat by recycling solar heat (from the summer) to store thermal energy in the ground for later use in autumn and winter. A demand response function will also be used to avoid operation of the heat pumps during electricity peak hours.

In Stanford, USA, Stanford University conducted the Stanford Energy System Innovations (SESI) project (Stanford University, 2014). This four-year project was initiated in 2011 and was operational in 2015. It involved nearly US$500 million in investment and included a new central energy facility, the conversion of an old steam district heating system to a modern hot water system with low temperatures, the closure of an old fossil CHP plant and short term thermal storage for both heat and cold. After its implementation, carbon dioxide emissions decreased by 65%. The heart of the SESI project is heat recovery by capturing excess heat from the district cooling system to generate hot water for the district heating system. This core feature was identified upon recognising that heating and cooling appeared simultaneously for three-quarters of the year (Stagner, 2016). An analysis has shown that the new system is robust and can be flexible through future application of carbon-aware scheduling (de Chalendar, Glynn, & Benson, 2019). Recently, an extension of the district cooling capacity was approved to cope with expected future heat waves (Stanford News, 2020).

In Oshawa, Canada, Ontario Tech included a geothermal source to obtain low-temperature heat storage for the district heating and cooling system at its North campus (Dincer & Rosen, 2007).

In Germany, a special grant program was made available for energy transition projects in university campus areas. Project outcomes were reported for four campus areas of different sizes and building characteristics: Rheinisch-Westfälische Technische Hochschule Aa-

chen, TU Braunschweig, Campus Leuphana Universität Lüneburg and Wissenschaftspark Telegrafenberg Potsdam (Erhorn-Kluttig, Doster, & Erhorn, 2016). While the projects consider measures inside the buildings as well as in the generation facilities and networks, they primarily focus on planning and simulation rather than actual implementation.

In the USA, the International District Energy Association started special campus conferences in 1989 since campus systems correspond to a significant proportion of the US district heating and cooling activities. These conferences have attracted an increasing number of university participants over the years. At the 2020 conference in Denver, 1160 people participated in learning, sharing and networking related to the development of campus systems (Editorial, 2020).

Overall, the information presented in this section suggests that some university campus areas have become important testing grounds for new district heating and cooling technologies, with a focus on LTDH. The high proportion of cold networks in the overview above can be explained by that grants from research funds are more easily obtained if new heat distribution configurations are implemented and explored.

8.6 Major conclusions from transition strategies

The major conclusions concerning visions, strategies and planning measures within the local transition strategies are:

1. Lower distribution temperatures are important to include in transition strategies for the decarbonisation of district heating systems.
2. Cooperation with research organisations should be considered when new technologies must be implemented in local district heating systems.
3. University campus areas are often forerunners when new technologies for heat distribution are introduced.

8.7 Literature references in Chapter 8

Abildgaard, M. (2017). Data Centers and 4GDH in practice - the case of Viborg. Paper presented at the 3rd International Conference on Smart Energy Systems and 4th Generation District Heating, 12-13 September, Copenhagen. https://www.4dh.eu/images/VF_-_Data_Centers_and_4DH_in_practice.pdf

AEE INTEC (2020). [Personal communication with Harald Schrammel about the transition strategies in Gleisdorf].

de Chalendar, J. A., Glynn, P. W., & Benson, S. M. (2019). City-scale decarbonization experiments with integrated energy systems. Energy & Environmental Science, 12(5), 1695-1707. doi:10.1039/C8EE03706J

Dincer, I., & Rosen, M. (2007). A unique borehole thermal storage system at University of Ontario Institute of Technology. In H. Ö. Paksoy (Ed.), Thermal Energy Storage for Sustainable Energy Consumption: Fundamentals, Case Studies and Design (pp. 221-228): Springer.

Durandeux, S. (2019). Abaissement des pics de demande dans les reseaux thermiques. SVGW Fachtagung Fernwärme October 31. https://www.aquaetgas.ch/aktuell/branchen-news/201912-11-fernw%C3%A4rmetagung-bern-2019/

Editorial. (2020). Conference wrapup (concerning the annual campus conference). District Energy, 106, issue 2, 39-41.

Energiföretagen. (2019). Färdplan för fossilfri konkurrenskraft - uppvärmningsbranschen (Roadmap for fossil-free competitiveness - the heating industry). Retrieved from https://www.energiforetagen.se/globalassets/energiforetagen/sa-tycker-vi/fardplaner-fossilfritt-sverige/ffs_fardplan-fossilfri-uppvarmning-med-undertecknare_191007.pdf

EPAPS. (2019). Paris-Saclay - Le réseau d'échange de chaleur et de froid. Retrieved from https://www.epaps.fr/wp-content/uploads/2019/11/brochure_reseau-min-1-compresse%CC%81.pdf

Erhorn-Kluttig, H., Doster, S., & Erhorn, H. (2016). Der energieeffiziente Universitätscampus: Pilotprojekte der Forschungsinitiative EnEff:Stadt. Stuttgart: Fraunhofer IRB Verlag. .

Frey, J., & Miller, A. (2017). The Munich renewable energy strategy. https://www.ca-eed.eu/content/download/3956/file/DistrictHeatingVision2040-StadtwerkeMuenchenSWM.pdf/attachment

Henke, Kröper, Spannig, Zeisberger, Ziegler, Braunmiller, . . . Piotrowski. (2015). EnEff:Wärme | LowEx-Systeme: Breitenanwendung von Niedertemperatur-Systemen für eine nachhaltige Wärmeversorgung. Retrieved from https://www.agfw-shop.de/agfw-fachliteratur/lowex-systeme-breitenanwendung-von-niedertemperatur-systemen.html

Häusermann, M., & Mast, M. (2017). Fallbeispiel - Anergienetz ETH Hönggerberg. Bericht Energie Schweiz. Retrieved from https://www.energieschweiz.ch/home.aspx?p=22949,22963,22985

Küng, M. (2019). Eröffnungsreferat - Basel Fernwärme. SVGW Fachtagung Fernwärme October 31. https://www.aquaetgas.ch/aktuell/branchen-news/201912-11-fernw%C3%A4rmetagung-bern-2019/

Laufkötter, S. (2019). European District Heating Industry Committed to Carbon-free District Heating Networks before 2050. Euroheat & Power (eng ed.), 16(3-4), 9-11.

Mlecnik, E., Hellinga, C., & Stoelinga, P. (2018). Energy Flexible Buildings, Case study: TU Delft campus, The Netherlands. IEA EBC Annex 67. Retrieved from http://annex67.org/media/1491/case-study-tu-delft-campus.pdf

Quiquerez, L., Lachal, B., Monnard, M., & Faessler, J. (2017). The role of district heating in achieving sustainable cities: comparative analysis of different heat scenarios for Geneva. Energy Procedia, 116, 78-90. doi:https://doi.org/10.1016/j.egypro.2017.05.057

Regierungsrat des Kantons Basel-Stadt. (2017). Energieverordnung 772.110, Verordnung zum Energiegesetz vom 29. August. Retrieved from https://www.gesetzessammlung.bs.ch/frontend/versions/pdf_file_with_annex/4540

RHC. (2019a). 100% renewable Energy Districts: 2050 Vision. Renewable Heating & Cooling - European Technology and Innovation Platform. Retrieved from https://www.euroheat.org/wp-content/uploads/2019/08/RHC-ETIP_District-and-DHC-Vision-2050.pdf

RHC. (2019b). 2050 vision for 100% renewable heating and cooling in Europe. Renewable Heating & Cooling - European Technology and Innovation Platform. Retrieved from https://www.rhc-platform.org/content/uploads/2019/10/RHC-VISION-2050-WEB.pdf

Secrétariat du Grand Conseil. (2013). Rapport du Conseil d'Etat au Grand Conseil sur la conception générale de l'énergie 2005-2009 et projet de conception générale de l'énergie 2013. RD986 Retrieved from http://ge.ch/grandconseil/data/texte/RD00986.pdf

Song, W. H., Wang, Y., Gillich, A., Ford, A., & Hewitt, M. (2019). Modelling development and analysis on the Balanced Energy Networks (BEN) in London. Applied Energy, 233-234, 114-125. doi:https://doi.org/10.1016/j.apenergy.2018.10.054

Stadt Gleisdorf. (2012). Aktionsplan für nachhaltige Energie (Action plan for sustainable energy). Retrieved from https://mycovenant.eumayors.eu/docs/seap/3264_1386074394.pdf

Stadtwerke München. (2017). Shaping the heating turnaround: District heating - 100 percent renewable. Retrieved from https://www.swm.de/dam/swm/dokumente/english/district-heating-shaping-turnaround.pdf

Stagner, J. C. (2016). Stanford University's „fourth-generation" district energy system. District Energy, 102, issue 4, 19-24.

Stanford News. (2020). Major expansion planned for Stanford's renewable energy system, October 6 [Press release]. Retrieved from https://news.stanford.edu/2020/10/06/major-sesi-expansion-planned/

Stanford University. (2014). Stanford Energy System Innovations - General Information. Retrieved from https://sustainable.stanford.edu/sites/default/files/Stanford%20SESI%20General%20Information%20Brochure%20%28rev%206%29.pdf

Stene, J., & Eggen, G. (1995). Heat pump system with distribution at intermediate temperature at the University of Bergen. Proceedings 19th International Congress of Refrigeration, volume 4b, 1262-1269. The Hague 1995.

Theis, A. (2019). Renewable District Heating for Munich. Paper presented at the Konferenz zum Thema Städte als Akteure der Energiewende. Paris 13. März. https://energie-fr-de.eu/de/veranstaltungen/leser/konferenz-zum-thema-staedte-als-akteure-der-energiewende.html

Tiefe Geothermie. (2018). Geothermiebohrung in Zwickau in Betrieb genommen [Press release]. Retrieved from https://www.tiefegeothermie.de/news/geothermiebohrung-in-zwickau-in-betrieb-genommen

9 CONCLUSIONS

Authors: Kristina Lygnerud and Sven Werner, Halmstad University

The conclusions of this guidebook are provided regarding technological developments, non-technical aspects and policy implications. Finally, recommendations are directed towards district heating providers that are expected to implement low-temperature heat distribution in the future. By considering the following conclusions, the implementation of LTDH can be facilitated. Hence, decarbonisation of the heat supply in the European Union, as described in Figure 1 (Section 1.1), can be supported over the coming years.

These conclusions can be considered simple advice and recipes for obtaining lower heat distribution temperatures. In the preface, this was expressed as the goal of this guidebook.

Since LTDH introduces new expressions and definitions into the district heating vocabulary, readers should consult the definition and abbreviation sections at the beginning of this guidebook if they discover expressions in this conclusion chapter that they are unfamiliar with.

9.1 Technological developments

In comparison with current heat distribution technologies, future low-temperature technologies require some modifications to both existing and new installations in buildings, substations, and heat distribution networks.

Regarding existing installations, only a minor proportion of the installations must be removed and exchanged. Refurbishments of existing buildings have already delivered – and will continue to deliver – heat demand reductions that will allow the possibility to use lower temperatures in existing installations. However, the *Legionella* issue must be considered in existing building installations with traditional technologies for provision of domestic hot water. The transition to using lower temperatures has begun in some buildings and should also be implemented in other buildings. Major reductions in heat distribution temperatures among existing networks can be obtained by identifying and eliminating malfunctions. Notably, existing district heating pipes can be reused when low-temperature heat distribution will be applied.

Concerning new installations, a higher degree of freedom is available. As presented in this guidebook, more diversified heat distribution configurations are now available. This is especially the case with cold networks, which offer the possibility to distribute heat with ultra-low temperatures for local upgrades in buildings. In the past, the focus was to recycle warm excess heat from various sources using district heating systems. In the future, cold and warm networks can complement each other. Hence, warm networks will be used when suitable warm heat sources are available, while cold networks can be used when heat with ultra-low temperature (including ambient heat) is the only remaining alternative.

New installations can also use redesigned components that are more suitable for LTDH. Notably, new buildings with low heat demand should not be equipped with smaller heat transfer areas than those that would have been installed in the past. Instead, new installations should have almost the same size as old installations. This simple design condition will facilitate the use of both heat pumps and LTDH in new buildings. By avoiding traditional installations with increased *Legionella* risk, domestic hot water can be prepared with lower supply temperatures. Redesigned heat exchangers should also be used in the substation installations. Previously, the focus was to obtain only low return temperatures in heat distribution networks. In the future, the focus should be to obtain both low supply and return temperatures. This can be achieved by reducing the temperature gap between district heating flows and customer flows in heat exchangers by using more efficient heat exchangers with longer thermal lengths.

Another technological development that will facilitate LTDH is the increased level of cross-over digitalisation used in buildings, substations, heat distribution networks and heat supply. Artificial intelligence will be used in monitoring methods for the rapid identification and elimination of malfunctions that increase network temperatures, smart load shifting by utilising the thermal mass of the buildings, continuous commissioning of control parameters and load forecasting for optimising the heat supply.

> In summary, tangible and appropriate technologies and methods are available for the implementation of low-temperature district heating. Early adopters have tested and implemented lower temperatures in both existing and new heat distribution networks. Thus, buildings can and should adopt the utilisation of lower temperatures in the future. Reductions in specific heat demand will also facilitate the use of lower temperatures. However, current technologies and methods can be further elaborated and refined by research and development.

9.2 Non-technical aspects

The competitiveness of combining LTDH with high-temperature district heating or using LTDH on a stand-alone basis has been discussed. It was identified that opportunities for low-temperature solutions are (i) the green value which is aligned with the future goals of fossil-free societies and (ii) the possibility to engage end users and prosumers in long term relationships. Both the green aspect and long customer relationships are non-technical opportunities that can improve the overall competitiveness of district heating in the decades to come.

Not only opportunities but also barriers for LTDH investments have been identified. For example, it has been shown that the payback period of such investments is long (8–15 years), which has led to the general opinion that LTDH is too expensive. In the context of LTDH being a future technology and the energy sector being built on infrastructure components with long operational and economic life spans, the payback argument should be weak. Another major barrier is that current business models with a focus on centralised heat supply and distribution dominate. When these are applied to the LTDH context, it makes the LTDH business case unattractive when compared to doing business as usual. In terms of standardisation, there is not yet any technical standard for LTDH installations, which slows implementation down which can be offset by efficient contracts.

While it remains difficult to predict the future, there are explicit goals to be reached by 2050. Notably, the future is fossil-free and circular. This means that there will be no fossil fuels nor any waste to incinerate for generating heat. Moreover, residuals from wood processing will most likely have alternative uses other than incineration. Thus, in the future, heat must come from other sources such as the sea, ground, sun, urban activity (e.g., sewage, infrastructure, datacentres, grocery stores, ice rinks) and industrial activity. LTDH is a very relevant technology for 2050. However, due to limited awareness about the conditions in 2050 and short term profit orientation, LTDH is a solution that is often foregone rather than exploited to mitigate the risk of limited heat supply in 2050.

> In summary, the largest barrier to undertaking LTDH investment is that it is not business as usual. One important factor that can explain the limited interest in future-proof LTDH technology is that the risk of limited heat supply in 2050, since current fossil fuels will not be available, has not yet become apparent for most end users and heat providers.

9.3 Policy implications

Current policies often have a conservative impact on the implementation of LTDH since most were implemented within a fossil fuel society. This creates a severe lock-in effect for lower temperatures. Hereby, a high demand occurs for revisions to current policies aimed at incentivising the implementation of LTDH. To facilitate the transition from the direct combustion of fossil fuels to the major use of renewable and recycled heat, it is important to promote competition that efficiently contributes to sustainable heat usage. Notably, policy revisions should provide a level playing field in the heat market.

Carbon pricing

The current lack of appropriate economic drivers for decarbonisation has been the most significant barrier to the decarbonisation of district heating systems. National governments can either apply carbon pricing to punish the use of fossil fuels or provide market support for alternative heating technologies that mitigate climate change. Carbon pricing is a more general and efficient option since the market will choose alternative technologies. Market support is a less efficient option since the alternative technologies to be supported must be chosen by governments, which lack perfect foresight regarding the heat market.

Carbon pricing should correspond to the forecasted damage cost from fossil fuel use. Sweden, Finland and Switzerland are the only countries in the world that apply appropriate carbon prices via high carbon dioxide taxes that nearly correspond to the forecasted damage costs from climate change. All other countries have not been sufficiently active in providing appropriate economic drivers for the mitigation of climate change.

However, as discussed in the introduction, many European countries have now considered the implementation of domestic carbon dioxide taxes as a complement to the European emission trading system. These national tax tools will be required when the countries must fulfil the new stronger EU ambition of considerably lowering carbon dioxide emissions by 2030. Through this recent and eagerly awaited policy change, more obvious business cases will appear for the decarbonisation of European district heating systems.

The economic benefits of lower heat distribution temperatures will be approximately five times higher when using renewables and recycled heat when compared to the traditional use of fuels that can easily create high temperatures. However, this economic driver for LTDH can, together with carbon pricing, be an important combined economic driver for decarbonisation. This situation is illustrated in Figure 91 to show the magnitudes of expected carbon prices and LTDH benefits in relation to current and future heat supply costs in district heating systems.

The information presented in Figure 91 is based on an assumed current delivery cost of 70 euro/MWh, while the current specific carbon dioxide emissions are 100 kg/MWh. By assuming an additional carbon price of 100 euro/ton, the additional LCOH will be 10 euro/MWh. The total current LCOH then becomes 80 euro/MWh for the first alternative in this figure.

In the second alternative, the LCOH for future heat supply with full decarbonisation is assumed to be 84 euro/MWh without using lower temperatures. This will provide a negative decarbonisation benefit with 4 euro/MWh for the second alternative since 84 euro/MWh is more costly than 80 euro/MWh for the first alternative.

In the third alternative, decarbonisation will also be combined with low-temperature heat distribution. Here, the LTDH benefit can be estimated as 15 euro/MWh. This estimation is copied from the end of Chapter 2, with a general possible total temperature reduction of 30 °C and a future cost reduction gradient of 0.5 euro/MWh °C. Hence, future heat delivery cost can be reduced by 18%. In this third alternative, the remaining decarbonisation benefit becomes 11 euro/MWh

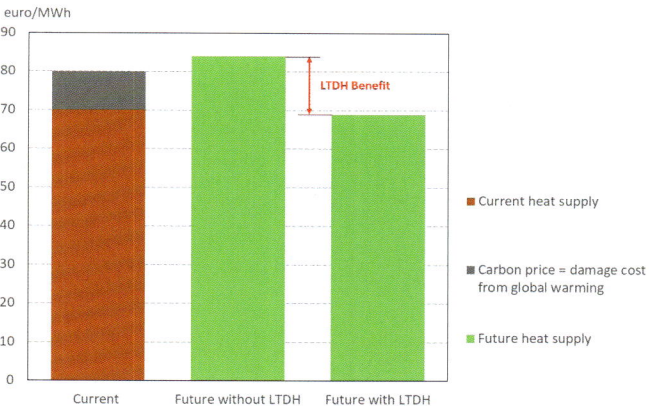

Figure 91. The LCOH for three alternatives: The current situation with an additional carbon price, the future situation with a decarbonised heat supply and without LTDH, and the future situation with LTDH. The difference in LCOH for the two latter alternatives is illustrated as the LTDH benefit.

since 69 euro/MWh is lower than 80 euro/MWh for the first alternative. Without the carbon price, the decarbonisation benefit would be only 1 euro/MWh. This low decarbonisation benefit alone would not be beneficial enough to implement the decarbonisation. Hence, the combination of both carbon pricing and lower heat distribution temperatures are essential to obtain a sufficiently strong combined incentive for decarbonisation.

The annual heat delivery in European district heating systems for 2050 has been forecasted at 950 TWh in Figure 1. The annual LTDH benefit then becomes 14 billion euro if LTDH heating were to be fully implemented. This annual benefit represents a net present value of over 200 billion euro.

Institutional rules

Various institutional rules (e.g., laws, regulations, standards etc.) regarding the heat market were originally set in the context of a fossil fuel society. Notably, many of these rules are now barriers to low-temperature heat sources and heat distribution. Therefore, these institutional rules must be properly revised.

One obvious example is the various national legislations concerning *Legionella*. Such legislations were mainly designed for two technical solutions (buffer tanks and hot water circulation) that are both very sensitive for *Legionella* growth. Higher temperature demands were previously chosen as a priority for the mitigation of *Legionella* risk via these methods. This option was appropriate when it was easy to create higher temperatures using fossil fuels. One important reflection is that the entire global HVAC community have continued to use the same traditional hot water preparation methods used before the initial discovery of the *Legionella* problem in Philadelphia, USA in 1976. Notably, new and safer preparation methods have not been implemented to systematically avoid *Legionella* risk. Instead, the national temperature demands for domestic hot water were only increased to reduce *Legionella* risk. One important policy action could involve using the national *Legionella* legislations to declare that other alternative preparation methods with considerably lower *Legionella* risk should not have the same temperature demand as the two less appropriate traditional technical solutions.

A second example is the traditional habit of using rather high design temperatures of approximately 60–70 °C when sizing radiators and other heat-emitting areas for space heating. This traditional design habit involves providing small radiators for new buildings with low heat demand. An excellent rule appears in Switzerland, where the 2009 national standard suggests 40 °C for radiators and 30 °C for floor heating as suitable design temperatures for new buildings. However, the upper permitted temperatures for these heating devices are 50 and 35 °C, respectively.

A third example considers when to ban the use of fossil fuels in new and existing buildings. In Norway, a 2018 regulation introduced a general ban on the use of domestic fuel oil for heating in buildings after January 1, 2020. However, some types of buildings were given a longer transition period until January 1, 2025. In the Austrian building sector, the government has announced the phasing out of all oil- and coal-fired heating systems for new construction by 2020 and all buildings by 2035. Moreover, the use of natural gas for heating will be restricted in new Austrian buildings from 2025 onward. Additionally, the German government is planning a ban on oil-based heating in buildings from 2026. Similar future bans on the use of natural gas in buildings will further provide improved decarbonisation conditions. Within the EU, a common ban on the use of any fossil fuels in buildings can be introduced into a revised Ecodesign directive.

Beyond these three rule revisions, other important policy drivers can also be considered. For example, green footprints could be requested from all energy providers in the national heat markets. This can be accomplished by requesting obligatory decarbonisation plans from all companies that provide energy input or heat into the heat market, such as distributors of fuel oil, natural gas, and biomass as well as distributors of heat and electricity. In this context, heat recycling can be made more attractive by defining this heat as being free of both primary energy supply and carbon dioxide emissions. The responsibility for these two aspects should be taken solely by the primary user based on the polluter pays principle.

Market support
Concerning governmental market support as an economic driver for decarbonisation, the use of LTDH can be made mandatory or beneficial for obtaining support for the decarbonisation of district heating systems. Notably, very few specific requests for LTDH have been identified in national supports for decarbonisation.

> In summary, while the economic benefit of low-temperature district heating can reduce the LCOH from future district heating systems, the benefit in current systems remains limited. Hence, this benefit alone is not currently strong enough to push the transition towards more decarbonised district heating systems. Carbon pricing or other efficient policy drivers must be used as strong parallel economic drivers for incentivising decarbonisation. Old institutional rules must also be properly revised for better alignment with low-temperature district heating.

9.4 Recommendations
The final recommendations concerning low-temperature heat distribution are directed towards district heating providers. These providers must initiate actions related to visions, strategies and planning measures for providing decarbonisation combined with LTDH.

Regarding long term visions, the recommendation is to involve owners and management to develop and establish ambitious decarbonisation paths for their district heating systems. The decarbonisation paths should be expressed with decarbonisation goals for specific years and coordinated with related activities (e.g., spatial energy planning). It is vital to communicate these decarbonisation paths to customers, staff, suppliers and the general public to gain awareness and support.

On the topic of short and long term strategies, the recommendation is to identify the actions necessary to achieve the adopted decarbonisation paths. An initial activity can involve studying how other district heating providers have implemented actions to fulfil their decarbonisation paths. For this context, it is vital to realise that lower heat distribution temperatures in both existing and new district heating systems are financially important for decarbonisation. However, it is also very important to realise that the major economic benefits of LTDH are not completely evident in traditional systems before they have been decarbonised with low-temperature heat sources. It should be communicated to the company staff that the main conclusion from district heating providers that have already lowered their heat distribution temperatures is: 'It was easier to implement than initially feared!' Thus, it is important to begin the learning process for identifying temperature barriers in substations and customer heating systems to achieve lower distribution temperatures in existing systems. It is also critical to establish workflows, technical solutions, communication strategies and business models to eliminate these barriers. New heat distribution configurations should also be assessed for new systems to safeguard low-temperature heat distribution.

Regarding short term planning measures, the initial recommendation is to pay attention to existing substations and customer heating systems. Low-hanging fruit should be identified with regard to high heat distribution temperatures from malfunctioning substations and high temperature demand in customer heating systems. Over the years, the company staff will recognise the low costs of eliminating malfunctions in existing substations and customer heating systems. Moreover, they will also learn how to avoid these malfunctions in new systems. They can also establish monitoring systems based on current heat delivery measurements and implement data-mining and intelligent algorithms in these systems for the rapid identification of malfunctions. Beyond substations, it is important to identify major possibilities for obtaining lower temperatures in existing heat distribution networks by tracking excessive circulation flows.

Regarding long term planning measures, district heating providers should follow the identified strategies to achieve their decarbonisation goals.

Regarding the learning process for establishing LTDH, providers should attempt to cooperate with research organisations such as universities or research institutes since these organisations are trained to develop new technologies and methods. The company staff should also participate in various education and training activities related to LTDH and appropriate installations in both buildings and networks.

> In summary, while it is easier to realise LTDH than many people fear, the actual implementation must be properly organised. In this work, long-term visions must express the future direction for decarbonisation, short- and long-term strategies should identify what to do, and both short- and long-term planning measures should outline how to perform the required work.

9.5 Main conclusion of this guidebook

Along with the necessary and expected national carbon pricing schemes, low-temperature district heating is a major economic driver for the decarbonisation of European district heating systems. From a technical viewpoint, it is possible to implement low-temperature district heating since several early adopters have provided clear evidence of its suitability. However, the hurdles to start the transition are old habits and lock-in effects from application of current technology together with a lack of understanding of how to efficiently link stakeholders to each other.

10 ANNEXES

Author: Sven Werner, Halmstad University.

In this final chapter, detailed information has been gathered concerning identified network configurations in early implementations, the short net list of detailed demonstration case descriptions compiled in this project, the longer gross list of identified inspiration initiatives for obtaining lower temperatures, the location index list for all activities mentioned in this guidebook, and brief presentations of participants in this project and some corresponding dissemination activities.

10.1 Typical configurations for low-temperature heat distribution networks

In this section, five typical network configurations for low-temperature heat distribution are presented together with some associated variants. These configurations have been identified from projects performed by early adopters of low-temperature heat distribution.

The five configurations are classified into two main groups: **warm** and **cold** district heating systems. In warm systems, the supply temperatures are high enough to deliver heat for typical heat demands in buildings without any additional heat supply in the buildings. In cold systems, ultra-low supply temperatures are applied. Some additional heat supply is then required in the customer buildings to meet the typical temperature demands.

In low-energy buildings, the typical thermal load with the highest temperature demand is the preparation of domestic hot water, requiring a supply temperature of at least 50 °C. Hence, supply sources below 50 °C can be classified as ultra-low supply temperatures in cold district heating systems. Hereby, supply temperatures above this temperature are used in warm district heating systems. The distinction between warm and cold networks has been previously defined in the initial definition pages and discussed in Sections 1.2 and 4.11.

The five configuration groups are presented in Table 23 together with their group affiliations and main features. These configurations are later presented in more detail in this section, including configuration layouts, characteristics, typical temperature levels, some implemented installation examples, and advantages/disadvantages, in the following sub-sections. Temperature levels for supply and return temperatures are later expressed as annual time-averaged values. Some variation can appear in daily and hourly values.

Warm – Classic configuration

In this first configuration group, the traditional network configuration is reused from the earlier second and third technology generations. The general Classic configuration is presented in Figure 92, while the corresponding characteristics are summarised in Table 24.

This configuration was originally implemented when the heat supply was based on primary energy use from fossil fuels, from which high temperatures from fuel combustion were readily available. Less attention was directed towards low-temperature heat distribution.

Table 23. Overview of the presented network configurations for low-temperature heat distribution with corresponding group affiliations and main features.

Main group	Configuration group	Main features
Warm	Classic	The traditional network configuration used in earlier technology generations
Warm	Modified Classic	Modification of the traditional configuration in order to obtain the lowest possible supply temperature without additional heating in buildings
Warm	Multi-Level	Configurations with at least two supply pipes that contain different supply temperatures
Cold	Ultra-Low	Configurations that use ultra-low supply temperatures for heat deliveries to buildings with additional heating
Cold/ Warm	CHC	Configurations that combine heat and cold deliveries by Combined Heating and Cooling (CHC); the heat can either be distributed in cold or warm networks

A circulation flow is required to keep a suitable supply temperature when continuous delivery flow is not available. The general conclusion is that it is not possible to utilise supply temperatures lower than 60–65 °C with this traditional network configuration. This limitation is linked to the need safeguard installed hot water tanks and hot water circulation from *Legionella* growth.

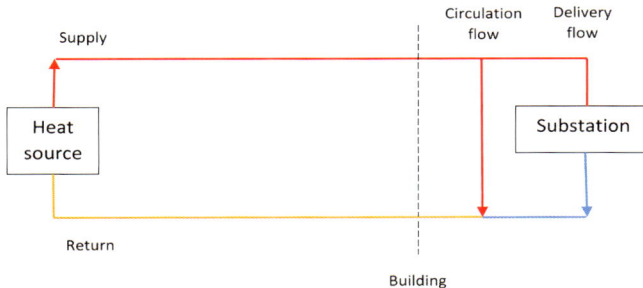

Figure 92. General layout for the Warm – Classic configuration

Two different variants can be implemented in this network configuration to reduce the temperature level. The first variant involves connecting a secondary network or a substation with a **supply-to-supply connection,** as shown in Figure 93. The return pipes from these local networks or substations are reconnected to the supply pipe to avoid undesired temperature pollution of the return flow. This cascading variant will generally provide lower return temperatures in the heat distribution network but requires a circulation pump in the supply pipe. One major disadvantage is that the connection must be close to a major supply pipe servicing many substations beyond the supply-to-supply connection. This variant is less suitable in the outskirts of the networks.

Examples of installations of this configuration include absorption chillers, industrial heat demands, etc. The Bromölla district heating network in Sweden recovers heat from a pulp and paper mill. The supply temperature is first delivered to a local industrial facility while the return pipe from this local industry facility is used as supply pipe for the municipal district heating system.

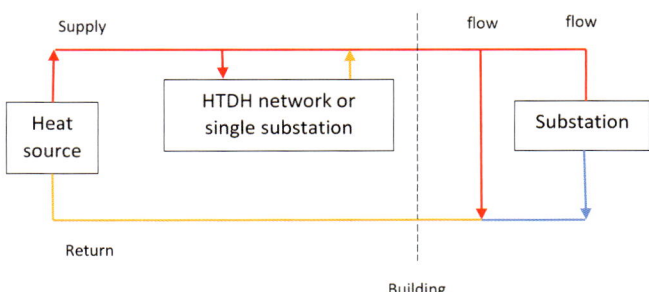

Figure 93. General layout for the supply-to-supply variant of the Warm – Classic configuration

The second variant of the Classic configuration involves connecting a local network or single substation with a **return-to-return connection,** as shown in Figure 94. The return temperature from the network is used as supply temperature for local networks or for substations that have low-temperature heat demands. This cascading variant will provide lower return temperatures in the heat distribution network but also requires a circulation pump in the return pipe. One ma-

Table 24. Characteristics for the Warm – Classic configuration

Characteristics	
Classic district heating technology with two parallel pipes that can be used with lowest possible temperatures	
Typical temperature levels [°C] in low temperature operation	
Supply temperatures	60-65
Return temperatures	30-35
Implemented installation examples	
Gleisdorf, Graz-Reininghaus, Salzburg-Lehen, Villach-Landskron, Kortrijk, Braunschweig-Rautheim, Chemnitz-Brühl, Crailsheim-Hirtenwiesen, Freiburg-Gutleutmatten, Munich-Ackermannbogen, Munich-Freiham, Wüstenrot-Weihenbronn, Aarhus-Lystrup, Boulogne-sur-Mer-EcoLiane, Västerås, Linköping-Ullstämma, Lund-Brunnshög, and Slough-Greenwatt Way	
Advantages	**Disadvantages**
• Current mainstream district heating technology utilises commercially available components that are installed using well-known methods. • The technology is familiar and often used when established district heating providers implement low-temperature heat distribution.	• *Legionella* risk is counteracted by high supply temperatures since hot water circulation and hot water tanks are often used in buildings. • Circulation and delivery flows are mixed in the return pipe, leading to higher return temperatures. • Relatively short thermal lengths are typically used in heat exchangers since the focus has been on low return temperatures.

jor disadvantage is that the connection must be close to a major return pipe servicing many substations before the return-to-return connection. This variant is also less suitable in the outskirts of networks. Sometimes a third pipe from the supply pipe is used when the temperature demand is occasionally higher than the return temperature in the network. This variant can be considered as a simple multi-level network.

Examples of installations of this configuration include Hamburg-Wilhelmsburg, Höje Tåstrup-Sönderby, Brescia-Via Carlo Venturi, and Nottingham-Sneinton. In addition, most ground heating systems for artificial snow melting in Sweden also use this network configuration.

Warm – Modified Classic configuration

This second configuration group involves modifications of the traditional configuration to obtain the lowest possible supply temperatures without any additional heating in the connected buildings. An example of a Modified Classic configuration is shown in Figure 95, while the corresponding characteristics are summarised in Table 25.

This Modified Classic configuration was developed by introducing three enhancements to the Classic configuration. First, the *Legionella* risk is reduced by avoiding the use of hot water circulation and hot water tanks in the buildings, enabling the use of lower supply temperatures. Flat substations are used in multi-family buildings to avoid hot water circulation inside buildings, and hot water is always generated in heat ex-

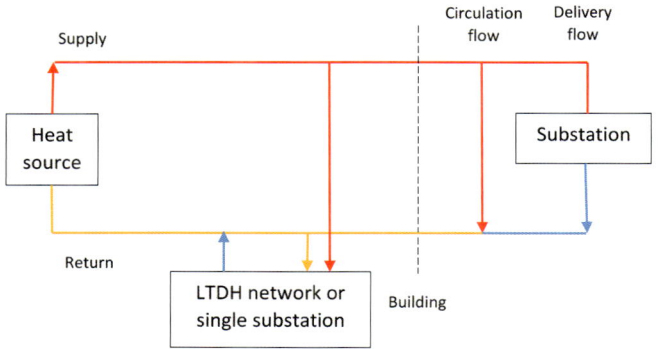

Figure 94. *General layout for the return-to-return variant of the Warm – Classic configuration*

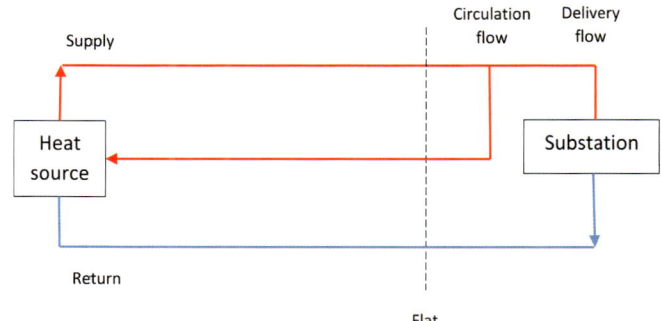

Figure 95. *General layout for the Warm – Modified Classic configuration*

Table 25. *Characteristics for the Warm – Modified Classic configuration*

Characteristics	
This modified classic configuration contains three parallel distribution pipes: one supply pipe, one return pipe for the delivery flow, and a second smaller return pipe for the circulation flow.	
Typical temperature levels [°C]	
Supply temperatures	50-55
Return temperatures	20-25
Planned installation example	
Halmstad-Ranagård (planned for 2022-23)	

Advantages	Disadvantages
• Absence of hot water circulation and storage tanks in buildings reduces the *Legionella* risk considerably. • Return pipe is not be polluted by uncooled supply temperatures and all circulation flows are controlled due to separation of the delivery and circulation flows. • Longer thermal lengths in heat exchangers allow lower supply temperatures.	• Current substations without hot water circulation must be redesigned since the valve for control of hot water preparation must maintain a minimum flow to provide a continuous uncontrolled circulation flow. • Heat exchangers with longer thermal lengths and service pipes with triple pipe casings are not yet commercially available.

changers to avoid hot water tanks. Second, circulation and return flows are separated by a smaller third pipe that takes care of the circulation flow, leading to lower temperatures in the return pipe. Third, longer thermal lengths are used in the substation heat exchangers, enabling the use of lower supply temperatures. This new network configuration has been presented and analysed in (Averfalk & Werner, 2018) and (Averfalk, Ottermo, & Werner, 2019). The main conclusion is that the temperature levels will be about 10 °C lower than the lowest possible level in the Classic configuration.

Warm – Multi-level configuration
This third configuration group contains designs with at least two supply pipes that contain different supply temperatures. The general configuration layout is shown in Figure 96, while the corresponding characteristics are summarised in Table 26.

In this configuration, substations are connected between suitable temperature levels depending on the customer's temperature demands. One substation can be connected between the high and medium temperatures, while another substation can be connected between the medium and low temperatures; a third substation can be connected between the high and low temperatures. Hence, the basic cascading principle is utilised by the return flow from one substation providing the supply flow to another substation.

The **first** variant of this network configuration utilises simply one supply pipe for space heating, another supply pipe for hot water preparation, and a third pipe for the return flow. This three-pipe variant was originally implemented in Berlin at the Charlottenburg and Neukölln networks in 1919 and 1921, respectively, and has been extensively used in the Berlin district heating system for many years.

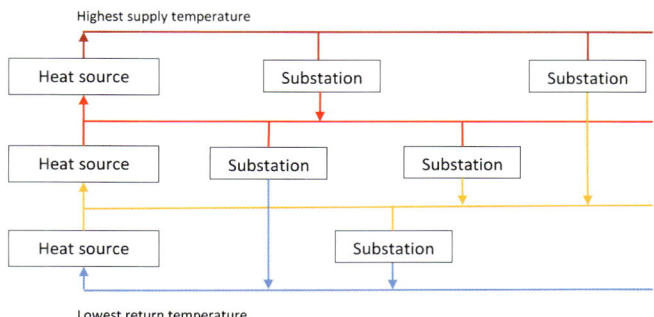

Figure 96. General layout for the Warm – Multi-Level configuration

Another **second** variant of the Multi-level configuration simply uses four pipes with one pair of pipes for space heating and another pair for preparation of domestic hot water.

Another **third** variant has been implemented at the Värtan supply site in Stockholm. The new residential development area, Hjorthagen, has somewhat lower network temperatures than ordinary networks; the reason for this is that the Hjorthagen area is located close to the Värtan site. The lower supply temperatures are obtained from the intermediate supply temperature within the Värtan site after the large heat pumps and the flue

Table 26. Characteristics for the Warm – Multi-level configuration

Characteristics	
This configuration uses more than one parallel supply pipe in pipe trenches; lower supply temperatures can then be provided from substations with high return temperatures.	
Typical temperature levels [°C]	
Supply temperatures	50-120
Return temperatures	30-80
Implemented installation examples	
Naters-Reka Feriendorf, Stuttgart-NeckarPark, Semhach geothermal-based network in Chevilly-Larue and L'Hay Les Roses (south of Paris), Paris-Clichy Batignolles, Paris-Nord Est, Dongen-Plan Beljaart, and Stockholm-Hjorthagen	
Advantages	**Disadvantages**
• Delivered supply temperatures are customised for each customer's temperature demand. • Lower return temperatures are obtained compared to those in the Classic configuration.	• Providing the required flow demand at each supply temperature level is a challenge, which must be addressed by a balancing function.

gas condensation unit in the biomass CHP plant. This configuration increases the heat recovery from these two groups of units. If the intermediate temperature becomes too low, the supply temperature to the Hjorthagen can be increased by blending in flow with the ordinary supply temperature.

Cold – Ultra-low configuration

This fourth configuration group contains designs that use ultra-low supply temperatures for heat deliveries. The general configuration layout is shown in Figure 97, while the corresponding characteristics are summarised in Table 27.

The Ultra-low configuration is based on simple utilisation of heat sources with temperatures that are lower than the traditional customer temperature demands. Instead of increasing the temperature centrally, the original heat source temperature is distributed in the network and each customer is responsible for meeting their own temperature demands. Typical heat sources are the sea, lake, ground, and mine waters together with low-temperature cooling waters from industrial processes or chillers.

This configuration can be used for a wide range of distributed supply temperatures. If the distributed supply temperature is relatively low, the additional heating in the buildings can be supplied by a heat pump using the heat distribution network as a heat source. If the distributed supply temperature is relatively high, a traditional electric hot water heater can be used for final heating of the domestic hot water, since the distributed supply temperature can be used to preheat the domestic hot water and meet the entire space heating demand.

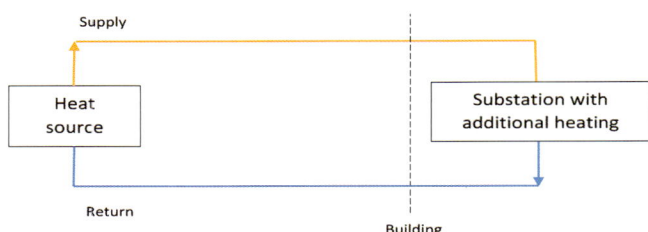

Figure 97. General layout for the Cold – Ultra-Low configuration

Cold/Warm – CHC configuration

This fifth configuration group contains designs that combine heat and cold deliveries by Combined Heating and Cooling (CHC), where the heat can either be distributed in cold or warm networks. This configuration takes advantage of the synergy of having both heating and cooling demands in the same area and becomes more beneficial when cooling demands are high compared to heating demands.

Table 27. Characteristics for the Cold – Ultra-Low configuration

Characteristics	
The obtained temperature from low-temperature heat sources is not upgraded centrally, but locally in each substation.	
Typical temperature levels [°C]	
Supply temperatures	25-50
Return temperatures	10-30
Implemented installation examples	
Vienna-Waldmühle-Roaun, Leuven-Janseniushof, Zürich-FGZ, Bochum-Werne, Hamburg-Harburg, Hameln-Ohrberg, Nümbrecht, Troisdorf, Wüstenrot-Agrothermie, Zwickau-Hochschule, Bjerringbro-Grundfos, Birkeröd-Norfors, Hörsholm-Norfors, Silkeborg-Balle, Aarhus-Geding, Hague-Duindrop, Hengelo, and Bristol-Owen square	

Advantages	Disadvantages
• The heat distribution system does not have to meet the temperature demand from the customer with the highest temperature demand. • In areas with low heat densities, the distribution heat losses will be considerably lower. • This configuration can be implemented everywhere, since ambient heat can always be used.	• The installation cost for many small heat pumps can be higher than those for corresponding central heat pumps in the network because of economy of scale. • The temperature differences between supply and return temperatures are often small. This leads to higher flow demands that require larger and more expensive pipes. • There is limited benefit in areas with high heat densities since traditional distribution heat losses are relatively low in these areas.

The excess heat from cooling can be used for heating, while the excess cold from heating can be used for cooling. Large buildings can also take advantage of this synergy without connecting to district heating and cooling systems.

The general configuration layout for **cold networks** is shown in Figure 98, while the corresponding characteristics are summarised in Table 28. The two heat pump function can be replaced with a reversible heat pump if heating and cooling demands do not coincide. An important condition is that each decentralised connected device must have its own distribution pump.

For this general configuration, flow demands for heating and cooling can partly be balanced directly in the distribution networks. During the winter when heating demands are highest, the supply temperatures decrease and must be provided with an external heat supply. Inversely, during the summer when cooling demands are the highest, the supply temperatures increase and must be provided by an external cold supply (heat removal). The short term balance between heating and cooling can be managed by using the heat distribution network as a short term heat storage. The annual heat balance between heating and cooling can be managed with a large low-temperature aquifer or borehole heat storage.

Several variants are possible within this configuration group with cold networks, since this general design can utilise a wide range of distributed supply temperatures.

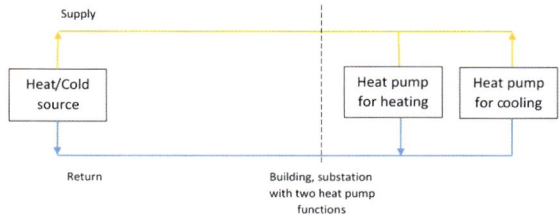

Figure 98. General layout for the Cold – CHC configuration

The **first variant** is possible when relatively high supply temperatures exist. In this design, heat delivery is possible without heat pumps for heating; instead, heat exchangers can be used for transferring heat to the low-temperature demands in the buildings.

The **second variant** is possible when relatively low supply temperatures exist. This design leads to low return temperatures, creating the opportunity to connect cooling substations directly to the return pipe. This solution can also be accomplished in ordinary district cooling systems when the return flow is utilised as a heat source for heat pumps, such as in Geneva-Lac Nation and Örebro-Hospital.

The **third variant** contains neither specific supply nor return pipes. Instead, all buildings are connected to the same pipe that contains a constant circulation flow in an integrated loop with intermediate temperatures. All heat pumps for heating transfer heat from the heat distribution system, while all heat pumps for cooling add heat to the heat distribution network. The purpose with this third variant is to eliminate local heat or cold sur-

Table 28. Characteristics for the Cold – CHC configuration

Characteristics	
This configuration offers both district heating and district cooling in cold networks.	
Typical temperature levels [°C] for cold networks	
Supply temperatures	10-45
Return temperatures	5-25
Implemented installation examples	
Vienna-Krieau, Kelowna-UBC, Richmond-Alexandra, Geneva-Lac Nation, Rotkreutz-Suurstoffi, Visp-West, Zürich-ETH Campus Hönggerberg, Paris-Saclay, Heerlen-Mijnwater, Bergen University, Ulstein-Fjordvarme, Lund-Medicon Village, Örebro-Hospital, and London-South Bank University	

Advantages	Disadvantages
• Direct heat recovery is possible for low-temperature heat obtained from cooling processes. • This configuration is advantageous when cooling demands appear at the same time as heating demands.	• The temperature differences between supply and return temperatures are often small; this leads to higher flow demands that require larger and more expensive pipes.

pluses or deficits in the network. This configuration variant can also be balanced with external heat and cold supplies, as well as with thermal storages.

The general configuration layout for **warm networks** is shown in Figure 99, while the corresponding characteristics are summarised in Table 29. Centralised heat pumps deliver heat to the district heating network and cold to the district cooling network.

The general configuration is based on centralised single heat pumps that are powerful enough to generate high supply temperatures for traditional district heating systems and low supply temperatures for traditional district cooling systems simultaneously. The heat is supplied from the heat pump condensers, while the cold is supplied from the heat pump evaporators. The short term balance between heating and cooling can be managed by separate heat and cold storages.

This synergy in warm networks has been implemented in some Swedish district heating systems that also operate district cooling systems. In 2019, 20% of the cold input to Swedish district cooling systems came from cold supplies using this configuration (Burstein, 2020).

In Helsinki, the Katri Vala underground heat pump plant utilises heat from both the district cooling system and the purified sewage water system, while simultaneously providing cold to the district cooling system (Lemola, 2016).

As previously discussed in Section 8.5, this is the core configuration that was implemented at Stanford University in 2015.

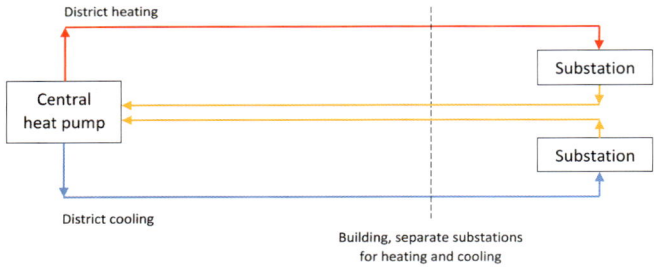

Figure 99. General layout for the Warm – CHC configuration

Table 29. Characteristics for the Warm – CHC configuration

Characteristics	
This configuration supplies heat to warm networks and cold to cold networks.	
Typical temperature levels for warm networks [°C]	
Supply temperatures	60-80
Return temperatures	40-50
Implemented installation examples	
Stockholm, Helsinki, and Stanford University	

Advantages	Disadvantages
• This configuration can be used in conjunction with existing district heating and cooling networks. • Single centralised heat pumps are used for both heating and cooling.	• More electricity is used for upgrading the heat in the heat pumps from the cooling levels to the higher temperature levels in warm networks.

10.2 Net list of completed detailed descriptions of cases

This list consists of all detailed descriptions of the forty demonstration cases from nine countries that were written within the frame of the IEA DHC Annex TS2 project. Shaded fields mark the fifteen cases with detailed descriptions in Chapter 7. The definitions for the six demo-classes were provided in Section 7.1.

Table 30. List of TS2 cases that aim for lower heat distribution temperatures.

Country	City	Title	Demo-class
AT	Gleisdorf	Transition of the district heating system of Gleisdorf	
DE	Darmstadt	Energy efficient university campus	
DE	Wüstenrot	District Heating Weihenbronn	
NO	Bergen	Cold district heating at Bergen University	
NO	Trondheim	District heating ring at the Norwegian University	
DE	Chemnitz	Quarter Brühl	
DE	Mannheim	Conversion of a military facility Benjamin Franklin Village.	
NL	Heerlen	Heerlen, Parkstad Limburg, Mijnwater	
CH	Zürich	FGZ Zürich – Ultra-low temperature district heating for the residential district	
AT	Salzburg-Lehen	Low temperature sub-system of Salzburgs district heating network	
AT	Graz-Reininghaus	New city quarter supplied by low temperature district heating based on waste heat utilisation with large scale heat pumps	
AT	Wörgl	Low temperature secondary network for 20 affordable row houses	
AT	Vienna	Viertel 2 plus	

Country	City	Title	Demo-class
AT	Villach	Landskron Energy Island	
DE	Braunschweig	Excess heat recovery from data centre	
DE	Bamberg	Energy supply concepts of the conversion area "Lagarde"	
DE	Wüstenrot	District Heating "Ortsmitte Wüstenrot" (EnVisaGePlus)	
IR	Dublin	Tallaght District Heating Scheme	
SE	Brunnshög-Lund	Brunnshög- Lund	
SE	Linköping	Ullstämma	
SE	Stockholm	Hjorthagen	
SE	Västerås	Five low-temperature district heating networks	
UK	Nottingham	REMOURBAN - H2020 Smart cities and communities project	
DE	Kassel	Geo-solar district heating in Feldlager Kassel	
DE	Moosburg	District Heating 4.0: Moosburg an der Isar	
DE	Recklinghausen	District Heating 4.0 network in Recklinghausen-Hillerheide	
DE	Aachen	Development of district heating systems for a new settlement	
DE	Neuburg	District heating supply with low-calorific waste heat sources and matrix control.	
DE	Oldenburg	Energetic Neighbourhood Quarter - Transition from high to low-temperature network	

Country	City	Title	Demo-class
DK	Bjerringbro	A pilot test of low and ultra-low-temperature district heating	🏢
DK	Copenhagen	Return temperature optimization in city areas	🏢
DK	Viborg	Optimised pump control	🏢
DK	Copenhagen	Improved heating of domestic hot water tanks	🏢
DK	Frederiksberg	Automatic return temperature limitation for space heating	🏢
DK	Viborg	Viborg district heating network / Motivation Tariff	🏢
DK	Viborg	Danish Clean Water	🏢
SE	Borlänge	Tjärna Ängar	🏢
SE	Göteborg	Backa Röd	🏢
DE	Kassel	Research facility District LAB	⚗️
SE	Mariehamn & Stockholm	EnerStore LowTemp	⚗️

10.3 Gross list of low-temperature inspiration initiatives

This longer gross list consists of 165 inspiration initiatives for obtaining lower temperatures in either buildings or heat distribution systems. The first column considers the country code, location, and project name. The second column reports the appropriate demo-classes that are defined in Section 7.1. The third column presents the weblink that was available in 2020. Note that weblinks can disappear or be moved. However, weblinks are not available for all initiatives. The fourth column contains the identified network configuration according to the definitions in Section 10.1. The fifth column reports the status for each initiative and considers the 2020 situation. The status ranges widely from early proposals and simulations to actual implemented systems.

Table 31. Gross list of identified inspiration initiatives targeting lower heat distribution temperatures.

Country code and location	Demo-class	Website link	Network configuration	Status 2020
AT, Gleisdorf, system	A. Existing area & existing system	https://thermaflex.greenenergylab.at/e4a_demonstrator/demo-4/	1C Rehabilitation	Planned
AT, Graz, Hummel Kaserne	E. Simulation	https://www.acr.ac.at/fileadmin/documents/ACR-Wissen/GET_Next_Generation_Heat_08_2017.pdf	1B Return to return	Proposed
AT, Graz, Reininghaus	C. New area & new system	http://www.reininghaus-findet-stadt.at/infrastruktur/energie/	1 Classic	Implemented
AT, Güssing, Aktivpark	E. Simulation	https://www.acr.ac.at/fileadmin/documents/ACR-Wissen/GET_Next_Generation_Heat_08_2017.pdf	1 Classic	Proposed
AT, Salzburg, Lehen	C. New area & new system	http://task52.iea-shc.org/Data/Sites/1/publications/IEA-SHC%20Task%2052%20STC2-Best%20practice%20summary%20report_2017-08-31_FINAL.pdf	1 Classic	Implemented
AT, T2LowEx, Austria	D. Building	https://www.aee-intec.at/t2lowex-transformation-von-konventionellen-waermenetzen-in-richtung-niedertemperaturnetze-durch-sekundaerseitige-massnahmen-p211	Secondary	Implemented
AT, Vienna, Seestadt Aspern	E. Simulation	https://www.acr.ac.at/fileadmin/documents/ACR-Wissen/GET_Next_Generation_Heat_08_2017.pdf	1B Return to return	Proposed
AT, Vienna, Waldmühle-Roaun	C. New area & new system	https://www.klimafonds.gv.at/wp-content/uploads/sites/6/20170412Wien-Energie-GmbHSolare-Groanlagen-2013EBB-366774KR13ST4K11070.pdf	4 Ultra-Low	Implemented

Country code and location	Demo-class	Website link	Network configuration	Status 2020
AT, Vienna, Krieau	C. New area & new system	https://energie-krieau.at/	5 CHC	Implemented
AT, Villach, Landskron	C. New area & new system	https://www.klimafonds.gv.at/wp-content/uploads/sites/6/Anlagensteckbrief-Energieinsel-Landskron.pdf	1 Classic	Implemented
AT, Wörgl, 20 houses	C. New area & new system	https://thermaflex.com/en/download/file/670	1 Classic	Implemented
AT, Wörgl, Winkelweg Siedlung	E. Simulation	https://www.acr.ac.at/fileadmin/documents/ACR-Wissen/GET_Next_Generation_Heat_08_2017.pdf	1B Return to return	Proposed
BE, Kortrijk, De Venning and Mijlpalen	C. New area & new system	https://guidetodistrictheating.eu/kortrijk/	1 Classic	Implemented
BE, Leuven, Janseniushof	C. New area & new system	https://www.bouwkroniek.be/article/eerste-koudenet-in-vlaanderen-verwarmt-honderd-leuvense-woningen.17343	4 Ultra-Low	Implemented
CA, Hamilton ON, McMaster Innovation Park	C. New area & new system	http://www.districtenergy-digital.org/districtenergy/2017q1?pg=9#pg9	1 Classic	Implemented
CA, Kelowna BC, UBC Okanagan Campus	B. Existing area & new system	https://facilities.ok.ubc.ca/geoexchange/des-operation/	5 CHC	Implemented
CA, Okotoks AB, Drake Landing	C. New area & new system	https://www.iea-dhc.org/fileadmin/documents/Annex_X/IEA_Annex_X_Final_Report_2014_-_Toward_4th_Generation_District_Heating.pdf	1 Classic	Implemented
CA, Oshawa ON, Ontario Tech University	B. Existing area & new system	https://sites.ontariotechu.ca/gogreen/initiatives/on-campus/energy.php	1 Classic	Implemented

Country code and location	Demo-class	Website link	Network configuration	Status 2020
CA, Richmond BC, Alexandra	C. New area & new system	http://www.luluislandenergy.ca/alexandra-district-energy-utility/	5 CHC	Implemented
		https://higherlogicdownload.s3.amazonaws.com/DISTRICTENERGY/UploadedImages/7d8cbd67-3d1f-4e6d-a9f5-c730372a054b/IDEA_Innovation_Award_2020_ADEU_Phase_4_Presentation_Format.pdf		
CA, Whistler BC, Cheakamus Crossing	C. New area & new system	https://www.whistler.ca/services/water-and-wastewater/district-energy-system	4 Ultra-Low	Implemented
CH, Geneva, Lac Nation	B. Existing area & new system	https://thermdis.eawag.ch/static/templates/files/ThermischeNetze_EtudeDeCas_ReseauThermiqueGeneveLacNations.pdf	5 CHC	Implemented
CH, Naters, Reka Feriendorf	C. New area & new system	http://task52.iea-shc.org/Data/Sites/1/publications/IEA-SHC%20Task%2052%20STC2-Best%20practice%20summary%20report_2017-08-31_FINAL.pdf	3 Multi-Level	Implemented
CH, Oberwald, Furka tunnel	B. Existing area & new system	http://www.energieregiongoms.ch/index.php/projekte/item/20-warmwasser-aus-dem-furkatunnel	4 Ultra-Low	Implemented
CH, Rotkreutz (Zug), Suurstoffi	C. New area & new system	https://www.energie-apero-schwyz.ch/media/files/abicht_160411.pdf	5 CHC	Implemented
CH, Visp, Visp-West	C. New area & new system	https://www.visp.ch/dokumente/artikel/16infoschreibenanergienetzvispwest.pdf	5 CHC	Implemented
CH, Zürich, ETH Hönggerberg	B. Existing area & new system	https://www.ethz.ch/en/the-eth-zurich/sustainability/campus/environment/energy/anergy-grid.html	5 CHC	Implemented
CH, Zürich, FGZ Friesenberg	B. Existing area & new system	http://www.fgzzh.ch/index.cfm?Nav=31&ID=151	5 CHC	Implemented
CR, Topusko,	A. Existing area & existing system	https://www.rewardheat.eu/en/Demonstration-Networks/Topusko	1B Return to return	Planned

Country code and location	Demo-class	Website link	Network configuration	Status 2020
DE, Aachen, Karl-Kuck-Strasse	C. New area & new system	https://docplayer.org/111127869-Erlaeuterungsbericht-zum-bebauungsplan-karl-kuck-strasse-im-stadtbezirk-aachen-brand.html	4 Ultra-Low	Planned
DE, Bamberg, Lagarde campus	B. Existing area & new system	https://www.stadtwerke-bamberg.de/unternehmen/beteiligungen/lagarde/	3 Multi-Level	Planned
DE, Bochum, Werne	C. New area & new system	http://www.bine.info/fileadmin/content/Presse/Projektinfos_2013/PM_13_2013/ProjektInfo_1313_engl_internetx.pdf	4 Ultra-Low	Implemented
DE, Braunschweig, Rautheim	C. New area & new system	https://www.reuseheat.eu/brunswick/	1 Classic	Implemented
DE, Chemnitz, Brühl	B. Existing area & new system	http://proceedings.ises.org/paper/swc2017/swc2017-0032-LalShrestha.pdf	1 Classic	Implemented
DE, Crailsheim, Hirtenwiesen	B. Existing area & new system	https://www.solarthermalworld.org/content/solar-district-heating-crailsheim-seasonal-borehole-storage	1 Classic	Implemented
DE, Darmstadt, Campus Lichtwiese	A. Existing area & existing system	https://www.enargus.de/pub/bscw.cgi/?op=enargus.eps2&v=10&q=lichtwiese&id=1051508	1C Rehabilitation	Proposed
DE, Freiburg, Gutleutmatten	C. New area & new system	http://task52.iea-shc.org/Data/Sites/1/publications/IEA-SHC%20Task%2052%20STC2-Best%20practice%20summary%20report_2017-08-31_FINAL.pdf	1 Classic	Implemented
DE, Hamburg, Bergedorfer Tor	C. New area & new system	https://www.enercity-contracting.de/pool/downloads/medienberichte/20181120_PM_Wegweisendes_Energiekonzept_Bergedorfer_Tor.pdf	3 Multi-Level	Planned
DE, Hamburg, Eisenbahnverein Harburg	B. Existing area & new system	https://www.ebv-harburg.de/unternehmen/infofilme/	4 Ultra-Low	Implemented
DE, Hamburg, Holsten	C. New area & new system	https://www.rewardheat.eu/en/Demonstration-Networks/Hamburg	2 Modified Classic	Planned

Country code and location	Demo-class	Website link	Network configuration	Status 2020
DE, Hamburg, Jenfelder Au	C. New area & new system	http://www.ewa-online.eu/id-17-symposium-proceedings.html	4 Ultra-Low	Proposed
DE, Hamburg, Wilhelmsburg	B. Existing area & new system	https://www.iba-hamburg.de/projekte/energiebunker/projekt/energiebunker.html	1B Return to return	Implemented
DE, Hameln, Solarsiedlung Am Ohrberg	C. New area & new system	http://www.fvee.de/fileadmin/publikationen/Themenhefte/th1997/th1997_02_03.pdf	4 Ultra-Low	Implemented
DE, Herne, Shamrockpark	B. Existing area & new system	https://project-clue.eu/demo-sites/shamrockpark-germany/	5 CHC	Planned
DE, Kassel, IEE Fraunhofer	F. Demolab	https://www.iee.fraunhofer.de/en/laboratories/District_LAB.html	3 Multi-Level	Planned
DE, Kassel, Zum Feldlager	E. Simulation	http://vbn.aau.dk/files/233873747/paper_137.pdf	4 Ultra-Low	Proposed
DE, Ludwigsburg, Sonnenberg LowEx	C. New area & new system	https://www.iea-dhc.org/fileadmin/documents/Annex_TS1/150420_IEA_DHC_Annex_TS1_case_studies_Brochure.pdf	4 Ultra-Low	Planned
DE, Ludwigsburg, Solar-Heat-Grid	A. Existing area & existing system	https://www.swlb.de/de/Energie/Nachhaltigkeit/SolarHeatGrid-/ALT-Kommunales-Klimaschutz-Modellprojekt-SolarHeatGrid.html	1 Classic	Planned
DE, Mannheim, Benjamin-Franklin	E. Simulation	https://www.mannheim.de/sites/default/files/page/22427/energiestudie.pdf	1 Classic	Proposed
DE, Moosburg, LTDH	B. Existing area & new system	https://www.iee.fraunhofer.de/de/projekte/suche/laufende/Machbarkeitsstudie_Moosburg.html	3 Multi-Level	Planned
DE, Munich, Ackermannbogen	C. New area & new system	https://www.iea-dhc.org/fileadmin/documents/Annex_X/IEA_Annex_X_Final_Report_2014_-_Toward_4th_Generation_District_Heating.pdf	1 Classic	Implemented
DE, Munich, Freiham	C. New area & new system	http://iea-gia.org/wp-content/uploads/2017/11/2-03-GRAME-Geothermal-Energy-in-the-Bavarian-Molasse-Basin.pdf	1 Classic	Implemented

Country code and location	Demo-class	Website link	Network configuration	Status 2020
DE, Neuburg, LTDH	C. New area & new system	https://sdec.is/wp-content/uploads/2019/11/kallert_anna.pdf	4 Ultra-Low	Planned
DE, Nümbrecht, Sohnius-Weide	C. New area & new system	https://www.gwn24.de/klimaumwelt/nahwaerme/	4 Ultra-Low	Implemented
DE, Nürnberg, Windsbach	C. New area & new system	https://www.tempo-dhc.eu/ http://www.4dh.eu/images/Leoni_Capretti_2018.pdf http://www.waerme-natuerlich.de/Nahwaerme-Neubaugebiet-Badstrasse/Downloads/	1 Classic	Planned
DE, Oldenburg, ENaQ	E. Simulation	https://www.enaq-fliegerhorst.de/	4 Ultra-Low	Planned
DE, Radolfzell, Kreuzbühl	C. New area & new system	https://www.radolfzell.de/kreuzbuehl	5 CHC	Planned
DE, Recklinghausen, Hillerheide	C. New area & new system	https://www.isek-hillerheide.de/energetische-sanierung	4 Ultra-Low	Planned
DE, Rosenheim, Heat Dispatch Centre	B. Existing area & new system	https://www.sciencedirect.com/science/article/pii/S1876610218305083	1 Classic	Proposed
DE, Stuttgart, NeckarPark	C. New area & new system	https://www.sciencedirect.com/science/article/pii/S1876610218305071	3 Multi-Level	Implemented
DE, Troisdorf, Friedrich-Wilhems-Hütte	C. New area & new system	https://docplayer.org/78291825-Praxisbeispiel-geothermie-in-troisdorf.html	4 Ultra-Low	Implemented
DE, Wüstenrot, Agrothermie	C. New area & new system	https://www.iea-dhc.org/fileadmin/documents/Annex_TS1/150420_IEA_DHC_Annex_TS1_case_studies_Brochure.pdf	4 Ultra-Low	Implemented
DE, Wüstenrot, Ortsmitte	B. Existing area & new system	http://www.envisage-wuestenrot.de/	1 Classic	Implemented
DE, Wüstenrot, Weihenbronn	A. Existing area & existing system	http://www.envisage-wuestenrot.de/teilprojekte/nahwarmenetz	1 Classic	Implemented

Country code and location	Demo-class	Website link	Network configuration	Status 2020
DE, Zwickau, Westsächsische Hochshule	B. Existing area & new system	http://floez-sachsen.de/dokument/Prof._Dr._Matthias_Hoffmann%2C_WHZ%2C_Thermische_Nutzung_von_Grubenw%C3%A4ssern_%E2%80%93_Projekt_Zwickau_1355474839_751.pdf	4 Ultra-Low	Implemented
DK, Albertslund, System	A. Existing area & existing system	https://fjernvarme.albertslund.dk/fjernvarme/lavtemperaturfjernvarme/	1C Rehabilitation	Planned
DK, Albertslund, Syd	A. Existing area & existing system	https://www.gate21.dk/nyhed/boliger-i-albertslund-syd-goeres-klar-til-fremtidens-varmeforsyning/	1C Rehabilitation	Planned
DK, Bjerringbro, Grundfos	A. Existing area & existing system	https://www.rm.dk/api/NewESDHBlock/DownloadFile?agendaPath=%5C%5CRMAPPS0221.onerm.dk%5Ccms01-ext%5CESDH%20Data%5CRM_Internet%5CDagsordener%5CUdvalg_vedroerende_bae%202016%5C20-04-2016%5CAaben_dagsorden&appendixId=134489	4 Ultra-Low	Implemented
DK, Copenhagen, Nordhavn	A. Existing area & existing system	http://www.energylabnordhavn.com/	Unknown	Planned
DK, Copenhagen, Building with 23 apartments	D. Building	Not available	Secondary	Implemented
DK, Copenhagen, Improved DHW tank	D. Building	Not available	Secondary	Implemented
DK, Frederiksberg, Building with 20 apartments	D. Building	Not available	Secondary	Implemented
DK, Hilleröd, Ullerödbyen	E. Simulation	http://www.fbbb.dk/Files/Filer/Jens_Lunding_-_Hilleroed_Kommune_varmeforsyning_29-10_2009.pdf	1 Classic	Proposed
DK, Höje Tåstrup, Sönderby	B. Existing area & new system	https://www.iea-dhc.org/fileadmin/documents/Annex_TS1/150420_IEA_DHC_Annex_TS1_case_studies_Brochure.pdf	1B Return to return	Implemented

Country code and location	Demo-class	Website link	Network configuration	Status 2020
DK, Höje Tåstrup, Österby	B. Existing area & new system	http://www.cooldh.eu/demo-sites-and-innovations-in-cool-dh/osterby-hoje-taastrup/	4 Ultra-Low	Planned
DK, Karise near Köge, Permatopia	C. New area & new system	https://www.permatopia.dk/andelsselskab/	Unknown	Planned
DK, Middelfart, System	A. Existing area & existing system	https://www.iea-dhc.org/fileadmin/documents/Annex_TS1/IEA_DHC_Annex_TS1_Subtask__D_Report_V8.pdf	1C Rehabilitation	Implemented
DK, Norfors-Birkeröd	B. Existing area & new system	http://www.norfors.dk/da-DK/Fjern-varme/F%C3%B8lg-Fjernvarmeprojek-tet-i-Birker%C3%B8d.aspx	4 Ultra-Low	Implemented
DK, Norfors-Hörsholm	C. New area & new system	https://energiforskning.dk/sites/energ-iteknologi.dk/files/slutrapporter/171026_slutrapport_64015-0053_-_ultra-lavtem-peraturfjernvarme_i_boligblokke_fi-nal_1_26102017_10411_1_29112017_14411.pdf	4 Ultra-Low	Implemented
DK, Norfors-Nivå, Teglbakken	C. New area & new system	https://energiforskning.dk/da/project/ul-tra-lavtemperaturfjernvarme-i-nybyggeri	4 Ultra-Low	Planned
DK, Odder, Five single-family houses	A. Existing area & existing system	https://energiforskning.dk/da/project/fjernvarmeunit-med-elpatron-til-ul-tra-lavtempeatur-fjernvarme	4 Ultra-Low	Implemented
DK, Silkeborg, Balle bygade	C. New area & new system	https://www.danskfjernvarme.dk/vi-den-om/f-u-konto-subsection/rapporter/2016-05-kold-fjernvarme	4 Ultra-Low	Implemented
DK, Tilst, Skjoldhøjparken	A. Existing area & existing system	https://www.danskfjernvarme.dk/groen-energi/projekter/lavtemperaturfjernvarme	1 Classic	Implemented
DK, Viborg, Optimised pump	D. Building	Not available	Secondary	Implemented
DK, Viborg, Motivation tariff	A. Existing area & existing system	Not available	1C Rehabilitation	Implemented
DK, Viborg, Danish clean water	A. Existing area & existing system	Not available	Secondary	Implemented

Country code and location	Demo-class	Website link	Network configuration	Status 2020
DK, Vinge, 59 single-family houses	C. New area & new system	http://www.relatedproject.eu/demonstrations/	4 Ultra-Low	Planned
DK, Aarhus, Geding	A. Existing area & existing system	https://www.sweco.dk/vi-tilbyder/cases/energi/fjernvarmekonvertering-fra-traditionel-til-ultra-lavtemperaturfjernvarme/	4 Ultra-Low	Implemented
DK, Aarhus, Lystrup	C. New area & new system	https://www.iea-dhc.org/fileadmin/documents/Annex_X/IEA_Annex_X_Final_Report_2014_-_Toward_4th_Generation_District_Heating.pdf	1 Classic	Implemented
EE, Kopli, System	A. Existing area & existing system	http://www.4dh.eu/images/Igor_Krupenski_2018.pdf	1 Classic	Proposed
EE, Maardu, System	A. Existing area & existing system	http://www.4dh.eu/images/Aleksandr_Hlebnikov_Aalborg_4thDH.pcf	1C Rehabilitation	Planned
EE, Tartu, TARKON-TUGLASE district heating network area	A. Existing area & existing system	http://www.relatedproject.eu/demonstrations/	4 Ultra-Low	Proposed
ES, Iurreta, Basque country, Government campus	B. Existing area & new system	http://www.relatedproject.eu/demonstrations/	4 Ultra-Low	Proposed
FR, Boulogne-sur-Mer, EcoLiane	A. Existing area & existing system	https://guidetodistrictheating.eu/south-dublin/	1 Classic	Implemented
FR, Chevilly-Larue and L'Hay Les Roses, SEMHACH	A. Existing area & existing system	www.semhach.fr	3 Multi-Level	Implemented
FR, La Seyne-sur-mer, System	A. Existing area & existing system	https://www.dalkia.fr/fr/thalassothermie_Seyne-sur-Mer	4 Ultra-Low	Implemented
FR, Marseille, Massileo	B. Existing area & new system	https://france3-regions.francetvinfo.fr/provence-alpes-cote-d-azur/bouches-du-rhone/marseille/marseille-lancement-second-projet-thalassothermie-1347229.html	5 CHC	Proposed

Country code and location	Demo-class	Website link	Network configuration	Status 2020
FR, Nice, Grand Arenas	B. Existing area & new system	https://www.reuseheat.eu/nice/	5 CHC	Proposed
FR, Nice, Meridia	B. Existing area & new system	https://reseau-meridia.idex.fr/web/p	5 CHC	Proposed
FR, Paris, Clichy Batignolles	C. New area & new system	https://www.engie.fr/actualites/geothermie-clichy-batignolles/	3 Multi-Level	Implemented
FR, Paris, Nord Est	C. New area & new system	https://tractebel-engie.com/en/references/paris-geothermal-heat-and-cold	3 Multi-Level	Implemented
FR, Paris, Saclay	C. New area & new system	https://www.construction21.org/articles/h/Recovering-upcycling-and-exchanging-energy-making-the-energy-transition-with-heating-and-cooling-networks-a-success.html	5 CHC	Implemented
IE, Dublin, Tallaght	B. Existing area & new system	https://guidetodistrictheating.eu/south-dublin/	1 Classic	Planned
IT, Brescia, Via Carlo Venturi	A. Existing area & existing system	https://www.tempo-dhc.eu/ http://www.4dh.eu/images/Leoni_Capretti_2018.pdf	1B Return to return	Planned
IT, Milano, Via Balilla	B. Existing area & new system	https://www.rewardheat.eu/en/Demonstration-Networks/Milan	4 Ultra-Low	Planned
IT, Ospitaletto, Cogeme	A. Existing area & existing system	https://www.euroheat.org/wp-content/uploads/2020/03/20200624_Life4HeatRecovery_Ospitaletto.pdf	5 CHC	Implemented
IT, Padova, Abano Terme	B. Existing area & new system	http://vbn.aau.dk/files/233986714/paper_451.pdf	4 Ultra-Low	Proposed
NL, Brunssum, Mijnwater	B. Existing area & new system	https://www.construction21.org/articles/h/Underground-power-plants-for-5th-generation-district-heating-and-cooling-in-the-city-of-Brunssum.html	5 CHC	Planned

Country code and location	Demo-class	Website link	Network configuration	Status 2020
NL, Delft, TU Campus	A. Existing area & existing system	http://campusdevelopment.tudelft.nl/en/project/heating-network-transition-programme/	Unknown	Planned
NL, Dongen, Plan Beljaart	C. New area & new system	https://tijdovanderzee.com/2017/06/23/warmtepompendrama-in-nieuwbouwwijk-dongen/	3 Multi-Level	Implemented
NL, Hague, Duindrop	B. Existing area & new system	https://www.c40.org/case_studies/the-hague-uses-seawater-to-heat-homes	4 Ultra-Low	Implemented
NL, Heerlen, Mijnwater	A. Existing area & existing system	https://www.mijnwater.com/?lang=en	5 CHC	Implemented
NL, Hengelo, Akzo Nobel	C. New area & new system	http://www.wiefm.eu/wp-content/uploads/2017/11/Warmtenet-Hengelo-Totstandkoming.pdf	4 Ultra-Low	Implemented
NL, Roosendaal, DER	C. New area & new system	http://energieder.nl/	4 Ultra-Low	Implemented
NL, Rotterdam, Maastad Ziekenhuis	A. Existing area & existing system	https://www.mijnwater.com/wp-content/uploads/2019/08/20190729_LIFE4HeatRecovery-first-newsletter.pdf	1 Classic	Planned
NO, Bergen, Old university system	A. Existing area & existing system	Not available	5 CHC	Implemented
NO, Kristiansand, Agder, Kasernehagen	D. Building	Not available	Secondary	Implemented
NO, Trondheim, Bröset	E. Simulation	Not available	Unknown	Proposed
NO, Trondheim, NTNU Campus	A. Existing area & existing system	https://www.ntnu.no/campusutvikling/2018/campus-smarte-energiloesninger	1C Rehabilitation	Implemented
NO, Ulstein, Fjordvarme	B. Existing area & new system	https://www.iea-dhc.org/fileadmin/documents/Annex_TS1/150420_IEA_DHC_Annex_TS1_case_studies_Brochure.pdf	5 CHC	Implemented

Country code and location	Demo-class	Website link	Network configuration	Status 2020
RS, Belgrade, Beoelektrane	A. Existing area & existing system	http://www.relatedproject.eu/demonstrations/	1 Classic	Proposed
SE, Borlänge, Tjärna Ängar	D. Building	https://www.e2b2.se/forskningsprojekt-i-e2b2/renovering/varsam-energieffektiv-renovering-tjaerna-aengar/	Secondary	Implemented
SE, Borås, System	A. Existing area & existing system	Not available	1C Rehabilitation	Implemented
SE, Bromölla, System	A. Existing area & existing system	Not available	1A Supply to supply	Implemented
SE, Ellös, Ellös	B. Existing area & new system	https://www.orust.se/amnesomrade/byggaboochmiljo/energiochuppvarmning/fjarrvarme.4.32c676ab13fc39bc3232dfa.html	1 Classic	Implemented
SE, Göteborg, Backa Röd	D. Building	http://www.sundolitt.se/download.aspx?object_id=DFCE3D9B961A4AB0B-7D1A46A132CA632.pdf	Secondary	Implemented
SE, Göteborg, Chalmers Living Lab	D. Building	https://hll.livinglab.chalmers.se/	1B Return to return	Implemented
SE, Halmstad, Maratonvägen	D. Building	http://www.bebostad.se/library/1778/go-daex-maratonvaegen-2014.pdf	Secondary	Implemented
SE, Halmstad, Ranagård	C. New area & new system	https://www.hem.se/inspiration/low-temp/	2 Modified Classic	Planned
SE, Helsingborg, Råå	C. New area & new system	https://www.rewardheat.eu/en/Demonstration-Networks/Helsingborg	4 Ultra-Low	Planned
SE, Kiruna, Garaget/Kapellet	C. New area & new system	https://www.ltu.se/research/subjects/VA-teknik/Dag-Nat/Nyheter-och-aktuellt/Kan-man-anvanda-fjarrvarme-for-att-lagga-VA-ledningar-ovanfor-tjalgransen-1.154796?l=en	1 Classic	Planned

Country code and location	Demo-class	Website link	Network configuration	Status 2020
SE, Kungsbacka, Vallda Heberg	C. New area & new system	http://task52.iea-shc.org/Data/Sites/1/publications/IEA-SHC%20Task%2052%20STC2-Best%20practice%20summary%20report_2017-08-31_FINAL.pdf	1 Classic	Implemented
SE, Kungsör, Södra Kungsladugården	C. New area & new system	https://www.malarenergi.se/om-malarenergi/vara-anlaggningar/varmeverken/	1 Classic	Implemented
SE, Kungälv, Nya Grinden	C. New area & new system	http://www.kungalvenergi.se/fjarrvarme/om-var-fjarrvarme/	1 Classic	Implemented
SE, Linköping, Ullstämma	C. New area & new system	Not available	1 Classic	Implemented
SE, Linköping, Contemporary	E. Simulation	https://www.sciencedirect.com/science/article/pii/S0360544217322004	2 Modified Classic	Proposed
SE, Linköping, Future	E. Simulation	https://www.sciencedirect.com/science/article/pii/S0360544217322004	2 Modified Classic	Proposed
SE, Lund, Brunnshög	C. New area & new system	http://www.cooldh.eu/demo-sites-and-innovations-in-cool-dh/brunnshog-in-lund/	1 Classic	Planned
SE, Lund, Linero	D. Building	http://www.cityfied.eu/Demo-Sites/Lund/Lund.kl	Secondary	Implemented
SE, Lund, Medicon Village	B. Existing area & new system	http://ectogrid.com/	5 CHC	Implemented
SE, Piteå, Haraholmen	C. New area & new system	http://www.diva-portal.org/smash/get/diva2:1219404/FULLTEXT01.pdf	1 Classic	Implemented
SE, Sigtuna, Stadsängar	C. New area & new system	https://www.eon.se/artiklar/vi-aer-med-och-bygger-sveriges-mest-hallbara-stadsdel.html	1 Classic	Planned
SE, Slöinge, SIA Glass	B. Existing area & new system	https://via.tt.se/pressmeddelande/glass-fabrik-varmer-djursjukhus-med-en-ergilosning-fran-eon?publisherId=1035173&releaseId=3274836	5 CHC	Planned
SE, Stockholm, Hjorthagen	A. Existing area & existing system	https://www.stockholmexergi.se/nyheter/kallare-fjarrvarme-ger-klimatvinst/	3 Multi-Level	Implemented

Country code and location	Demo-class	Website link	Network configuration	Status 2020
SE, Varberg, Västerport	C. New area & new system	Not available	2 Modified Classic	Proposed
SE, Västerås, Bergsgrottan	C. New area & new system	Not available	1 Classic	Implemented
SE, Västerås, Bjärby	C. New area & new system	Not available	1 Classic	Implemented
SE, Västerås, Gotö källa	C. New area & new system	Not available	1 Classic	Planned
SE, Västerås, Kaptenen	C. New area & new system	Not available	1 Classic	Implemented
SE, Västerås, Kartbladet	C. New area & new system	Not available	1 Classic	Planned
SE, Västerås, Kassel	C. New area & new system	Not available	1 Classic	Implemented
SE, Västerås, Lahti	C. New area & new system	Not available	1 Classic	Planned
SE, Västerås, Lillhamra	C. New area & new system	Not available	1 Classic	Planned
SE, Västerås, Spinnakern	C. New area & new system	Not available	1 Classic	Implemented
SE, Växjö, READY-project	D. Building	http://www.smartcity-ready.eu/	Secondary	Implemented
SE, Örebro, Sjukhuset	B. Existing area & new system	https://bioenergyinternational.com/app/uploads/2018/11/Helen_Carlstrom.pdf	5 CHC	Implemented
SF, Hyvinkää, Building fair	C. New area & new system	https://www.iea-dhc.org/fileadmin/documents/Annex_TS1/150420_IEA_DHC_Annex_TS1_case_studies_Brochure.pdf	1 Classic	Implemented
UK, Aberdeen, Torry	B. Existing area & new system	https://www.nweurope.eu/projects/project-search/heatnet-transition-strategies-for-delivering-low-carbon-district-heat/news/aberdeen-pilot-works-to-extend-the-dh-network-have-started/	1 Classic	Planned

Country code and location	Demo-class	Website link	Network configuration	Status 2020
UK, Bristol, Owen square	B. Existing area & new system	https://www.owensquare.coop/	4 Ultra-Low	Implemented
UK, Glasgow, Clyde Gateway	B. Existing area & new system	http://www.clydegateway.com/europe-an-funded-programme-assists-low-carbon-energy-ambitions/	4 Ultra-Low	Planned
UK, London, London South Bank Univ.	B. Existing area & new system	http://www.lsbu.ac.uk/research/centres-groups/sites/ben-project	5 CHC	Implemented
UK, Nottingham, Sneinton	A. Existing area & existing system	https://www.ntu.ac.uk/research/groups-and-centres/projects/remourban-city-demonstrator-project	1B Return to return	Implemented
UK, Nottingham, Bullwell	B. Existing area & new system	https://www.construction21.org/articles/h/a-publier-le-23-d2grids-and-nottingham-city-council-at-forefront-of-exciting-future-for-mine-energy-in-uk.html	4 Ultra-Low	Planned
UK, Plymouth, Civic Centre	B. Existing area & new system	https://guidetodistrictheating.eu/plymouth/	5 CHC	Planned
UK, Slough, Greenwatt Way	C. New area & new system	https://www.iea-dhc.org/fileadmin/documents/Annex_X/IEA_Annex_X_Final_Report_2014_-_Toward_4th_Generation_District_Heating.pdf	1 Classic	Implemented
USA, Stanford, Stanford University	B. Existing area & new system	https://sustainable.stanford.edu/sites/default/files/Stanford%20SESI%20General%20Information%20Brochure%20%28rev%206%29.pdf	5 CHC	Implemented

10.4 Location index

In this location index, all locations mentioned in this guidebook are listed with information about what page/pages that the location is mentioned.

Aachen, 131, 164, 180, 185
Aalborg, 44
Aarhus, 131, 173, 176, 190
Aberdeen, 195
Albertslund, 121, 188
Amstetten, 93
Antwerp, 129
Bamberg, 131, 134, 135, 136, 141, 151, 156, 185
Basel, 161, 162, 163
Belgrade, 193
Bergen, 95, 164, 177, 179, 192
Birkeröd, 176, 189
Bjerringbro, 140, 152, 157, 176, 180, 188
Bochum, 176, 185
Borlänge, 54, 181, 193
Borås, 43, 45, 81, 193
Boulogne-sur-Mer, 173, 190
Braunschweig, 131, 141, 148, 156, 165, 173, 179, 185
Brescia, 174, 191
Bristol, 176, 196
Bromölla, 173, 193
Brunssum, 191
Chemnitz, 173, 179, 185
Chevilly-Larue, 175, 190
Copenhagen, 180, 188
Crailsheim, 173, 185
Darmstadt, 100, 112, 113, 114, 115, 117, 120, 131, 134, 141, 156, 179, 185
Delft, 164, 192
Dongen, 175, 192
Dublin, 131, 134, 180, 190
Edmonton, 128
Ellös, 193
Enköping, 43
Frederiksberg, 141, 153, 157, 181, 188
Freiburg, 131, 173, 185
Geneva, 131, 132, 160, 161, 162, 163, 177, 184
Glasgow, 196
Gleisdorf, 44, 82, 131, 140, 141, 142, 156, 160, 162, 163, 179, 182
Graz, 84, 140, 147, 156, 155, 179, 182
Güssing, 182
Göteborg, 44, 181, 193
Hague, 176, 192
Halmstad, 131, 174, 193
Hamburg, 174, 176, 185, 186
Hameln, 176, 186
Hamilton, 183
Heerlen, 141, 143, 144, 156, 177, 179, 192
Helsingborg, 193
Helsinki, 128, 178
Hengelo, 192
Herne, 186
Hilleröd, 188
Hjortshøj, 96
Hyvinkää, 195
Höje Tåstrup, 131, 174, 188
Hörsholm, 176, 189
Iurreta, 190
Karise, 189
Kassel, 77, 131, 135, 141, 150, 154, 155, 156, 157, 180, 181, 186
Kelowna, 177, 183
Kiruna, 193
Klagenfurt, 86
Kopli, 190
Kortrijk, 173, 183
Kristiansand, 192
Kungsbacka, 121, 194
Kungsör, 194
Kungälv, 194
L'Hay Les Roses, 175, 190
La Seyne-sur-mer, 190
Leuven, 176, 183
Linköping, 131, 173, 180, 194
London, 164, 177, 196
Ludwigsburg, 186
Lund, 131, 141, 149, 156, 173, 177, 180, 194
Lüneburg, 164
Maardu, 190
Mannheim, 131, 179, 186
Mariehamn, 181
Marseille, 190
Middelfart, 43, 45, 82, 189
Milano, 191
Moosburg, 180, 186
Munich, 31, 131, 160, 161, 162, 163, 173, 186
Naters, 175, 184
Neuburg, 131, 180, 187
Nice, 191
Nivå, 189
Nottingham, 121, 174, 180, 196
Nümbrecht, 176, 187
Nürnberg, 187
Oberwald, 184
Odder, 189

Okotoks, 131, 183
Oldenburg, 131, 180, 187
Oshawa, 164, 183
Ospitaletto, 191,
Padova, 191,
Paris, 164, 175, 177, 191
Piteå, 194
Plymouth, 196
Potsdam, 165
Radolfzell, 187
Recklinghausen, 131, 180, 187
Richmond, 177, 184
Roosendaal, 192
Rosenheim, 187
Rotkreutz, 131, 184
Rotterdam, 192
Salzburg, 84, 86, 131, 136, 141, 146, 156, 173, 179, 182
Sigtuna, 194
Silkeborg, 176, 189
Slough, 131, 173, 196
Slöinge, 194
Stanford, 164, 178, 196
Stockholm, 175, 178, 180, 181, 194
Stuttgart, 175, 187
Tartu, 190
Tilst, 189
Topusko, 184
Troisdorf, 176, 187
Trondheim, 131, 179, 192
Ulstein, 177, 192
Varberg, 195
Whistler, 184
Viborg, 44, 81, 141, 152, 157, 160, 161, 162, 163, 180, 181, 189
Vienna, 84, 86, 131, 176, 177, 179, 182, 183
Villach, 131, 173, 179, 183
Vinge, 190
Visp, 131, 177, 184
Wüstenrot, 141, 142, 156, 173, 176, 179, 180, 187
Västerås, 127, 131, 135, 173, 180, 195
Växjö, 195
Wörgl, 122, 179, 183
Zwickau, 164, 176, 188
Zürich, 95, 131, 132, 133, 134, 137, 141, 145, 156, 164, 176, 177, 179, 184
Örebro, 177, 195

10.5 Participant organisations in the TS2 annex

In this final section, the project participant organisations are presented together with their own recent projects associated with low-temperature district heating. Written articles and conference presentations that have been given in conjunction to this TS2 project are also listed.

Within the TS2 project, each active participant was involved in research projects concerning low-temperature district heating in either a national or an international context; the experiences and results from these projects were aggregated into this TS2 project. In total, approximately 1.2 million euro was available for the core author group for participation in the project.

The TS2 abbreviation signifies that this was the second task sharing (TS) project within the IEA-DHC technology collaboration programme focused on district heating and cooling. The task sharing cooperation model involves each participant securing its own financial resources for its research participation. The first task sharing project within the IEA-DHC programme was titled 'Low Temperature District heating for Future Energy Systems' and was coordinated by Fraunhofer IEE.

The core author group in the TS2 project consisted of participants from five countries and ten organisations:

- From Austria, the participants were the Austrian Institute of Technology (AIT), Institute of Sustainable Technologies (AEE INTEC), Austroflex, and TU Wien. AIT, AEE INTEC and TU Wien cooperated in the national project T2LowEx, financed by FFG. AIT participated also in the Austrian project NextGenerationHeat and in the European TEMPO project that was performed under the Horizon 2020 programme. AEE INTEC coordinates the Thermaflex project funded by the Austrian climate and energy fund.
- From Denmark, the participants were the Danish Technical University together with Danfoss, financed by the Danish EUDP programme. Other relevant DTU research projects were 'Return Temperature Optimisation of Radiators for the EUDP programme' and a PhD project concerning optimisation of radiators for obtaining lower temperature levels in heat distribution networks. The latter project was co-financed with VITO in Belgium.

- From Germany, the participants were Fraunhofer IEE, TU Darmstadt, and Kassel University. Fraunhofer IEE had previous projects such as Smart Thermal Subgrid within the German EnEff:Wärme programme, the Zum Feldlager project in Kassel, and several feasibility studies within the German Wärmenetzsysteme 4.0 programme. TU Darmstadt have had two research projects with the German EnEff:Stadt programme concerning future development of their Lichtwiese campus. Kassel University also participated in the Smart Thermal Subgrid and Zum Feldlager projects. They have also had strategy development projects with the local district heating provider in Kassel.
- From Norway, the participant was the Norwegian University of Science and Technology (NTNU). At NTNU, the national research project 'Understanding behaviour of district heating systems integrating distributed sources' was financed by the Norwegian Research Council.
- From Sweden, Halmstad University coordinated the TS2 project. Their participation was financed by the Swedish Energy Agency. Halmstad University has participated in projects such as Future District Heating in Sweden and in the TEMPO and Reuseheat projects within the European Horizon 2020 programme.

Beyond the core author group, the following participants were involved in project meetings and discussions:

- From Germany, Stuttgart University of Applied Sciences participated.
- From Ireland, Codema participated and coordinated the international HeatNet NW project that was financed by one of the European regional Interreg programmes.
- From United Kingdom, Nottingham Trent University, Leeds University, and Carbon Alternatives participated. Nottingham Trent University participated in the REMOURBAN lighthouse project within a smart city call from the European Horizon 2020 programme.

Project participants have delivered the following presentations associated to the TS2 project at various conferences:

Andrés, M., Regidor, M., Macía, A., Vasallo, A., & Lygnerud, K. (2018). Assessment methodology for urban excess heat recovery solutions in energy-efficient District Heating Networks. Energy Procedia, 149, 39-48. doi:https://doi.org/10.1016/j.egypro.2018.08.167

Benakopoulos, T., Salenbien R., Tunzi, M., & Svendsen, S. (2020). Faults detection and low operating temperatures in radiator system by using data from existing digital heat cost allocators in a multifamily building. Paper presented at the 6th Smart Energy Systems and 4DH conference, Aalborg.

Best, I., Braas, H., Orozaliev, J., Jordan, U., & Vajen, K. (2019). Systematic investigation of building energy efficiency standard and hot water preparation systems' influence on the heat load profile of districts. Paper presented at the 5th International Conference on Smart Energy Systems, Copenhagen.

Best, I., Orozaliev, J., & Vajen, K. (2018). Impact of Different Design Guidelines on the Total Distribution Costs of 4th Generation District Heating Networks. Energy Procedia, 149, 151-160. doi:https://doi.org/10.1016/j.egypro.2018.08.179

Geyer, R. (2019). Potentials and effects of temperature reductions in heat networks. Paper presented at the Sustainable District Energy Conference, Reykjavik. https://sdec.is/wp-content/uploads/2019/11/roman_geyer.pdf

Geyer, R. (2020). Reduced system temperatures in heating networks – Energy-economic assessments of the effects. Paper presented at the 6th International Conference on Smart Energy Systems, Aalborg.

Li, H., & Nord, N. (2018). Transition to the 4th generation district heating - possibilities, bottlenecks, and challenges. Energy Procedia, 149, 483-498. doi:https://doi.org/10.1016/j.egypro.2018.08.213

Oltmanns, J. (2019). Decreasing the temperature of an existing district heating network. Paper presented at the 5th Smart Energy Systems and 4DH conference, Copenhagen.

Oltmanns, J., Freystein, M., Dammel, F., & Stephan, P. (2018). Improving the operation of a district heating and a district cooling network. Energy Procedia, 149, 539-548. doi:https://doi.org/10.1016/j.egypro.2018.08.218

Schmidt, D. (2018). Low Temperature District Heating for Future Energy Systems. Energy Procedia, 149, 595-604. doi:https://doi.org/10.1016/j.egypro.2018.08.224

Schmidt, D. (2019). Implementation of low temperature district heating systems – Successful case studies of IEA DHC ANNEX TS2. Paper presented at the 5th Smart Energy Systems and 4DH conference, Copenhagen.

Project participants wrote the following scientific articles related to low-temperature district heating issues in international journals:

Averfalk, H., & Werner, S. (2020). Economic benefits of fourth generation district heating. Energy, 193, 116727. https://doi.org/10.1016/j.energy.2019.116727

Benakopoulos, T., Salenbien, R., Vanhoudt, D., & Svendsen, S. (2019). Improved Control of Radiator Heating Systems with thermostatic radiator valves without pre-setting function. Energies, 12(3215). https://doi.org/10.3390/en12173215

Benakopoulos, T., Salenbien, R., Vanhoudt, D., Tunzi, M., & Svendsen, S. (2021). Low return temperature from domestic hot-water system based on instantaneous heat exchanger with chemical-based disinfection solution. Energy, 215, 119211. https://doi.org/10.1016/j.energy.2020.119211

Best, I., Braas, H., Orozaliev, J., Jordan, U., & Vajen, K. (2020). Systematic investigation of building energy efficiency standard and hot water preparation systems' influence on the heat load profile of districts. Energy, 197, 117169. https://doi.org/10.1016/j.energy.2020.117169

Braas, H., Jordan, U., Best, I., Orozaliev, J., & Vajen, K. (2020). District heating load profiles for domestic hot water preparation with realistic simultaneity using DHWcalc and TRNSYS. Energy, 201, 117552. https://doi.org/10.1016/j.energy.2020.117552

Geyer, R., Krail, J., Leitner, B., Schmidt, R.-R., Leoni, P. (2021). Energy-economic assessment of reduced district heating system temperatures. Smart Energy, https://doi.org/10.1016/j.segy.2021.100011

Leoni, P., Geyer, R., & Schmidt, R.-R. (2020). Developing innovative business models for reducing return temperatures in district heating systems: Approach and first results. Energy, 195, 116963. https://doi.org/10.1016/j.energy.2020.116963

Lygnerud, K. (2019). Business model changes in district heating: The impact of the technology shift from the third to the fourth generation. Energies, 12(9). doi:10.3390/en12091778

Oltmanns, J., Sauerwein, D., Dammel, F., Stephan, P., & Kuhn, C. (2020). Potential for waste heat utilization of hot-water-cooled data centers: A case study. Energy Science & Engineering, 8(5), 1793-1810. doi:10.1002/ese3.633

10.6 Literature references in Chapter 10

Averfalk, H., Ottermo, F., & Werner, S. (2019). Pipe Sizing for Novel Heat Distribution Technology. Energies, 12(7), 1276.

Averfalk, H., & Werner, S. (2018). Novel low temperature heat distribution technology. Energy, 145, 526-539. doi:https://doi.org/10.1016/j.energy.2017.12.157

Burstein, N. (2020). Fjärrkylestatistik 2019 (District cooling statistics 2019). Retrieved from https://www.energiforetagen.se/statistik/fjarrkylestatistik/

Lemola, J. (2016). Helsinki Expands Smart Eco-Friendly District Cooling. Retrieved from https://www.globenewswire.com/news-release/2016/06/30/1301579/0/en/Helsinki-Expands-Smart-Eco-Friendly-District-Cooling.html

IMPRINT

Contact:
Fraunhofer Institute
for Energy Economics and Energy System Technology IEE
Königstor 59
34119 Kassel
Germany
Phone +49 561 804 1871
dietrich.schmidt@iee.fraunhofer.de
www.iee.fraunhofer.de

on behalf of the editors Kristina Lygnerud and Sven Werner within IEA DHC Annex TS2.

Bibliographic information of the German National Library:
The German National Library has listed this publication in its Deutsche Nationalbibliografie; detailed bibliographic data is available on the internet at www.dnb.de.

ISBN 978-3-8396-1745-8

Print and finishing:
RCOM Print GmbH, Würzburg-Rimpar

Layout and typesetting:
Ulf Cadenbach, www.ulfcadenbach.de

The book was printed with chlorine- and acid-free paper.

© Fraunhofer Verlag, 2021
Nobelstrasse 12
70569 Stuttgart
Germany
verlag@fraunhofer.de
www.verlag.fraunhofer.de

is a constituent entity of the Fraunhofer-Gesellschaft, and as such has no separate legal status.

Fraunhofer-Gesellschaft zur Förderung
der angewandten Forschung e.V.
Hansastrasse 27 c
80686 München
Germany
www.fraunhofer.de

Many of the designations used by manufacturers and sellers to distinguish their products are claimed as trademarks. The quotation of those designations in whatever way does not imply the conclusion that the use of those designations is legal without the consent of the owner of the trademark.